Chemical Analysis Handbook

Chemical Analysis Handbook

Edited by Natalia Wood

Larsen & Keller
www.larsen-keller.com

Chemical Analysis Handbook
Edited by Natalia Wood
ISBN: 979-8-88836-065-1 (Hardback)

Larsen & Keller
Published by Larsen and Keller Education,
5 Penn Plaza,
19th Floor,
New York, NY 10001, USA

Cataloging-in-Publication Data

Chemical analysis handbook / edited by Natalia Wood.
p. cm.
Includes bibliographical references and index.
ISBN 979-8-88836-065-1
1. Analytical chemistry. 2. Chemistry. I. Wood, Natalia.
QD75.22 .C44 2023
543--dc23

For more information regarding Larsen and Keller Education and its products, please visit the publisher's website www.larsen-keller.com

Table of Contents

Preface

This book aims to highlight the current researches and provides a platform to further the scope of innovations in this area. This book is a product of the combined efforts of many researchers and scientists, after going through thorough studies and analysis from different parts of the world. The objective of this book is to provide the readers with the latest information of the field.

Chemical analysis refers to the study of the chemical composition and structure of substances. It is divided into two categories, which include qualitative and quantitative analysis. Qualitative analysis is used for determining whether or not a substance is present in a sample. Quantitative analysis is utilized to determine the amount of a substance present in a sample. Chemical analysis studies the chemical components of a material which aids in the identification of unknown materials or the verification that the material satisfies the necessary conditions. The key steps performed during chemical analysis include sampling, field sample pretreatment, laboratory treatment, laboratory assay, calculations and results presentation. Chemical analysis is critical for ensuring the chemical's productivity, safety, consistency, quality and control. This book explores all the important aspects of chemical analysis in the present day scenario. It is appropriate for students seeking detailed information in this area of chemistry as well as for experts.

I would like to express my sincere thanks to the authors for their dedicated efforts in the completion of this book. I acknowledge the efforts of the publisher for providing constant support. Lastly, I would like to thank my family for their support in all academic endeavors.

Editor

1

ESCA as a Tool for Exploration of Metals' Surface

Eleonora Bolli [1,2,*], **Saulius Kaciulis** [2] and **Alessio Mezzi** [2]

1 Department of Industrial Engineering, University of Rome "Tor Vergata", Via del Politectico 1, 00133 Rome, Italy
2 Institute for the Study of Nanostructured Materials, ISMN—CNR, Monterotondo Stazione, 00015 Rome, Italy; Saulius.kaciulis@cnr.it (S.K.); alessio.mezzi@cnr.it (A.M.)
* Correspondence: Eleonora.bolli@ismn.cnr.it

Abstract: The main principles and development of electron spectroscopy for chemical analysis (ESCA) are briefly reviewed. The role of ESCA techniques (X-ray photoelectron spectroscopy and Auger electron spectroscopy) in the investigation of metallic surfaces is discussed, evidencing their importance and analytical potentiality. An overview is given of a series of recent experimental cases of ESCA application for the characterization of different metals and metallic alloys, illustrating the main results and various phenomena, such as the formation of impurity defects, corrosion, migration of constituent elements in various alloys, clustering in liquid alloy, etc., that can occur on the surface and the interface of investigated materials. These materials comprise the collection coins of noble metals, some metal alloys and Ni-based superalloys, nitride coatings on stainless steel, composite material with TiAlV alloy, treated austenitic steels, and graphene interface with polycrystalline metal foils. The present review could be particularly recommended for the newcomers to the research field of surface analysis and its application for various metals, their treatments, and possible modifications in operating conditions.

Keywords: surface analysis; metals and alloys; metal coatings; XPS; AES; SPEM

---

1. Introduction

From the very beginning of surface science around 1960, coinciding with the discovery of electron spectroscopy for chemical analysis (ESCA) [1], a great part of the first surface analysis studies has been dedicated to various metals [2]. This is easily understandable because the metals are stable in ultrahigh vacuum (UHV), and their surface is relatively clean (or can be easily cleaned) and is not modified under soft X-rays or electron beam. Therefore, during the initial boom of surface analysis—namely when the main experimental techniques were developed, the main principles were established, and spectroscopic catalogues were created—great attention of this research was given to the surface of metals, and in particular, the transition metals and noble ones. In this period, new scientific journals dedicated to surface analysis were also born, such as *Journal of Electron Spectroscopy* and *Related Phenomena*, etc. Classical examples of metals' surface studies can be found already in the first volume of the first journal mentioned [3,4]. Even later, when surface analysis became a common tool in the labs of materials characterization and when the ESCA handbooks of all chemical elements were available [5,6], the application for the surface of metals remained an important research field, as it was emphasized in the first text books on surface analysis [7,8].

Currently, when a great variety of surface-sensitive electron spectroscopies (see [9]) are available everywhere in the world, the most used ones remain X-ray photoelectron spectroscopy (XPS) and Auger electron spectroscopy (AES) that were born with a common term of ESCA. Of course, at present time, surface analyses are aimed at more sophisticated materials than elemental metals, but even for elemental metals these techniques are widely used for the simple and reliable control of surface purity in the fields of materials research and technological applications. It is namely for these reasons that we decided to prepare a short review, illustrating the importance and analytical capabilities of surface analysis for the exploration of metals, including also their modifications induced by operating conditions. In this review, we present the most interesting cases of experimental research carried out in our lab during the last few decades. These cases comprise the surface defects on noble metals (collection coins), some metal alloys and superalloys, nitride coatings on steel, composite material with TiAlV alloy, treatments of austenitic steel, and graphene growth on polycrystalline metals. The common denominator of all these cases is the application of ESCA, i.e., XPS and AES techniques, for materials' characterization.

The main working equations of ESCA are very simple, and they are based on the principle of energy conservation. In the case of XPS, this equation is

$$BE = h\nu - KE - WF, \tag{1}$$

where BE is the binding energy of the elemental core level, hν is the photon energy of X-rays, KE is the kinetic energy of the emitted photoelectron, and WF is the work function of the spectrometer.

In Auger effect, the electrons from three different atomic levels are involved, resulting in the finally excited Auger electron with kinetic energy equal to the following:

$$KE = EL1 - EL2 - EL3 - WF, \tag{2}$$

where KE is the kinetic energy of emitted Auger electron; EL1 is the binding energy of the first atomic level, where the hole is created (by X-rays or electron beam); EL2 is the energy of the second level from which the electron is falling down to the lower level EL1; and EL3 is the energy of the third atomic level from which the Auger electron is ejected. Of course, in some chemical elements the last two levels of the Auger process can be located in the valence band, where a core–valence–valence (CVV) Auger peak is then observed. Typical examples of such elements with broad Auger CVV peaks are carbon and silicon. A schematic diagram of the final state in photoemission and Auger excitations is illustrated in Figure 1.

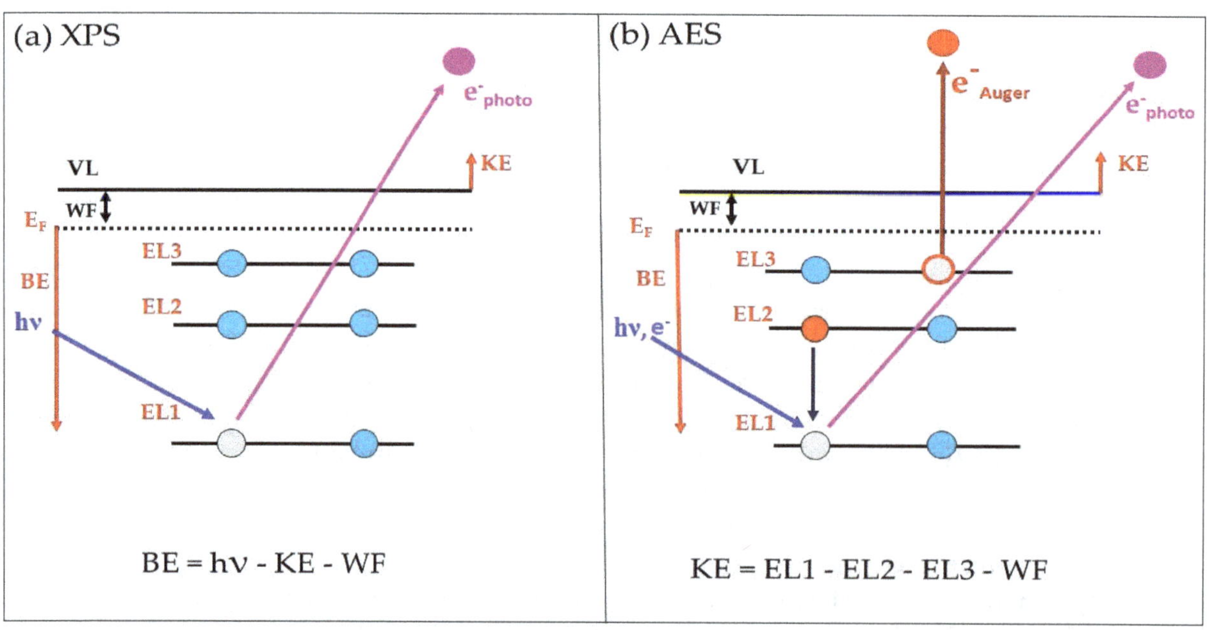

Figure 1. Schematic diagram of the final state in photoemission (**a**) and Auger (**b**) excitations.

From the photoemission and/or Auger spectra, it is possible to identify the chemical elements because the energy of these peaks is characteristic for every element. In the case of superposition of some peaks from different elements, other peaks of the same elements can be used for identification. In most cases, the XPS also permits the chemical state of constituent elements to be identified due to the chemical shift of the photoemission peaks [10]. Both the techniques are surface sensitive because their information depth is limited by the mean free path of the electrons in the solid, which depends on kinetic energy and is typically from 1 to about 10 nm. The detailed description of XPS and AES techniques can be found in numerous text books (e.g., [7–9]) and even online (e.g., [11]).

In the case of AES, the primary electron beam can be easily focused as with electron microscopy, and therefore it is possible not only to achieve a high lateral resolution in spectroscopy but also to acquire the chemical maps of the surface. This mode of operating is called Auger scanning microscopy (SAM). The first two generations of XPS spectrometers were equipped with standard soft X-ray sources (typically with Al and Mg anodes); because of this, the focusing of X-rays was impossible, and the lateral resolution of these instruments was limited to about 0.1–1 mm. Later on, with the introduction of monochromatized X-ray sources and electromagnetic input lenses, the lateral resolution of XPS was improved to about 1–3 microns, allowing also for the operation in XPS imaging mode, i.e., to acquire the surface chemical maps [12,13]. However, quite often this resolution can be too low for the investigation of the submicrometric features or patterns on the sample surface. A much higher lateral resolution of photoemission spectroscopy and imaging can be achieved by using the dedicated beamlines of synchrotron radiation. This technique, which is called scanning photoelectron microscopy (SPEM), enables us to investigate the surface chemical composition at a lateral resolution of about 100 nm [14,15]. It was successfully employed also for some of our experimental cases by using an ESCA microscopy beamline at the synchrotron Elettra in Trieste, Italy. The main features of XPS, AES, and SPEM techniques are summarized in the Table 1, including also the benefits and weak points of their practical applications.

Table 1. Comparison of the main features of three ESCA techniques.

Features	XPS	AES	SPEM
Probe	Soft X-rays	Electrons	Soft X-rays
Spectroscopic signal	Photoelectrons, Auger electrons	Auger electrons	Photoelectrons, Auger electrons
Detectable elements	Li and higher	Li and higher	Li and higher
Sampling depth	0.5–10 nm	0.5–10 nm	0.5–10 nm
Detection limit	1×10^{-4}	1×10^{-3}	1×10^{-4}
Information	Elemental, chemical	Elemental	Elemental, chemical
Quantification	OK	semi	OK
Lateral resolution	>3 µm	>30 nm	>50 nm
Advantages	Chemical bonding, no sample damage	High resolution of chemical imaging	Chemical bonding, chemical imaging
Disadvantages	Poor lateral resolution	Sample charging, beam-induced damage	Beam-induced damage

2. Experimental Techniques

Two different spectrometers were used for the XPS characterization of investigated materials: an aged Escalab MkII (VG Scientific Ltd., East Grinstead, UK) and a modern one, Escalab 250Xi (Thermo Fisher Scientific Ltd., East Grinstead, UK). In both the instruments, the spectroscopy was carried out by concentric hemispherical analyzers operating in a constant pass energy (20 or 40 eV) mode. The first one was equipped with a double-anode (Al/Mg Kα) X-ray source and electrostatic input lens, collecting the signal from the sample area of about 10 mm (large-area mode) and variable to about 0.3 mm (small-area mode). The photoemission signals were registered by a 5-channeltron detector. The second apparatus was equipped with a monochromatized Al Kα source and a combined system of electrostatic/electromagnetic input lenses. In the spectroscopy mode, this system allowed the diameter of analyzed sample area to vary from 900 to 20 μm, and the photoemission signals were registered by a 6-channeltron detector. In the imaging XPS mode, the best lateral resolution of chemical maps was of about 3 μm, and the signals were registered by a multichannel plate with 128 channels. The charging of insulating samples was suppressed by using a combination of two neutralizing floods: low energy electrons from an in-lens gun and low energy Ar+ ions from an external gun. For the sample surface cleaning and XPS depth profiling in both the Escalabs, rastered Ar+ ion guns were used, i.e., the EX05 model in MkII and the EX06 in 250Xi. The base UHV pressure in the analysis chambers of both spectrometers was always kept below 10^{-9} mbar.

The experiments of AES/SAM were carried out by using a LEG200 electron gun installed on the analysis chamber of Escalab MkII. This excitation source provided the primary beam of electrons with an energy up to 10 keV and a minimum beam diameter of 200 nm. For all the samples, the current of electron beam was kept very low (4–10 nA) in order to avoid any sample surface damage by the electron beam. Seeking to increase the signal-to-noise ratio, all the Auger spectra and chemical maps were acquired in a constant retard ratio (1:2) mode of the analyzer.

All experimental data were processed by the software Avantage v.5 (Thermo Fisher Scientific Ltd.). The peak fitting of photoemission spectra was performed by using the Shirley background, a Voigt peak-shape (mixed Gaussian-Lorentzian with variable ratio), and linked full widths at half maximum (FWHMs) for the same core level. Final calibration of the BE scale was done by fixing the main component of C 1s peak (aliphatic carbon) at 285.0 eV and controlling it, if the Fermi level in the valence band is positioned at BE = 0.0 eV.

High resolution SPEM experiments were performed at the ESCA microscopy beamline of the Elettra synchrotron [14,15]. By using Fresnel zone plate optics, the X-ray beam from the synchrotron source was focused to a microprobe with a diameter of about 150 nm on the sample, which was raster-scanned with respect to the microprobe. Photoelectrons were collected by the SPECS-PHOIBOS 100 hemispherical analyzer and registered by a 48-channel electron detector. All the samples were investigated in both imaging and spectroscopy modes with a 0.2 eV energy resolution by using 500–700 eV photon energy. The overall lateral resolution was below 50 nm. Before the measurements, the samples were cleaned by Ar^{+} ion sputtering at 2.0 keV energy. After the acquisition, the chemical maps were processed by the Igor v.6.3 software.

3. Experimental Cases of Different Materials

3.1. "Gold Corrosion" in Collection Coins

Can "gold corrosion" occur in gold coins? This question arose approximately two decades ago, when some owners of precious coins unexpectedly found the appearance of numerous stains on their gold coins. After many studies, even using the Pourbaix diagram, this enigma was successfully disclosed because of the application of surface analysis techniques. The chemical composition of these defects was determined, and their source was established.

The study of surface analysis was performed on gold and silver collection coins supplied by the Kunsthistorisches Museum in Vienna (historical Austrian Ducat) and Austrian Mint (coins and their blanks). XPS, AES, and SAM techniques were combined to get qualitative and quantitative information about the surface defects. Their stains, analysed by a stereomicroscope, were generally composed by a dark central area surrounded by a larger outer area, whose colour varied from red to dark blue [16]. The chemical composition of every single stain was determined by XPS. All elements in the spot were quickly identified by the assignment of the peaks found in the survey scan spectrum

(Figure 2), whereas their chemical state and atomic concentration were determined by processing the resolved spectra of the main peaks presented in Figure 3 [17].

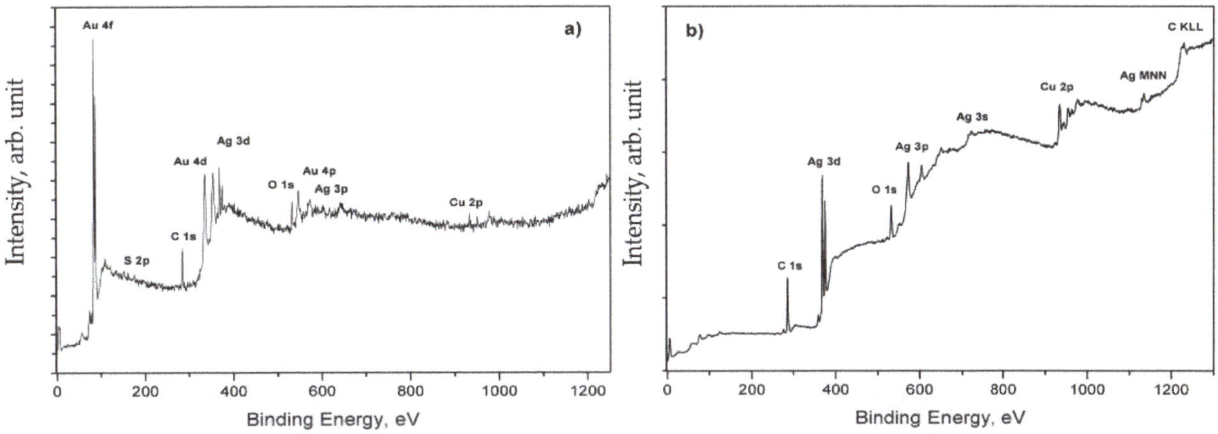

Figure 2. XPS survey scans of the gold (**a**) and silver (**b**) coins.

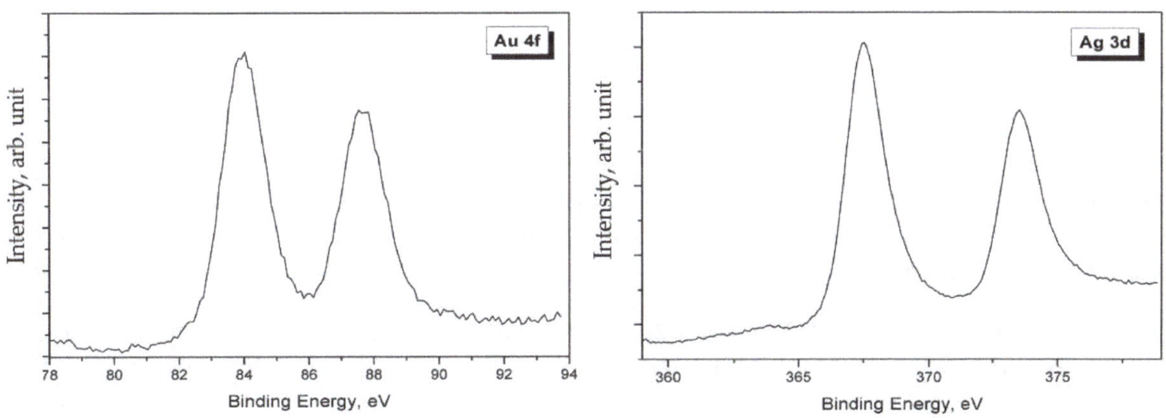

(**a**) (**b**)

Figure 3. High-resolution XPS spectra of Au 4f (**a**) and Ag 3d (**b**).

The obtained results promptly evidenced a strange composition of the stains on a pure (999.9) gold coin: A contamination with Ag and S was revealed. This was an astonishing finding for a pure gold coin, giving rise to the following questions: how and when had these impurities been added? The obtained results were confirmed by the multipoint AES analysis and SAM chemical maps acquired with a higher lateral resolution of approximately 200 nm, which are presented in Figure 4.

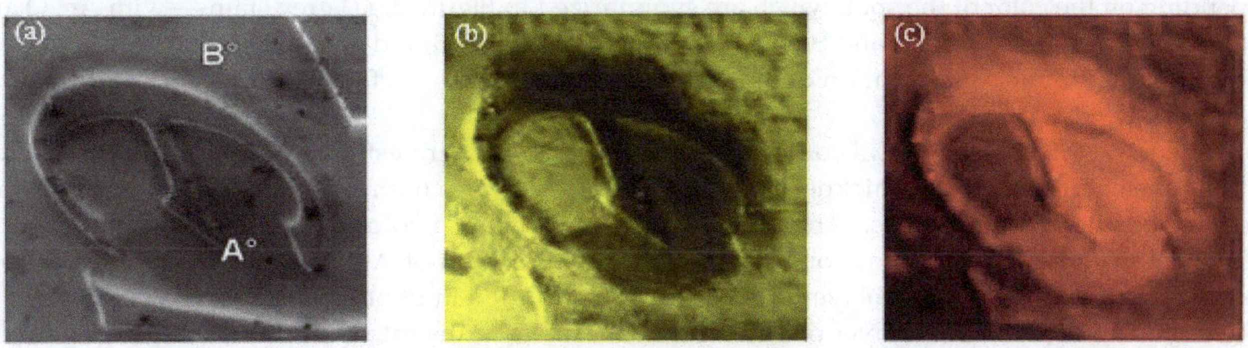

Figure 4. *Cont.*

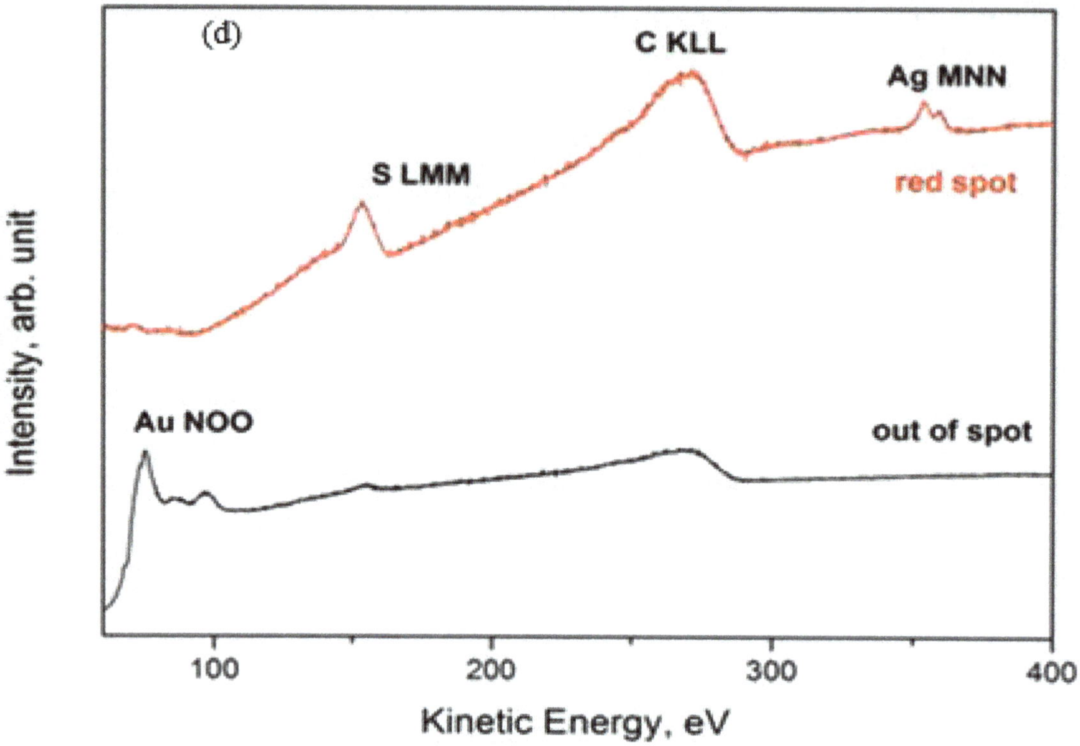

Figure 4. Auger electron spectroscopy (AES) investigation of the defect on gold coin: SEM image—size 1.8×1.8 mm^2 (**a**); Au NOO image (**b**); Ag MNN image (**c**) and AES spectra (**d**) acquired in the red spot (point A) and out of the spot (point B) [17].

The analysis of Ag 3d, Ag LMM, and S 2p spectra gave some indications on their chemical state. As it can be seen in Figure 4, the Ag 3d spectrum was characterized by the typical doublet of the spin-orbit splitting of the core level 3d (Ag $3d_{5/2}$–Ag $3d_{3/2}$), separated by 6.0 eV. The main Ag $3d_{5/2}$ peak was positioned at BE = 368.0 eV. However, it is well known that the Ag 3d signal is one of the few cases where the chemical shift is almost absent, i.e., it is impossible to identify the chemical state of Ag only from photoemission spectra. In these cases, it is necessary to calculate the modified Auger parameter α' by using a very simple formula: α' = BE (Ag $3d_{5/2}$) + KE (Ag LMM) [18]. The value of the Auger parameter can indicate the chemical state (metal, oxide, etc.) of the investigated element. In this case, it was α' = 725.2 – 725.3 eV, which is the typical value for Ag^+ in the silver sulfides, specifically in Ag_2S [5]. The analysis of the S 2p signal confirmed the presence of sulfides, since the S $2p_{3/2}$ peak was positioned at BE = 161.6–161.9 eV [5].

It is interesting to note that the XPS quantitative analysis identified four different scenarios, depending on the color of the spot, which are summarized in Figure 5: (1) grey stains—with Ag, O and S; (2) dark spots—with Ag, O, and S, but also with Au and Cu; (3) red spots—like the grey spots, but with different atomic concentration of the elements; and (4) clean surface—with Au, Cu, and a small amount of O.

Then, the different chemical composition of the stains was investigated by XPS depth profiling, which revealed the different thicknesses of the stains: from 5 to 6 nm for red ones to about 300 nm for dark blackish colored ones. Therefore, the variation of the color was principally related to a different thickness of contamination layer, where the thickness of Ag_2S was always limited to the first 3–5 nm and the second sublayer of metallic Ag continued in depth. These results suggest that a thin, almost transparent, overlayer of sulphide was formed by the interaction of metallic Ag with the sulfur-containing contaminants in air (like H_2S), whereas some bigger silver particles were mechanically embedded into the coin surface during the milling, rolling, or punching of the gold strips.

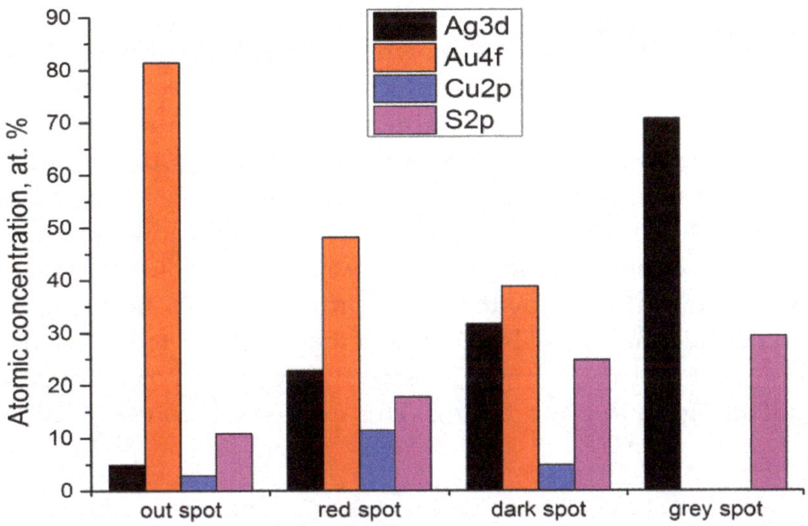

Figure 5. XPS elemental quantification of the different spots on the gold coin.

3.2. Hard Coatings of Nitrides

Hard coatings, based on transition metal nitrides or carbides, are characterized by excellent mechanical properties, suited for steel protection and fabrication of cutting tools. Their performance is continuously improved by the optimization of the fabrication processes, the development of new deposition technologies, and the production of composite materials with enhanced physical and chemical properties. An important contribution to the development of these coatings can be given by the use of surface analysis, which enables us to find the best production conditions and to improve their quality. In this section, the results obtained by the XPS and AES investigations of the TiN-Ti composite and a multilayer CrN–Cr coating are presented. As it can be seen in Figure 6, the deconvolution of the Ti 2p spectrum shows the presence of multiple contributions due to the different chemical states of Ti: the components 3 and 4 located at BE = 458.5 and 456.5 eV were assigned to the chemical states of Ti^{4+} and Ti^{3+} bound to oxygen; the component 1 positioned at BE = 454.1 eV was related to metallic Ti (0); finally, the component 2 positioned at BE = 455.0 eV was assigned to the bonds of Ti–N and Ti–C [19,20].

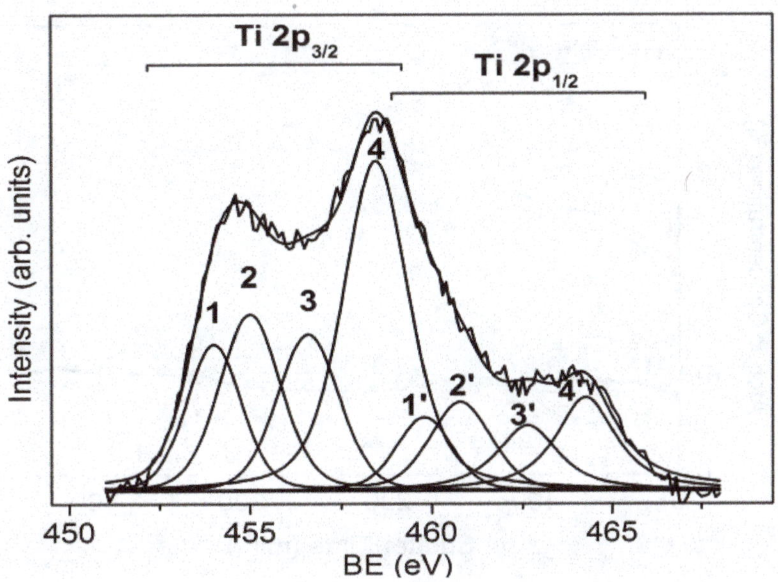

Figure 6. Ti 2p peak fitting of the Ti/TiN composite coating [19].

Naturally, the presence of oxides was caused by the oxidation of metallic Ti in air. After ion sputtering, they were almost removed as it is shown in Figure 7. Of course, the possible influence of the preferential sputtering of oxygen [21,22] to the reduction of oxides cannot be excluded, but in our case this effect was not considered, as this study aimed to determine the composition in the volume of Ti nitride after removal of the native oxides overlayer. By using XPS depth profiling, i.e., alternating cycles of ion sputtering and spectra acquisition, it is possible to investigate the changes of chemical composition until a depth of about 1 μm. From the depth profile shown in Figure 8, it is possible to observe how the content of oxides decreases in depth, whereas the contents of metallic Ti and nitrides increase. This trend was also confirmed by the depth profile of the N 1s signal, which was composed of two peaks positioned at BE = 397.0 and 399.5 eV and was assigned to the bonds of N–Ti and N–O in oxynitride compounds, most probably formed due to environmental contamination. The atomic ratio Ti/N = 3.5 was constant along the depth profiling. This excess of Ti content indicated the formation of composite TiN-Ti.

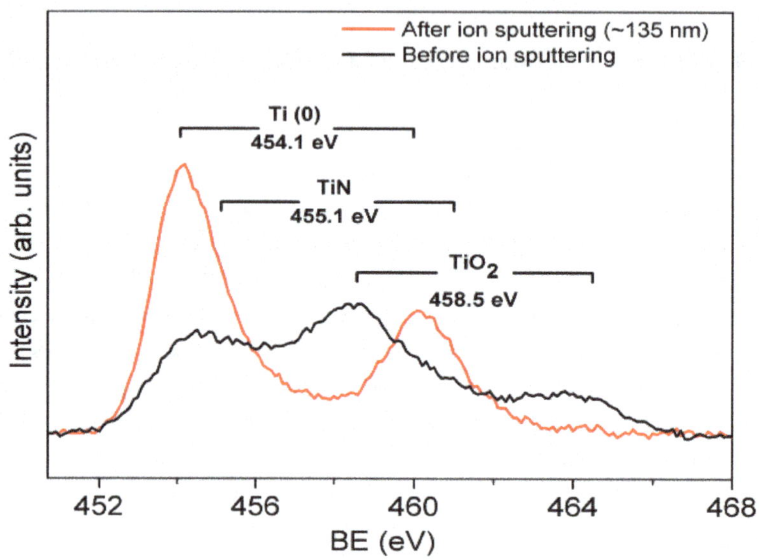

Figure 7. Comparison of the Ti2p signals acquired before and after Ar^+ ion sputtering [19].

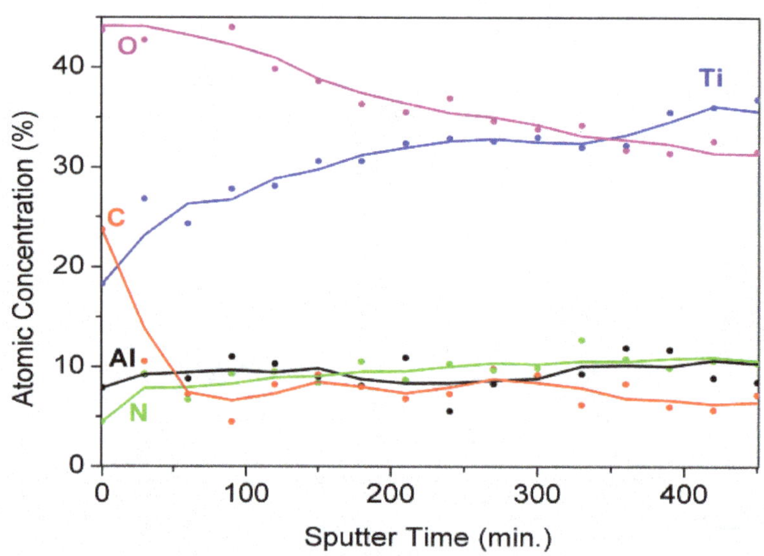

Figure 8. XPS depth profile of the Ti/TiN composite coating. The average sputtering rate was equal to 0.3 nm·min^{-1} [19].

Due to the limited depth of XPS depth profiling, the study of the multilayer coating CrN/Cr/CrN with the thicknesses of 1.5/1.0/1.5μm was carried out only for the top layer of this coating [23]. In this layer, the signal of Cr 2p (Figure 9) was composed of a typical Cr $2p_{3/2}$–$2p_{1/2}$ doublet, which was positioned at BE = 574.2 eV, and a large peak due to the contribution of multiplet splitting, centered at BE = 576.0 eV. The deconvolution of the N 1s spectrum evidenced the presence of two chemical species: chromium nitride at BE = 397.1 eV and a component of oxynitrides at BE = 398.6 eV, probably due to the presence of a low amount of oxygen in the deposition chamber. The obtained BE values of N 1s and Cr $2p_{3/2}$ (single component) indicated the formation of CrN, excluding the phase of Cr_2N characterized by a noticeably higher value of BE [5]. This supposition was confirmed also by the determined atomic ratio of Cr/N nearly at 1.0. The XPS depth profile, depicted in Figure 10, showed that after removal of the surface contamination, the composition of CrN coating remained almost constant. Since the total thickness of the coating (~4 μm) was too high for XPS depth profiling until the substrate, it was stopped after the removal of ~100 nm of CrN, and the cross section of the coating was further investigated. Due to the limited lateral resolution of XPS, the interfaces of CrN/Cr, Cr/CrN, and CrN/substrate were investigated by the AES/SAM technique.

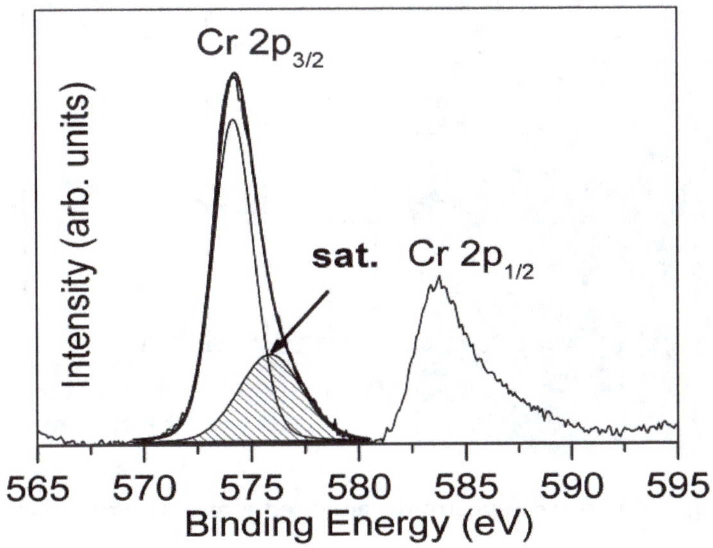

Figure 9. Peak fitting of Cr 2p spectrum acquired on the top of the multilayer coating [23].

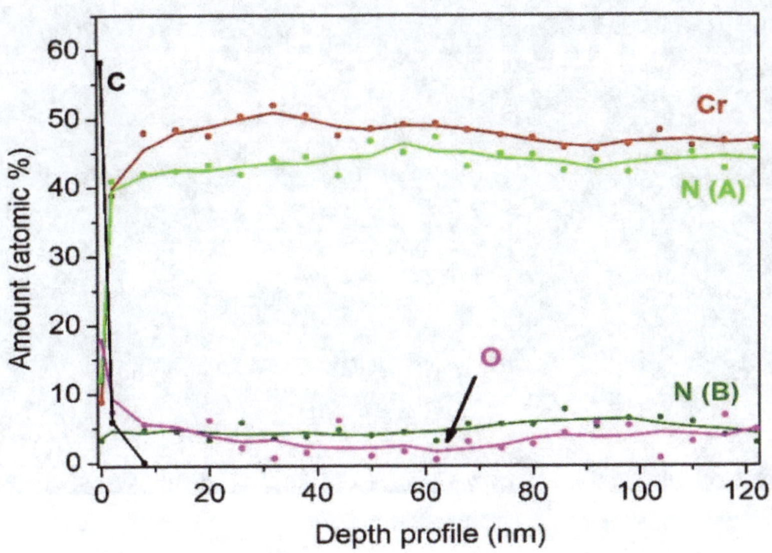

Figure 10. XPS depth profile of the first CrN layer in composite coating. The average sputtering rate was equal to 0.3 nm·min^{-1} [23].

Figure 11 shows the SEM image and the multipoint AES analyses carried out on the cross section of the sample. The AES spectra were acquired on different points, moving from the substrate (region 1) to the top of the coating (region 4). The substrate was characterized by the presence of Fe LMM peaks (KE = 594.0, 652.0 and 705.7 eV) and the low-intensity KLL peaks of C and O (see Figure 11b). In Regions 2 and 4, the peaks of Cr $L_3M_{23}M_{45}$ (KE = 530.6 eV) and N KLL (KE = 385.4 eV) were registered, whereas in Region 3, only a peak of Cr $L_3M_{23}M_{45}$ was present. In addition, the chemical maps were acquired by SAM, where the investigated area of the sample was represented by pixels of the peak-minus-background intensity of the selected Auger peak. The SAM images collected by using the peak-minus-background of the Cr $L_3M_{23}M_{45}$ and N KLL peaks are shown in Figure 12. The black points indicate the area without signal, whereas the lighter grayscale points indicate the area where the signals were detected. As it can be noticed, the layers are well-defined, and the interface is rather neat, suggesting the absence of diffusion phenomena during the deposition process. The coating thickness, estimated from the SEM/SAM images, was about 4.0 μm.

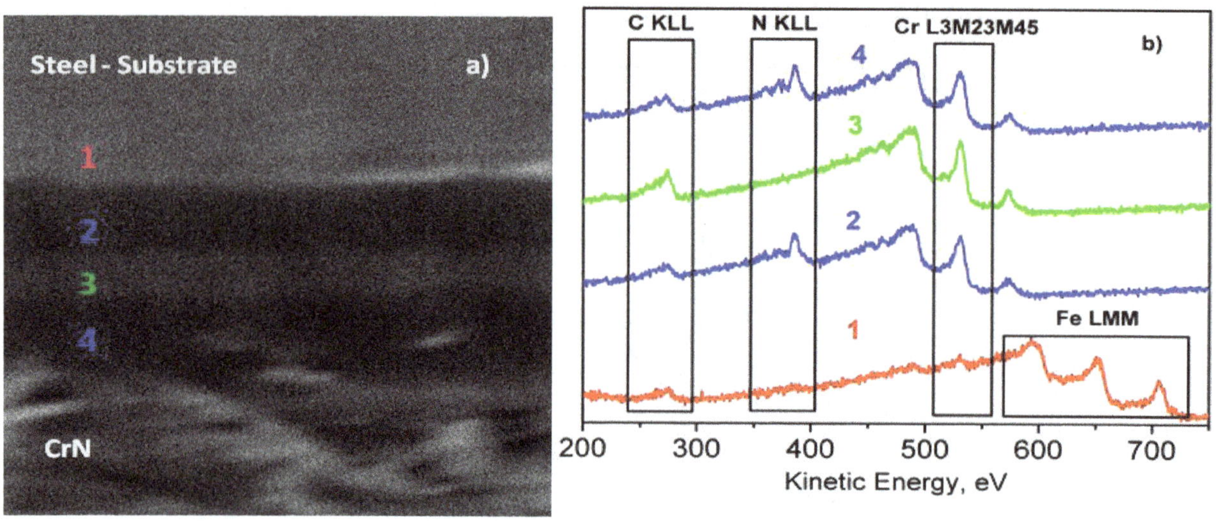

Figure 11. SEM image (**a**) and AES spectra (**b**) acquired along the cross section of the multilayer coating [23].

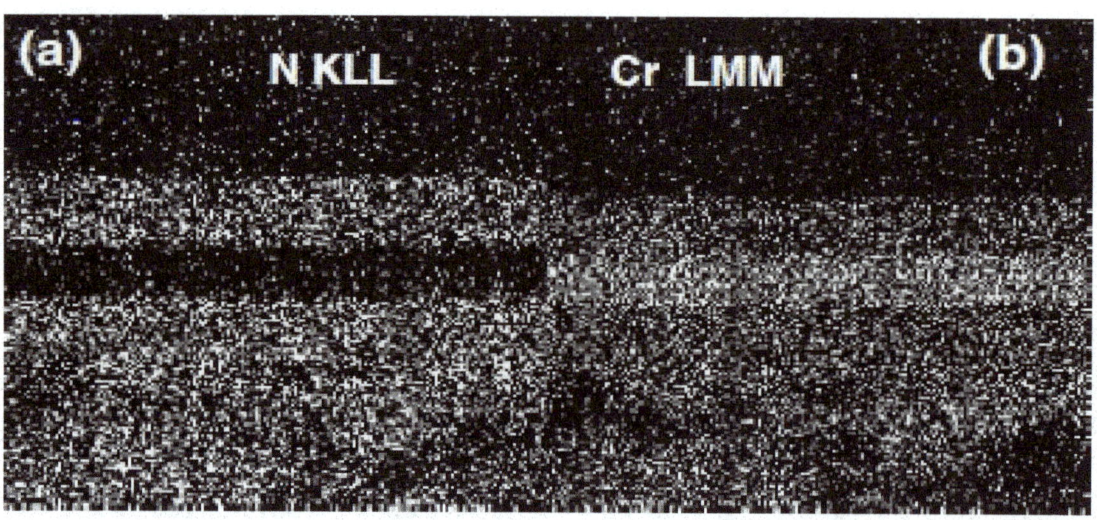

Figure 12. N (**a**) and Cr (**b**) chemical maps of the cross section of multilayer coating [23].

3.3. Microchemical Composition of Ni-Based Superalloys

Superalloys are a class of materials that find numerous applications in the metallurgical field, in particular when a high strength, superior oxidation, and corrosion resistance at temperatures above 700 °C are aquired. Many of superalloys properties are determined by their microstructure, and therefore it is quite important to predict the microstructural evolution during long-time operation, especially the coarsening and morphological changes of the γ' phase that take place at the operating temperature of 800–900 °C. These superalloys are composed of cuboidal γ' particles with submicrometric dimensions, embedded in the γ matrix. The chemical composition of two phases could be different, but most of the previous experimental studies have been dedicated to the morphology and microstructure of superalloys, e.g., [24,25] and the references therein. Practically, the data on the chemical composition of the two phases in various superalloys are absent in the literature. However, the coarsening of γ' particles strongly depends on the difference of chemical composition between a disordered matrix and cuboidal particles. Since this change must occur at the microscale, the surface investigations of the microchemical structure of a biphasic ($\gamma + \gamma'$) Ni-based CM186 superalloy were performed at a high lateral resolution by using the laboratory of scanning photoemission microscopy (SPEM) at the Elettra synchrotron (Trieste, Italy). This technique allows us to directly acquire the surface chemical maps of constituent elements and to determine the variation of their atomic concentrations, eventually induced by the creep tests. In order to prepare for SPEM investigations, the XPS spectra were collected and processed by using a standard XPS apparatus [26,27].

The spectral region, containing all 4f photoemission peaks of constituent elements together with the overlapping peaks of W 5p and Re 5p, is shown in Figure 13.

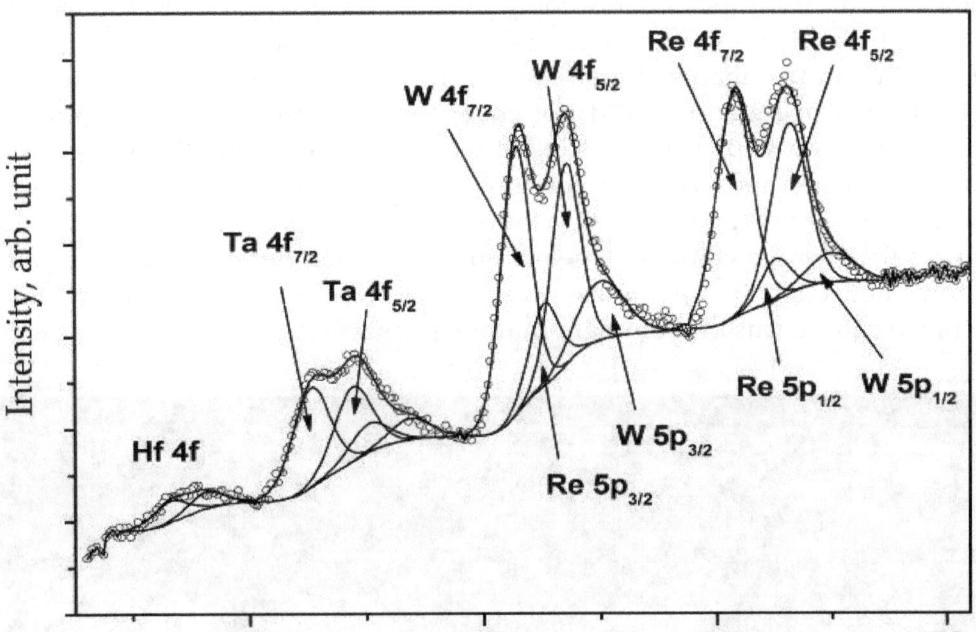

Figure 13. XPS spectrum of the 4f region acquired for the sample of CM186 superalloy [26].

The peak fitting analysis revealed that Re $4f_{7/2}$ and W $4f_{7/2}$ peaks were located at BE = 40.8 and 31.4 eV, corresponding to their metallic states, whereas the Ta $4f_{7/2}$ peak was characterized by two components at 22.6 eV and 25.1 eV, assigned to metallic and oxidized species [26], respectively. Finally, the peak of Hf $4f_{7/2}$ at BE = 16.5 eV was assigned to oxidized species [26].

The chemical maps were recorded in different zones of the samples before and after the creep test, shedding light on the compositional differences between γ and γ' phases. After the acquisition, each map was numerically processed in order to remove the contribution of surface morphology from the photoemission signals. It is worth noting that in SPEM the chemical images can be acquired without any chemical etching of the samples, which is the contrary of the standard microscopies (SEM, AFM, etc.) used for superalloys. Figure 14 shows some examples of obtained chemical maps. The Re 4f, W 4f, and Ta 4f images were acquired from the interdendritic zone on the as-received sample.

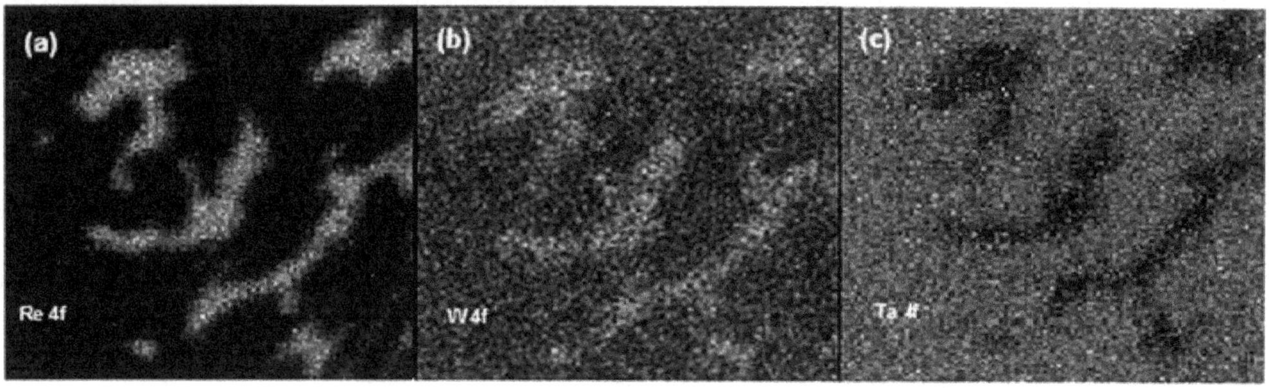

Figure 14. Re 4f (**a**), W 4f (**b**), and Ta 4f (**c**) chemical maps (6.4 × 6.4 μm^2) acquired on the interdendritic zone of the as-received sample [26].

The chemical maps of Re and Ta were complementary, namely, the bright zones in the Re map corresponds to the black zones in that of Ta and vice versa, whereas the tungsten was distributed homogeneously through the analyzed area, even if its content was slightly higher in the γ phase. The lateral distribution of Re and Ta did not change in the crept sample (Figure 15), since they were concentrated in γ and γ' phases, respectively. In comparison with the as-received sample, the distribution of W after creep appeared more uniform. The relative distribution of the main constituent elements between the γ and γ' phases in the as-received and crept samples is displayed in Figure 16. Each data point is the average value of 5 measurements carried out on different points of the same phase. As it can be noticed, both the phases were characterized by the same amount of Ni, while the concentration of Co and Re was predominant in the γ phase.

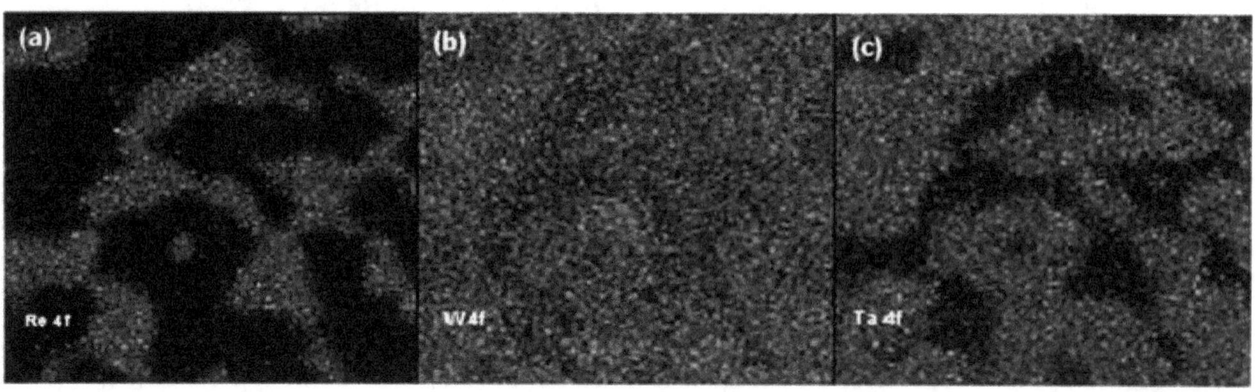

Figure 15. Re 4f (**a**), W 4f (**b**), and Ta 4f (**c**) chemical maps (5.1 × 5.1 μm^2) of the sample after the creep test [26].

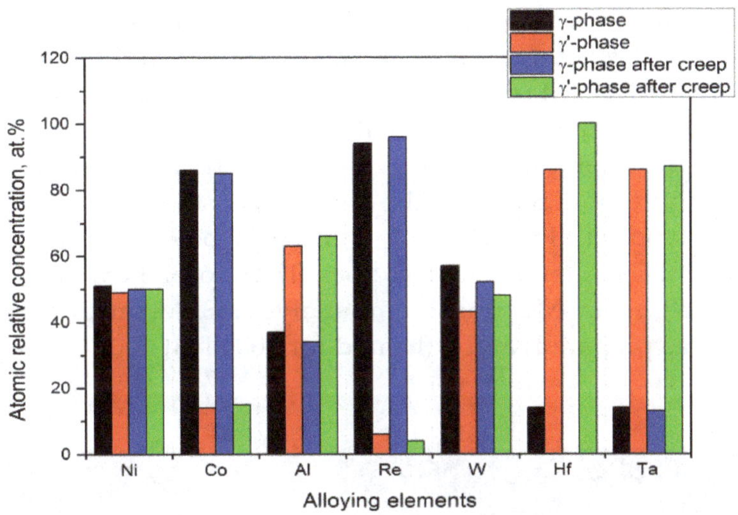

Figure 16. Elemental distribution between the γ and γ' phases before and after the creep test.

After the creep test, their distribution remained almost the same. On the contrary, the amount of Al and Ta was predominant in the γ' phase, remaining unchanged after the creep test. Significant differences were found for W and Hf, where the creep test induced a migration of these elements from the γ phase to γ' phase. The obtained results evidenced that this diffusion process is responsible for the weakening of the disordered matrix during the creep.

3.4. *Diffusion Phenomena in the Ti_6Al_4V/SiCf Composite*

There are only a few analytical techniques capable of investigating the diffusion mechanism of the elements in a solid-state sample. Among them, the surface analysis techniques represent the most powerful tool of the investigation, especially in the proximity of the interface between different materials. In this section, we illustrate the multitechnique approach applied for the investigation of a composite material consisting of a Ti_6Al_4V matrix and SiC fibers [28–32].

To avoid the formation of brittle compounds like Ti_5Si_3 at the interface matrix/fiber, each fiber was coated with a 3 μm thick graphite layer. However, at the high temperatures during the fabrication process and in-service life, some elemental diffusion could be induced, reducing the mechanical performance of the composite. Figure 17 shows the elemental distribution on the cross section of the sample.

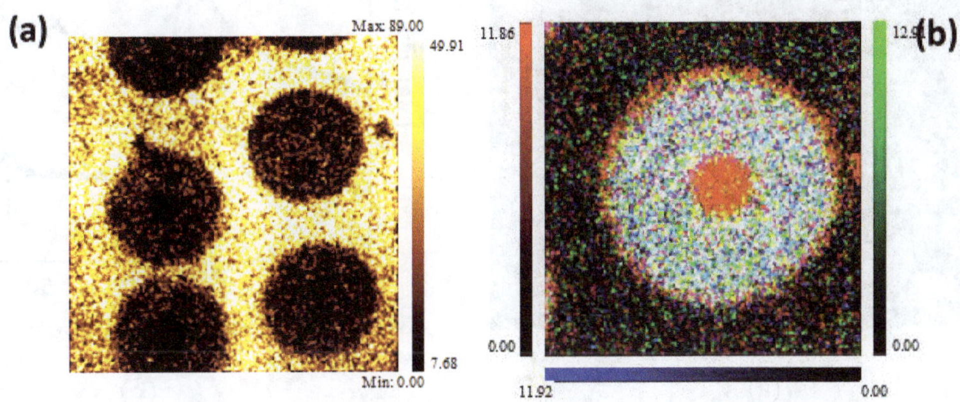

Figure 17. XPS images: (**a**) Ti 2p—400 × 400 μm^2; (**b**) Si 2p (blue), C 1s—190 × 190 μm^2, carbide (green) and graphite (red) [29].

The XPS chemical maps were acquired by collecting the intensity of the signals positioned at BE = 458.8 eV (Ti $2p_{3/2}$), BE = 529.0 eV (O 1s), BE = 99.9 eV (Si 2p), and the intensity of C 1s, where the contributions of graphite (BE = 284.6 eV) and carbide (BE = 283.0 eV) were separated. As it can be seen, the fibers were embedded in the Ti_6Al_4V matrix, which the surface contained oxidized Ti species due to the reaction with atmospheric oxygen. The layer of titanium oxides was promptly removed after a brief time of ion sputtering, reducing the Ti chemical state to metallic one. Unfortunately, the lateral resolution of the standard XPS imaging (>3 μm) was too low for us to investigate the diffusion processes that can occur at the interface matrix/fiber. Therefore, the investigation at a higher lateral resolution was performed by an AES multipoint analysis. SEM images and AES line scan spectra acquired for Samples 1 (as prepared) and 2 (heated for 1000 h at 600 °C) are displayed in Figures 18 and 19, respectively.

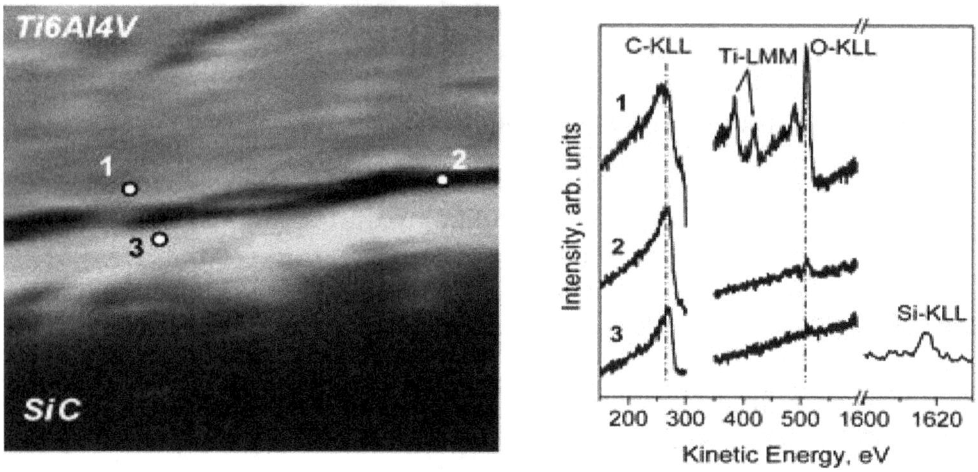

(a) (b)

Figure 18. SEM image 80 × 80 μm^2 (**a**) and AES spectra (**b**) acquired on the cross section of the sample 1 across the carbon layer; analysis points are labelled 1, 2, and 3 [29].

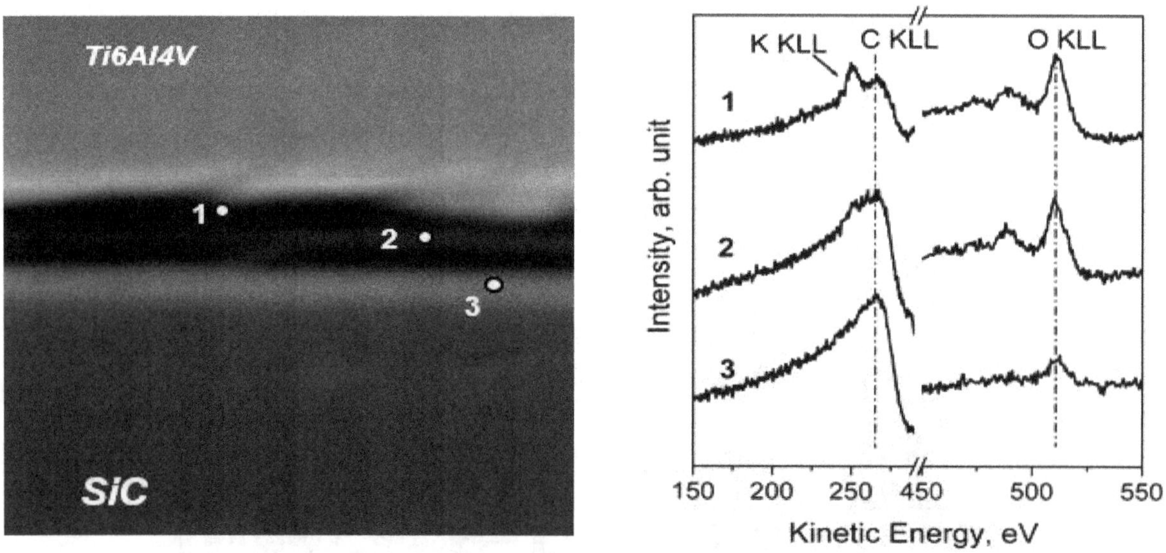

(a) (b)

Figure 19. SEM image 80 × 80 μm^2 (**a**) and AES spectra (**b**) acquired on the cross section of Sample 2 across the carbon layer; analysis points are labelled 1, 2 and 3 [29].

The obtained results revealed that the graphite layer acts as a good protection barrier, avoiding the diffusion of Si in the Ti matrix. However, as evidenced by the SEM analysis, the morphology of the graphite layer became irregular after a thermal treatment at 600 °C for 1000 h, despite its thickness remaining unchanged. This result can be explained by taking into consideration the reaction between carbon and atmospheric oxygen in producing CO. However, the carbon diffusion in the Ti matrix should also be considered. Since the samples have a curvy geometry, the resolution of standard XPS and AES was not sufficient to characterize the chemical species at the interface. To solve this problem, the interface between the graphite and the metallic alloy was investigated by covering the Ti_6Al_4V and Ti 99.99+ foils with a thin layer of graphite. In order to simulate the diffusion of carbon, the samples were heated in vacuum for 8 h at 500 °C (Figure 20). The XPS depth profiles demonstrated the diffusion of elemental carbon in the metallic matrix, forming a thin layer (about 10 nm) of carbides (Figure 21).

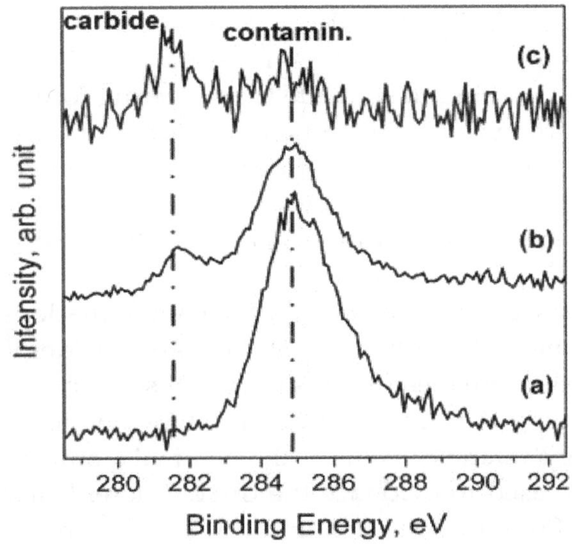

Figure 20. Comparison of C 1s spectra of the (**a**) as-prepared Ti 99.99+ sample; (**b**) Ti 99.99+ sample after 30 min of thermal process at 500 °C; (**c**) Ti 99.99+ sample after 140 min of ion sputtering at 1 keV [29].

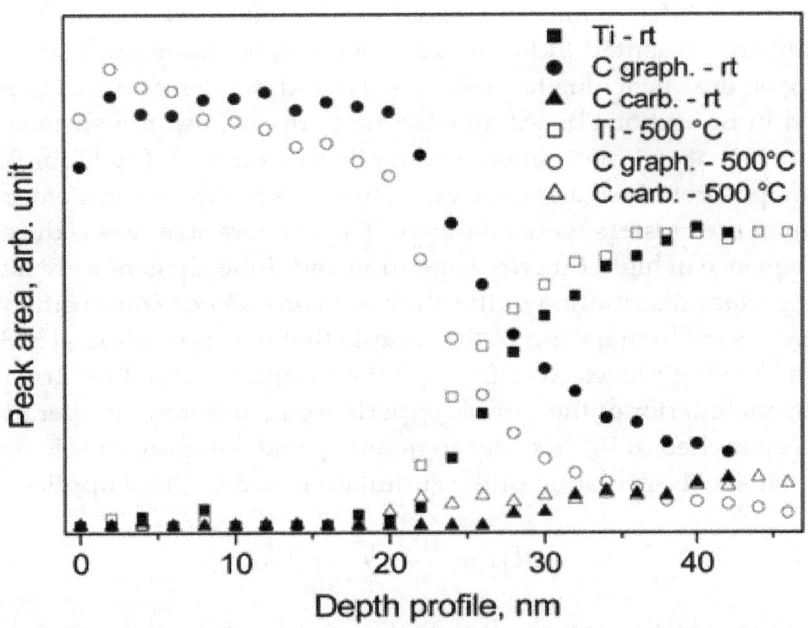

Figure 21. XPS depth profiles of the Ti 99.99+ sample covered with carbon at room temperature (rt) and after heating at 500 °C [29].

From the SPEM analyses [32], carried out on the composite samples, it was concluded that the formation of carbides included not only TiC, but also the interstitial–substitutional (i-s) pairs of C–Al and C–V, present in the α phase of the matrix near the fibers.

3.5. Microchemical Structure of the PbBi Liquid Alloy

The development of a new generation of nuclear reactors has involved many aspects of material science. One of them was the investigation of the microchemical inhomogeneities occurring at high temperature in a liquid Pb-Bi eutectic (LBE) alloy. LBE finds its application in the nuclear reactor as a coolant and spallation source of MYRRHA, an accelerator-driven system. Therefore, it is quite important to investigate any changes of the microchemical structure of LBE that may induce corrosion and embrittlement phenomena in the structural materials. The microstructure of the LBE alloy was evaluated using high-temperature X-ray diffraction (HT-XRD) [33], whereas the microchemical composition was investigated by SPEM [34–36].

In this section, we focus our attention on the surface analysis. Generally, by SPEM, only the solid samples can be analyzed; thus, in order to simulate the clustering formation, we used a rapid cooling (quenching) of liquid alloy starting from different temperatures and assumed that the microchemical composition of the liquid was preserved on the surface of the obtained solid LBE alloy. The selected temperatures for quenching were 126 °C (eutectic temperature) and 200, 300, 400, 518, and 700 °C. The surface chemical maps were acquired by measuring the intensity of the Pb $4f_{7/2}$ and Bi $4f_{7/2}$ peaks, positioned at BE = 137.0 and 156.0 eV, respectively.

Before collecting the maps, the carbon and oxygen contaminations were removed, operating a short cycle of Ar+ ion sputtering. Although the Pb and Bi native oxides were not completely removed, they were neglected because they are meaningless for this discussion. For convenience, the chemical maps were displayed, indicating the Pb/Bi atomic ratio (AR), which is more representative to the elemental distribution. Taking a reference value of the nominal atomic ratio of the eutectic alloy Pb/Bi = 0.8, three pixel colors were used to evidence the three different cases: (i) blue—lack of Pb with AR < 0.6, (ii) red—excess of Pb with AR > 1.0, and (iii) yellow—near a nominal ratio of 0.6 < AR < 1.0. After the acquisition, each image was processed by applying the following procedure of the Igor software: (1) elimination of morphology effects from Pb and Bi maps by using the correction (peak minus background)/background and (2) the superposition of obtained maps and conversion to the maps of atomic ratio AR. Obtained maps of the AR ($100 \times 100\ \mu m^2$ or $50 \times 50\ \mu m^2$) processed by MATLAB software are presented in Figure 22. As it can be noted, a strong inhomogeneity was observed. Depending on the quenching temperature, Pb and Bi atoms formed the clusters of different dimensions enriched in Bi and/or Pb. At an eutectic temperature, the surface of the sample was characterized by the presence of micrometer clusters enriched in Bi (~90% of Bi), immersed in the alloy with eutectic composition. Increasing the quenching temperature, the elemental distribution and atomic concentration in the clusters were changed. The clusters size was reduced to a few microns (1–5 µm) as the consequence of higher thermal agitation and these clusters were alternatively enriched in Pb and Bi, and the surface distribution of the alloy with an eutectic composition 0.6 < AR < 1.0 was also changed. The cross-section mapping of the sample that was quenched at 518 °C (see Figure 22f) demonstrates how the cooling process was freezing the sample surface in a structure quite similar to the liquid alloy, while the interior of the sample experienced a different temperature gradient, giving rise to the big clusters enriched in Bi. In order to quantify and compare the elemental distribution in different samples, a statistical calculation of the cumulative area CA was applied:

$$\mathrm{CA}(\mathrm{AR_i}) = \frac{100}{\mathrm{n}} \sum\nolimits_{\mathrm{n}} \mathrm{p_i}(\mathrm{AR_i}), \quad (3)$$

where n is the total number of selected pixels p_i that have AR_i in the chemical map.

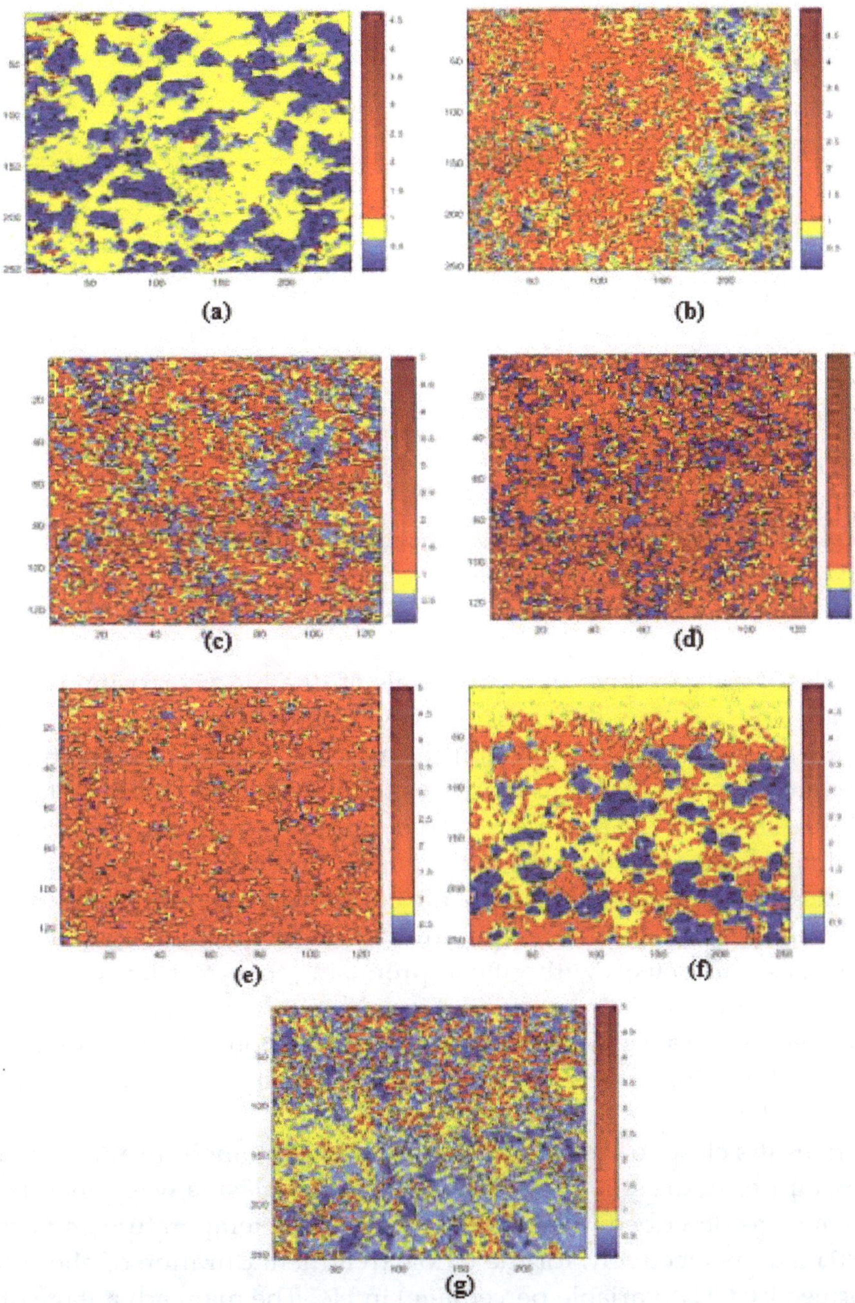

Figure 22. Scanning photoelectron microscopy (SPEM) images of the Pb/Bi atomic ratio (AR) distribution for the samples quenched at **(a)** 126 C, **(b)** 200 C, **(c)** 313 C, **(d)** 401 C, **(e)** 518 C, **(f)** 518 C (cross section of the sample), **(g)** 700 C.

Figure 23 shows the plot of cumulative area (CA) versus the quenching temperature (QT), where the curves were calculated for AR1, AR2, and AR3. At the melting temperature (126 °C), the CA value of AR2 was approximately 2.5%, indicating a very low concentration of Pb, whereas the CA values of AR1 and AR3 were almost similar at 52% and 45%, respectively. Increasing QT, the CA of AR2 was augmenting almost linearly until over 80% at QT = 600 °C, then suddenly falling down below 10% at QT = 700 °C. The comparison of these curves with the phase transition determined by HT-XRD investigations confirmed that the structural modification is also accompanied by the change of the number of clusters enriched in Pb (AR2). As regards the curves of AR1 and AR3, they substantially showed a complementary trend with respect to the AR2 one.

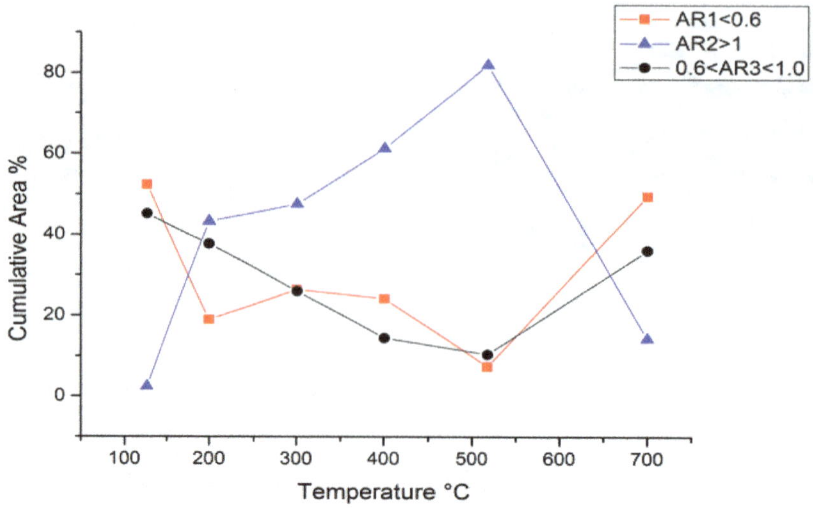

Figure 23. Cumulative area (CA) vs. quenching temperature for the three ranges of atomic ratio (AR): the values of AR1, AR2, and AR3.

3.6. Austenitic Steels

Austenitic stainless steels are known as materials with a high corrosion resistance in different environments. Because of their low hardness, however, they cannot be used in several industrial applications, unless after modifications through thermochemical surface treatments. Carburizing, nitriding, and carbo-nitriding are common examples of the heat treatments that are used to increase the hardness of stainless steels. These processes need to reach a temperature higher than 550 °C, which could cause the local microstructural changes in the austenitic steel phase, such as the precipitation of Cr carbides. Since these precipitates can reduce the corrosion resistance of the steel, it is necessary to adopt some heat treatment at a lower temperature. A good alternative is the kolstering process, which can harden the austenitic steels without compromising their resistance to corrosion. Although kolstering is a good low-temperature treatment, it is unfortunately very long lasting and expensive. It involves a pretreatment of the steel in an HCl atmosphere at about 250 °C to remove the Cr_2O_3 layer from the surface. Then, the stainless steel is treated at 450 °C in a gaseous atmosphere of CO, H_2, and N_2 for a duration about 30 h.

Very promising results close to those of kolstering were obtained through a plasma carburizing process at low temperature. In the study presented in [37], the plasma was generated by microwaves operating up to 200 mbar as described in detail in [38], while the temperature and pressure were set to about 420 °C and 80 mbar, respectively, for the whole treatment duration of about 6 h. The chamber gas mixture was formed by CH_4 (variable percentage) in H_2. The main advantage of this treatment is the reduction of the process time, and consequently this is more convenient also for process costs.

XPS and AES techniques allow for the study of the steel surface before and after these treatments. In particular, these techniques permit us to examine the chemical composition of the superficial hardened layer. In this way, it is possible to identify the best process condition, for example, by changing some parameters of the treatment. In this study, the percentage of CH_4, added to H_2 in the gas mixture was varied from 2% to 10%.

The results of microhardness tests and XRD measurements [1] have established that the sample treated with 2% of CH_4 was the one with the best results in terms of hardness (700 HV) and corrosion resistance, without the presence of any precipitates of Cr carbides. For a better understanding of these results, all the samples were investigated by surface analysis.

An AES line scan over the cross section, shown in Figure 24a, revealed the presence of an additional carbon layer with a thickness of about 2–3 μm (lighter zone) above the hardened layer of 20 μm. As it is possible to see from Figure 24b, the Auger signals of C KLL, O KLL, Cr LMM, and Fe LMM were detected along the whole cross section. In the first point, corresponding to the zone near the surface, the amount of carbon is the highest, and the concentrations of Fe and Cr are low. The ratio of the signals intensity (Fe LMM)/(C KLL) is equal to 0.5. Instead, at the point closest to the bulk, the amount of carbon returns to the nominal value of the alloy, and the ratio Fe/C is equal to 0.9. The intensity ratio of (Fe LMM)/(C KLL) for the entire line scan is shown in Figure 25.

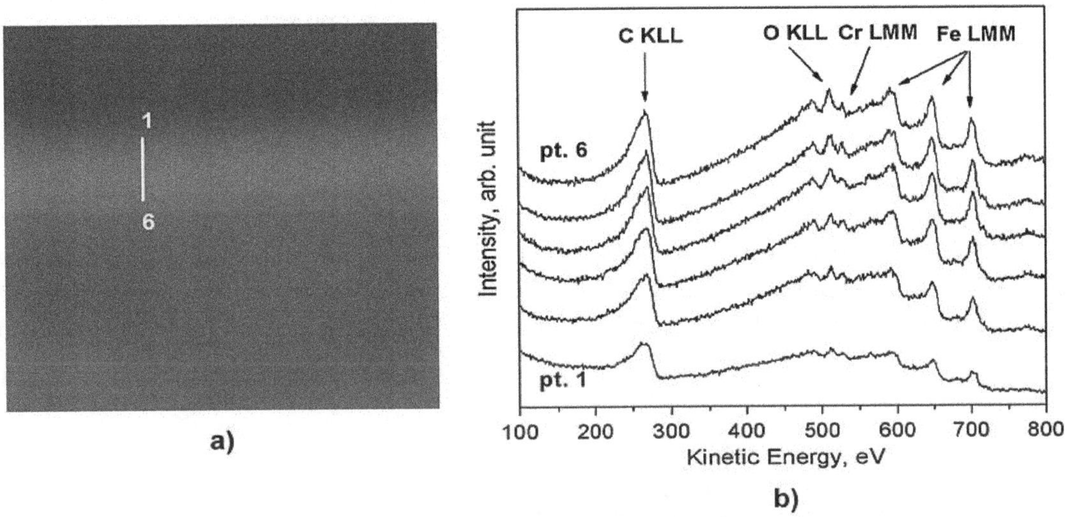

Figure 24. SEM image (24 × 24 μm^2) of the sample treated by plasma at CH_4 2% in H_2 (**a**), and Auger spectra (**b**) recorded in points 1–6 are marked in the SEM image [37].

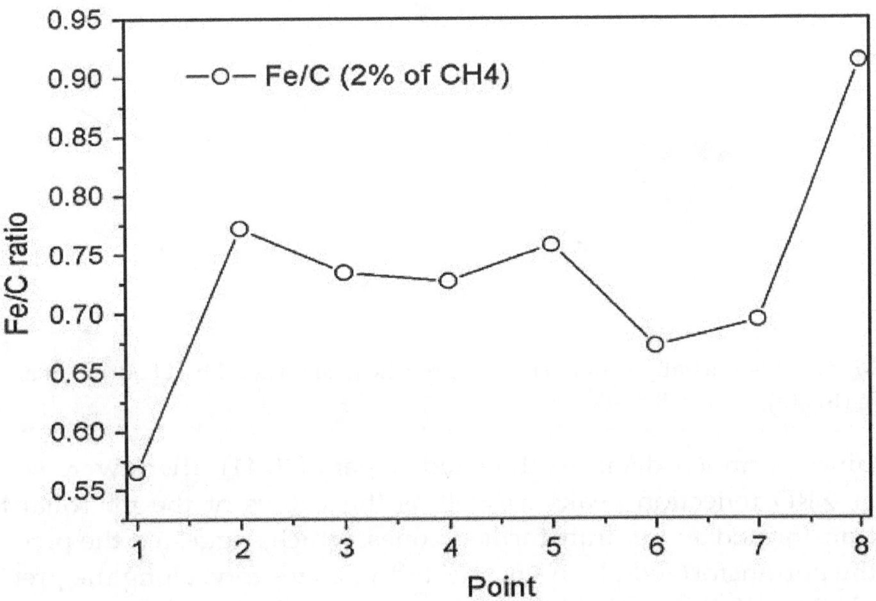

Figure 25. (Fe LMM)/(C KLL) intensity ratio calculated along the line scan. [37].

From the value of the D parameter, which is the distance between the most positive maximum and the most negative minimum of the first derivative of C KLL spectrum [39], it is possible to establish that the samples subjected to carburization with a gas mixture composed of 2% of CH_4 and H_2 have a ultrathin outer layer of graphitic nature (C–C bond with a majority of planar sp^2 hybridization). Because of the presence of this additional hard graphitic layer, which was not present in the other samples treated with higher percentages of CH_4, it is possible to conclude that 2% of CH_4 is the best gas mixture process condition. In this way, the hardened surface of the austenitic steel is comparable to the one obtained with the kolstering treatment.

Another interesting discovery of austenitic steels that has been reported in the papers [40,41], concerns the microstructural modification in the steel with a high content of N (about 0.8 wt.%) induced by heating. Although nitrogen stabilizes the austenitic phase and increases the corrosion resistance, it is important to note that N is soluble only for quantities less than 0.4 wt.% (both in the liquid and solid phase). After exceeding this value, the discontinuous precipitations of chromium nitride are formed in the steel in the range of temperature between 700 and 900 °C.

The transformation that occurs during heat treatments in that temperature range is the following:

$$\gamma_S \rightarrow \gamma + Cr_2N, \tag{4}$$

where γ_S is the N-supersaturated austenitic phase (initial phase of the steel), γ is the austenitic transformed phase which appears as a lamellar structure, and Cr_2N is the chromium nitride precipitates. A SEM image with the corresponding schematic structure of the austenitic steel is shown in Figure 26.

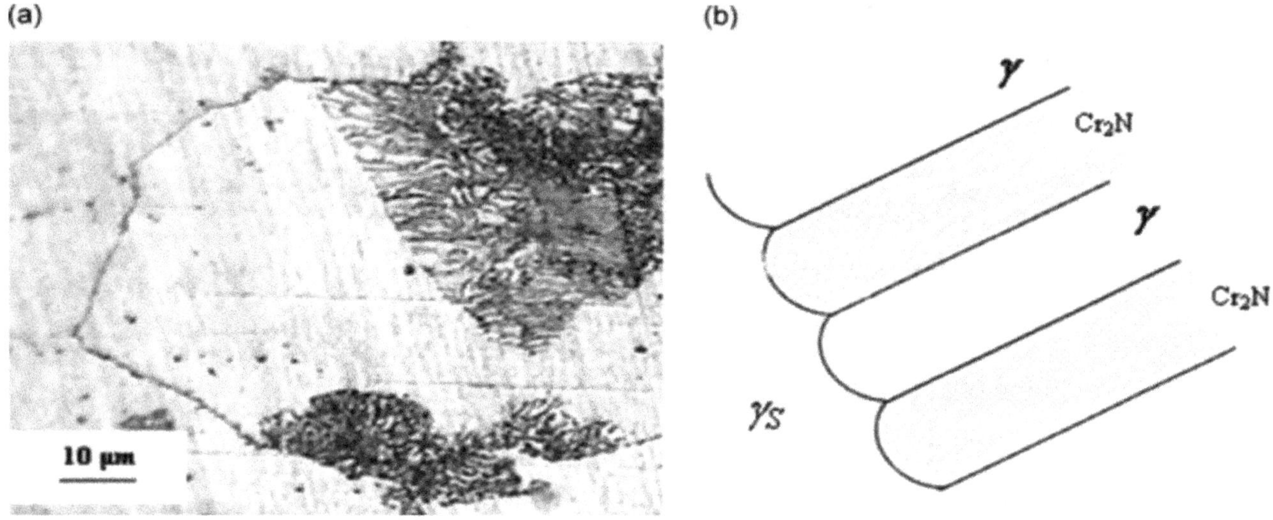

Figure 26. SEM image of partially transformed austenitic grain after 2 h at 850 °C (**a**) and a schematic of the structure (**b**) [40].

As it is explained in much detail in the cited papers [40,41], there were some experimental evidences, such as XRD reflection peaks as well as the values of the microhardness and lattice parameter in the transformed and untransformed zones, which suggested the presence of a net flow of nitrogen from the untransformed N-supersaturated γ_S zones to γ, along the precipitation process. Therefore, XPS and AES techniques could be used to establish final evidence of this phenomenon.

As it is possible to see from Figure 26a, the grains size of γ_S is about 100 μm, while the dimension of the transformed areas is much smaller, at about 10 μm. A traditional XPS apparatus is not adequate for us to study the chemical composition of the transformed zones with sufficient resolution because it can analyze only surface areas between 0.1 and 1 mm. For this reason, it was necessary to use a scanning photoelectron microscopy (SPEM) operating in both imaging and spectroscopy modes. Indeed, this type of analysis can use an X-ray microprobe with a diameter less than 100 nm.

By using the SPEM technique, it was possible to determine the chemical composition and spatial distribution of the elements in the lamellae and interlamellar spaces. In Figure 27, a spatially resolved XPS image of the transformed zone before and after the topographical correction is shown.

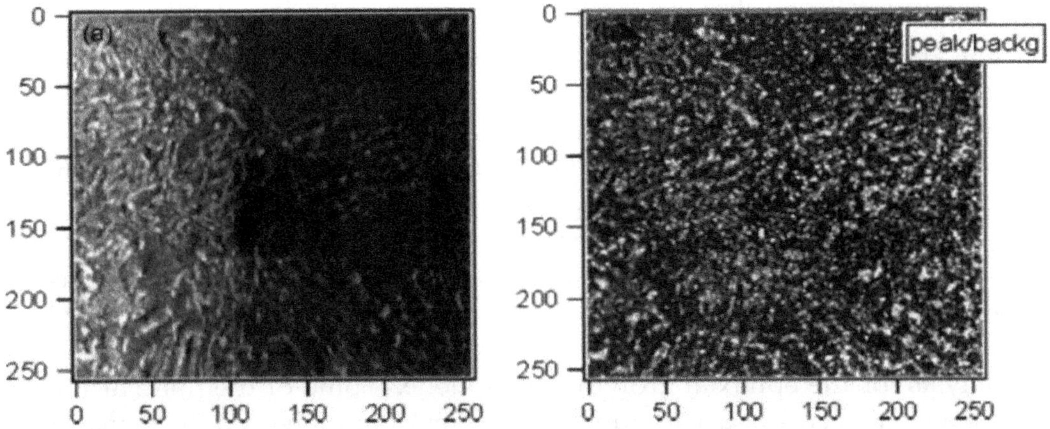

Figure 27. SPEM Cr 3p chemical images (51.2 × 51.2 μm^2) of the sample treated for 23 h at 850 °C: (**a**) as-acquired and (**b**) corrected for the sample topography by calculating the ratio peak/background for Cr 3p signal [40].

The information obtained from these images and from the microscopy was in good agreement with traditional XPS measurements. From the SPEM images of Cr 3p signal, it was found that in the transformed zone, Cr is concentrated in the lamellae, whereas it is uniformly distributed in low concentration in the untransformed region. In an opposite way to Cr distribution, a Fe-enrichment in the untransformed zone and impoverishment in the lamellae were revealed from Fe 3p images. These analyses indicate a migration of Cr, which is mainly accumulated in the Cr_2N precipitates across the interface between γ and γ_s.

Furthermore, from the Auger spectra shown in Figure 28, the Cr/N atomic ratio was calculated. It was found to be equal to 2.9 and 5.9 for the transformed and untransformed zones, respectively. This result confirms the nitrogen enrichment in the transformed zones during the heat treatment. Moreover, another phenomenon was also explained. In fact, from these analyses it was possible to hypothesize that the precipitation of Cr_2N takes place as long as the flow of nitrogen from the untransformed to the transformed area is present. Finally, when γ and γ_s zones have the same concentration of N, the precipitation process is stopped, even if not all the cells of the steel were transformed.

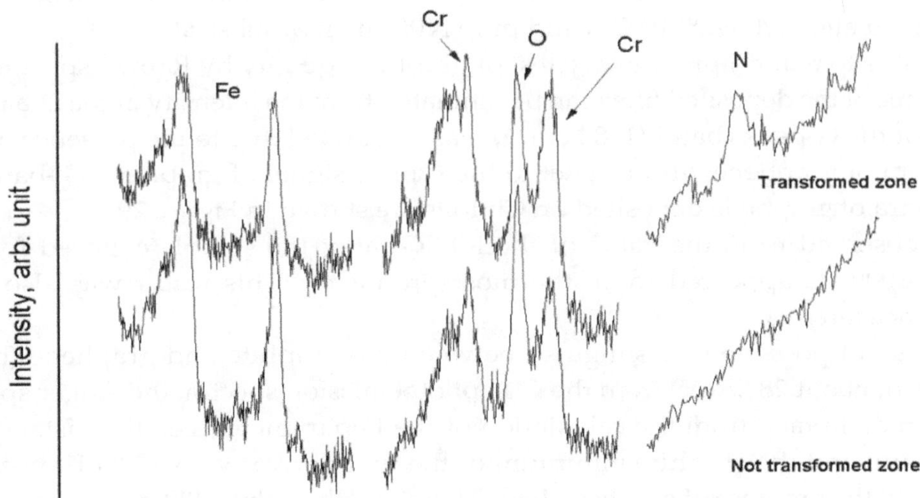

Figure 28. Auger spectra (Fe LMM, Cr LMM, O KLL, and N KLL) of untransformed and transformed zones of the steel sample after 23 h at 850 °C [40].

3.7. Graphene on Polycrystalline Metals

The last few decades of material science will be remembered as the years of the graphene revolution. In fact, although the theoretical predictions can be traced starting from the 19th century, the experimental evidences occurred only in the 2004. After that, throughout the scientific community there was a continuous race to discover the new fields of application in order to exploit the full potential of this 2D material. At the same time, it was essential to develop an industrial method of synthesis that could guarantee a large-scale production of graphene.

Recently, the research for the development of microelectronic devices, transparent conductive films, and in general different type of sensors with graphene has focused on the growth of graphene via the chemical vapor deposition (CVD) on polycrystalline metal substrates. These substrates act as excellent catalysts for the epitaxial growth of graphene. Some of them are also cheap and are easily removable, when it is necessary to transfer the layer of graphene on the device where it has to operate [42,43].

Numerous studies, extensively reviewed in [42–45], have been dedicated to the growth and characterization of graphene on various metals. Among many analytical techniques for the characterization of graphene, the mostly attractive ones are Raman spectroscopy, atomic force, and transmission electron microscopies, XPS and AES. However, in many papers including those on XPS, only the photoemission spectrum of C 1s has been used for graphene characterization, even if it does not allow the main peaks of graphene and graphite to be differentiated [44,46] without the angle-resolved analysis of low intensity σ and π bands accompanying the main peak [45]. A more useful and easier approach is the unequivocal identification of graphene from the analysis of C KVV spectrum combined with the main photoemission peaks of substrate and C 1s [46]. This approach, combined with Raman spectroscopy, allows us to obtain the information on the uniformity of graphene layer over a large area. In the same manner, these analyses permit us to determine the graphene thickness, which can often differ from the monolayer. The parameters, as well as the morphology and the thickness, depend on the type of growth mechanism of graphene. In the case of the CVD technique, two different growth mechanisms can take place: the decomposition of hydrocarbon gas at high temperature or the segregation of C atoms on the metal surface during the cooling phase.

For example, in the study in [47], the graphene was synthesized on the substrates of various polycrystalline metals. The growth was carried out by the CVD technique in a mixture of CH_4-H_2 gas at 1000 °C, with different times of exposure to the gas: 2, 4, and 6 min for the Cu substrate; and only 2 min for the Ni-Cu alloy (20 wt.% of Cu) and pure Ni film on Si substrate.

It was possible to make a preliminary test of graphene quality by Raman spectroscopy. At first, the disorder degree of the deposited films can be estimated from the intensity of the D-band (1350 cm^{-1}). Then, the ratio of the G-peaks band (1582 cm^{-1}) was calculated due to the presence of graphite or a multilayer system of graphene with respect to the typical signal of graphene G′-band (2700 cm^{-1}). The Raman spectra of graphene deposited on Cu foils are shown in Figure 29.

As it was observed from the value of the IG′/IG ratio, the sample exposed for 6 min to the gas mixture at 1000 °C appeared to be the most promising. This result was also confirmed by photoemission measurements.

Because it is not possible to distinguish between the graphite and graphene (both peaks are positioned at BE of about 284.5 eV) from the C 1s photoemission spectra, the Auger spectra of C KLL were also acquired. In fact, from the calculation of the D parameter, i.e., the distance between the absolute maximum and the absolute minimum of the first derivative of C KLL spectrum [39], it is possible to identify the presence of graphene [46]. Therefore, the value of D parameter was determined from the C KLL spectra induced by an X-ray source (XAES) and then it was compared with the same parameter obtained by using excitation with an electron gun (AES).

The typical spectra of C 1s and C KLL regions are shown in Figure 30, whereas all the results of the XPS characterization are summarized in the Tables 2–4.

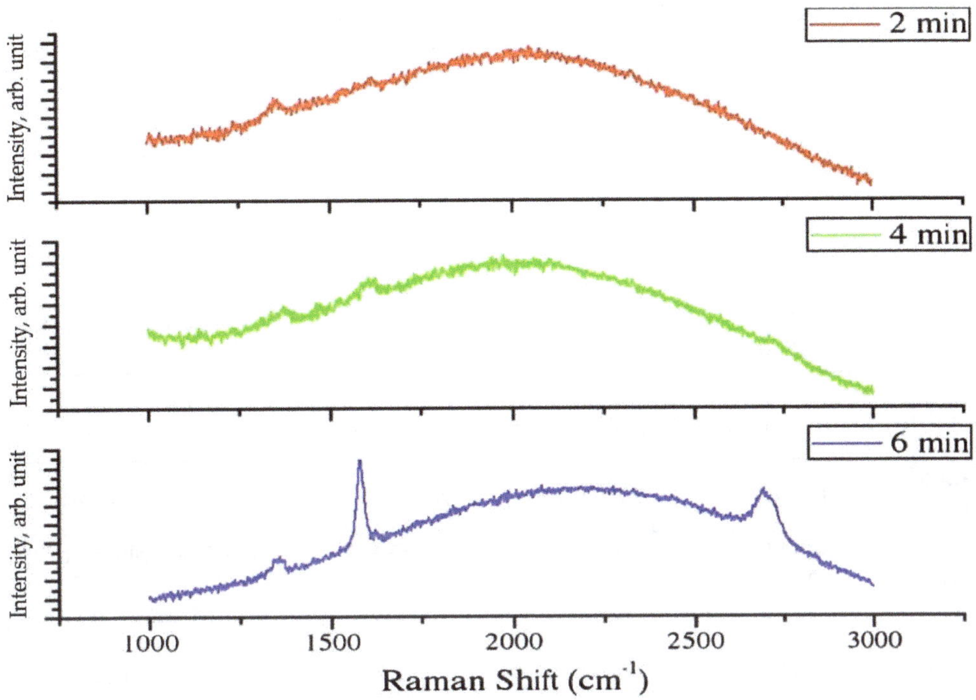

Figure 29. Raman spectra of graphene grown on electroformed Cu foils: exposure time of 2, 4, and 6 min [47].

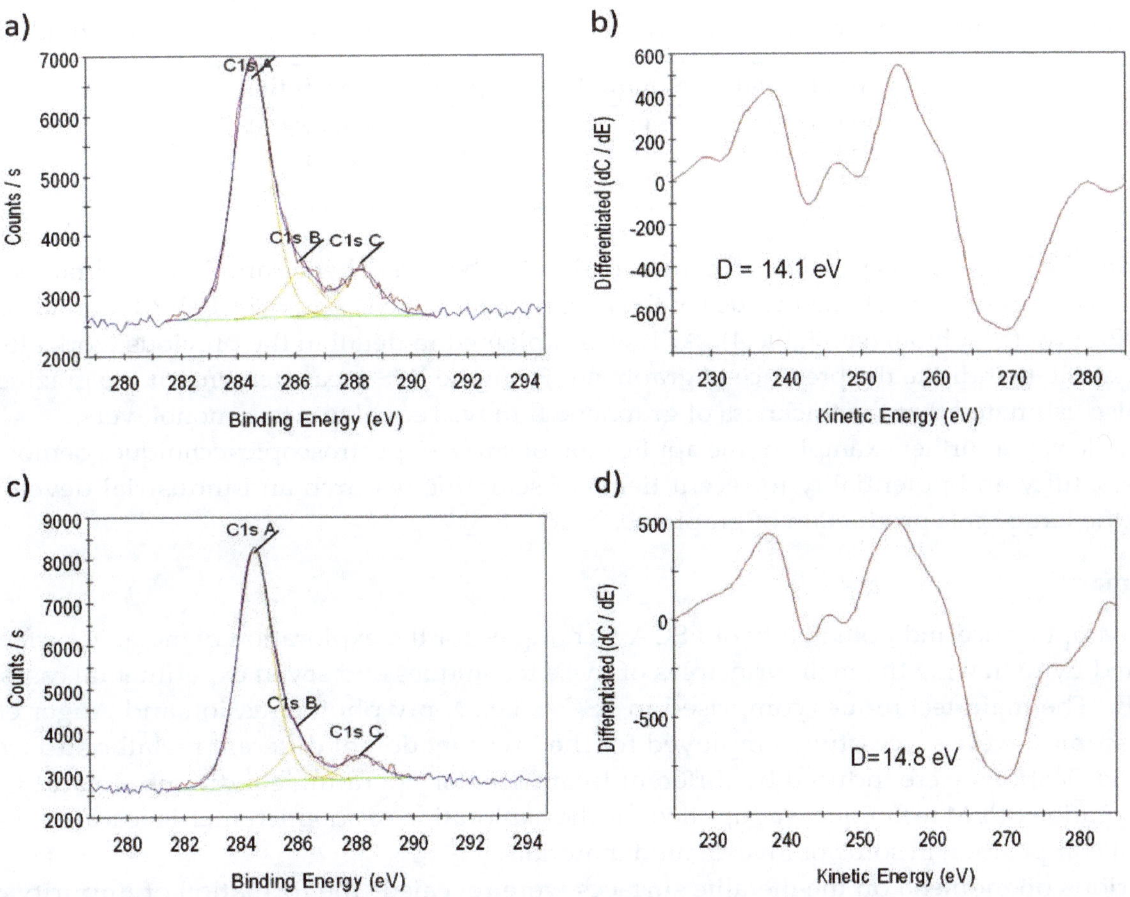

Figure 30. XPS spectra of C 1s for the samples 4 (**a**) and 6 (**c**) grown on Cu at 1000 °C. The corresponding C KLL spectra in the first derivative are shown in (**b**) and (**d**), respectively [47].

Table 2. Summary of XPS results for Sample 4 of graphene on Cu substrate [47].

Sample 4 Peak	BE, eV	FWHM, eV	Atomic %	Bond
C1s A	284.5	1.51	44.1	C–C (graphene or graphite)
C1s B	286.2	1.51	7.4	C–O
C1s C	288.3	1.51	7.7	Carboxyl O–C=O
$Cu2p_{3/2}$ A	932.5	1.71	9.3	Cu(0)
$Cu2p_{3/2}$ B	934.7	1.71	1.6	$Cu(OH)_2$
O1s A	530.7	1.87	6.1	C–O
O1s B	531.8	1.87	20.9	OH groups, C=O
Cl2p	199.0	1.41	2.8	-

Table 3. Summary of XPS results for Sample 6 of graphene on Cu substrate [47].

Sample 6 Peak	BE, eV	FWHM, eV	Atomic %	Bond
C1s A	284.5	1.42	51.8	C–C (graphene or graphite)
C1s B	286.0	1.42	7.2	C–O
C1s C	288.2	1.42	5.2	Carboxyl O–C=O
$Cu2p_{3/2}$ A	932.4	1.51	12.8	Cu(0)
$Cu2p_{3/2}$ B	934.6	1.51	1.2	$Cu(OH)_2$
O1s A	530.5	1.61	8.3	C–O
O1s B	531.8	1.61	12.3	OH groups, C=O
Cl2p	199.0	1.50	1.2	-

Table 4. D parameter (eV) determined by XPS and XAES for graphene samples on Cu substrate [47].

Experimental	Sample 4	Sample 6	Description
XPS at 90°	14.1	1.42	Diamond-like
XPS at 45°	13.3	14.2	Diamond-like
XAES (e^- beam)	22.1	21.5	Graphitic

From the Table 3, it is possible to conclude that the best graphene sample was obtained by the deposition of 6 min: The obtained values of the D parameter were D_{XAES} = 14.1 eV (diamond-like) and D_{AES} = 22.1 eV (graphitic) (see Table 4). As it was explained in detail in the previous work [46], these values definitely indicate the presence of graphene. From the XPS measurements at the grazing angle, it was also estimated that the thickness of graphene film was equal to a few monolayers.

In this way, a further example of the application of surface spectroscopic techniques demonstrated their versatility and potentiality in recent fields of scientific research and industrial development, such as the large-scale production of graphene.

4. Summary

The importance and potentiality of ESCA techniques for the exploration of metallic surfaces was illustrated by reviewing the main principles of these techniques and seven experimental cases of our research. The main techniques comprised in ESCA, i.e., X-ray photoemission and Auger electron spectroscopies, were successfully employed for the investigation of different metallic surfaces, and their modifications were induced by different treatments or operating conditions. In addition, the high resolution SPEM technique was applied for the exploration of submicrometric features of surface chemical composition in some of investigated materials.

Various phenomena on the metallic surfaces were revealed: the formation of impurity defects on collection coins, the microchemical composition and corrosion of stainless steel coated by Cr and Ti nitrides, modifications of microchemical composition in biphasic Ni-based superalloys, carbon diffusion at high temperature in the interface of Ti6Al4V/SiCf composite, microchemical inhomogeneity of liquid PbBi alloy, surface modification of austenitic steels by plasma carburizing, and nitrogen migration at high temperature, with an influence of polycrystalline metal substrates (Cu, Ni, and NiCu alloy) on the growth of graphene. One more recent example of an advantageous ESCA application for the study of Cr segregation in martensitic stainless steel is reported in the present issue of this journal [48].

Author Contributions: Conceptualization, E.B., S.K., and A.M.; investigation, E.B., S.K., and A.M.; writing and editing, E.B., S.K., and A.M.; All authors have read and agreed to the published version of the manuscript.

Acknowledgments: The authors are grateful to all the coauthors of reviewed papers for their contributions. Our special thanks are dedicated to Roberto Montanari (Tor Vergata University of Rome), who initiated the research in a major part of the reviewed cases, also to Luca Gregoratti and his team for hosting us at the ESCA microscopy beamline in the Elettra synchrotron and providing extensive experimental support.

References

1. Hagström, S.; Nordling, C.; Siegbahn, K. Electron spectroscopy for chemical analyses. *Phys. Lett.* **1964**, *9*, 235–236. [CrossRef]
2. Fahlman, A.; Hamrin, K.; Nordberg, R.; Nordling, C.; Siegbahn, K. Revision of Electron Binding Energies in Light Elements. *Phys. Rev. Lett.* **1965**, *14*, 127–129. [CrossRef]
3. Carlson, T.A.; Mcguire, G.E. Study of the x-ray photoelectron spectrum of tungsten—Tungsten oxide as a function of thickness of the surface oxide layer. *J. Electron Spectrosc. Relal. Phenom.* **1972**, *1*, 161–168. [CrossRef]
4. Schön, G.; Lundin, S.T. ESCA studies of oxygen adsorption on nickel. *J. Electron Spectrosc. Rel.Phenom.* **1972**, *1*, 105–109. [CrossRef]
5. Moulder, J.F.; Stickle, W.F.; Sobol, P.E.; Bomben, K.D. *Handbook of X-ray Photoelectron Spectroscopy*; Phys. Electronics Inc.: Eden Prairie, MN, USA, 1992.
6. Childs, K.D.; Carlson, B.A.; LaVanier, L.A.; Moulder, J.F.; Paul, D.F.; Stickle, W.F.; Watson, D.G. *Handbook of Auger Electron Spectroscopy*; Phys. Electronics Inc.: Eden Prairie, MN, USA, 1995.
7. Feuerbacher, B.; Fitton, B.; Willis, R.F. *Photoemission and the Electronic Properties of Surfaces*; J. Wiley & Sons: Chichester, UK, 1978.
8. Briggs, D.; Seah, M.P. *Practical Surface Analysis Auger and X-ray Photoelectron Spectroscopy*, 2nd ed.; J. Wiley & Sons: Chichester, UK, 1990.
9. De Crescenzi, M.; Piancastelli, M.N. *Electron Scattering and Rel. Spectroscopies*; World Scientific Publishing: Singapore, 1996.
10. Lindgren, I. Chemical shifts in X-ray and photo-electron spectroscopy: A historical review. *J. Electron Spectrosc. Rel. Phenom.* **2004**, *137*, 59–71. [CrossRef]
11. X-ray Photoelectron Spectroscopy. Available online: https://mmrc.caltech.edu/SS_XPS/XPS_PPT/XPS_Slides.pdf (accessed on 17 November 2020).
12. Walton, J. Quantitative surface chemical microscopy by X-ray photoelectron spectroscopy. *Spectrosc. Eur.* **2007**, *19/6*, 7–10.
13. McIntyre, N.S.; Davidson, R.D.; Kim, G.; Francis, J.T. New frontiers in X-ray photoelectron spectroscopy. *Vacuum* **2003**, *69*, 63–71. [CrossRef]
14. Gunther, S.; Kaulich, B.; Gregoratti, L.; Kiskinova, M. Photoelectron microscopy and applications in surface and materials science. *Prog. Surf. Sci.* **2002**, *70*, 187–260. [CrossRef]
15. Potts, A.W.; Morrison, G.R.; Gregoratti, L.; Barinov, A.; Kaulich, B.; Kiskinova, M. The exploitation of multichannel detection in scanning photoemission microscopy. *Surf. Rev. Lett.* **2002**, *9*, 705–708. [CrossRef]
16. Gusmano, G.; Montanari, R.; Kaciulis, S.; Montesperelli, G.; Denk, R. Gold corrosion: Red stains on a gold Austrian Ducat. *Appl. Phys. A* **2004**, *79*, 205–211. [CrossRef]
17. Gusmano, G.; Montanari, R.; Kaciulis, S.; Mezzi, A.; Montesperelli, G.; Rupprecht, L. Surface defects on collection coins of precious metals. *Surf. Interface Anal.* **2004**, *36*, 921–924. [CrossRef]
18. Moretti, G. Auger parameter and Wagner plot in the characterization of chemical states by X-ray photoelectron spectroscopy: A review. *J. Electron Spectrosc. Rel. Phenom.* **1998**, *95*, 95–144. [CrossRef]

19. Ingo, G.M.; Kaciulis, S.; Mezzi, A.; Valente, T.; Gusmano, G. Surface characterization of titanium nitride composite coatings fabricated by reactive plasma spraying. *Surf. Interface Anal.* **2004**, *36*, 1147–1150. [CrossRef]
20. Ingo, G.M.; Kaciulis, S.; Mezzi, A.; Valente, T.; Casadei, F.; Gusmano, G. Characterization of composite titanium nitride coatings prepared by reactive plasma spraying. *Electrochim. Acta.* **2005**, *50*, 4531–4537. [CrossRef]
21. Caballero, A.; Leinen, D.; Espinos, J.P.; Fernandez, A.; Gonzalez-Elipe, A.R. Use of XAS and chemical probes to study the structural damage induced in oxide ceramics by bombardment with low-energy ions. *Surf. Interface Anal.* **1994**, *21*, 418–424. [CrossRef]
22. Hashimoto, S.; Hirokawa, K.; Fukuda, Y.; Suzuki, K.; Suzuki, T.; Usuki, N.; Gennai, N.; Yoshida, S.; Koda, M.; Sezaki, H.; et al. Correction of peak shift and classification of change of X-ray photoelectron spectra of oxides as a result of ion sputtering. *Surf. Interface Anal.* **1992**, *18*, 799–806. [CrossRef]
23. Kaciulis, S.; Mezzi, A.; Montesperelli, G.; Lamastra, F.; Rapone, M.; Casadei, F.; Valente, T.; Gusmano, G. Multi-technique study of corrosion resistant CrN/Cr/CrN and CrN: C coatings. *Surf. Coat. Technol.* **2006**, *201*, 313–319. [CrossRef]
24. Monjati, H.; Jahazi, M.; Bahrami, R.; Yue, S. The influence of heat treatment conditions on characteristics in Udimet®720. *Mat. Sci. Eng. A.* **2004**, *373*, 286–293. [CrossRef]
25. Ges, A.M.; Fornaro, O.; Palacio, H.A. Coarsening behaviour of a Ni-base superalloy under different heat treatment conditions. *Mat. Sci. Eng. A* **2007**, *458*, 96–100. [CrossRef]
26. Kaciulis, S.; Mezzi, A.; Amati, M.; Montanari, R.; Angella, G.; Maldini, M. Relation between the microstructure and microchemistry in Ni-based superalloy. *Surf. Interface Anal.* **2012**, *44*, 982–985. [CrossRef]
27. Amati, M.; Gregoratti, L.; Balijepalli, S.K.; Kaciulis, S.; Mezzi, A.; Montanari, R.; Angella, G.; Donnini, R.; Maldini, M.; Ripamonti, D. Microchemical analysis through SPEM for a nickel superalloy after creep test. *Metall. Ital.* **2013**, *105*, 21–26.
28. Donnini, R.; Kaciulis, S.; Mezzi, A.; Montanari, R.; Tata, M.E.; Testani, C. MicrO–Chemical characterization of fibermatrix interface in TI6ALAV-SICF composite. Caratterizzazione microchimica dell'interfaccia fibra-matrice nel composito TI6AL4V-SICf. *Metall. Ital.* **2007**, *99*, 13–18.
29. Donnini, R.; Kaciulis, S.; Mezzi, A.; Montanari, R.; Testani, C. Composite of Ti6Al4V and SiC fibres: Evolution of fibre-matrix interface during heat treatments. *Surf. Interface Anal.* **2008**, *40*, 277–280. [CrossRef]
30. Deodati, P.; Donnini, R.; Kaciulis, S.; Mezzi, A.; Montanari, R.; Testani, C.; Ucciardello, N. Anelastic phenomena at the fibre-matrix interface of the Ti6Al4V-SiC f composite. *Key Eng. Mater.* **2010**, *425*, 263–270. [CrossRef]
31. Deodati, P.; Donnini, R.; Kaciulis, S.; Mezzi, A.; Montanari, R.; Testani, C.; Ucciardello, N. Microstructural characterization of Ti6Al4V-SiCf composite produced by new roll-bonding process. *Adv. Mat. Res.* **2010**, *89*, 715–720.
32. Kaciulis, S.; Mezzi, A.; Donnini, R.; Deodati, P.; Montanari, R.; Ucciardello, N.; Amati, M.; KazemianAbyaneh, M.; Testani, C. Microchemical characterisation of carbon-metal interface in Ti6Al4V-SiCf composites. *Surf. Interface Anal.* **2010**, *42*, 707–711. [CrossRef]
33. Montanari, R.; Varone, A. Physical phenomena leading to melting of metals. *Mater. Sci. Forum.* **2017**, *884*, 3–17. [CrossRef]
34. Mezzi, A.; Kaciulis, S.; Balijepalli, S.K.; Montanari, R.; Varone, A.; Amati, M.; Aleman, B. Microchemical inhomogeneity in eutectic Pb-Bi alloy quenched from melt. *Surf. Interface Anal.* **2014**, *46*, 877–881. [CrossRef]
35. Amati, M.; Balijepalli, S.K.; Mezzi, A.; Kaciulis, S.; Montanari, R.; Varone, A. Temperature dependent phenomena in liquid LBE alloy. *Mater. Sci. Forum.* **2017**, *884*, 41–52. [CrossRef]
36. Montanari, R.; Varone, A.; Gregoratti, L.; Kaciulis, S.; Mezzi, A. Lead-bismuth eutectic: Atomic and micro-scale melt evolution. *Materials* **2019**, *12*, 3158. [CrossRef]
37. Ciancaglioni, I.; Donnini, R.; Kaciulis, S.; Mezzi, A.; Montanari, R.; Ucciardello, N.; Verona-Rinati, G. Surface modification of austenitic steels by low-temperature carburization. *Surf. Interface Anal.* **2012**, *44*, 1001–1004. [CrossRef]
38. Bagalà, P.; Gusmano, G.; Montesperelli, G.; Montanari, R.; Ucciardello, N.; Verona-Rinati, G. Valutazione della resistenza alla corrosione di acciai inossidabili ad alto tenore di azoto sensibilizzati ad alta temperature. In *Giornate Nazionali sulla Corrosione e Protezione*; Monte Porzio Catone: Rome, Italy, 2011.

39. Kaciulis, S.; Mezzi, A. Surface investigation of carbon films: From diamond to graphite. *Surf. Interface Anal.* **2010**, *42*, 1082–1084.
40. Carosi, A.; Amati, M.; Gregoratti, L.; Kaciulis, S.; Mezzi, A.; Montanari, R.; Rovatti, L.; Ucciardello, N. Heating modification of an austenitic steel with high-nitrogen content. *Surf. Interface Anal.* **2010**, *42*, 726–729. [CrossRef]
41. Rovatti, L.; Montanari, R.; Ucciardello, N.; Mezzi, A.; Kaciulis, S.; Carosi, A. Discontinuous precipitation in a high-nitrogen austenitic steel. *Mater. Sci. Forum.* **2010**, *638*, 3597–3602. [CrossRef]
42. Mattevi, C.; Kim, H.; Chhowalla, M. A review of chemical vapour deposition of graphene on copper. *J. Mater. Chem.* **2011**, *21*, 3324–3334. [CrossRef]
43. Wintterlin, J.; Bocquet, M.-L. Graphene on metal surfaces. *Surf. Sci.* **2009**, *603*, 1841–1852. [CrossRef]
44. Batzill, M. The surface science of graphene: Metal interfaces, CVD synthesis, nanoribbons, chemical modifications, and defects. *Surf. Sci. Rep.* **2012**, *67*, 83–115. [CrossRef]
45. Algdal, J.; Balasubramanian, T.; Breitholtz, M.; Kihlgren, T.; Walldén, L. Thin graphite overlayers: Graphene and alkali metal intercalation. *Surf. Sci.* **2007**, *601*, 1167–1175. [CrossRef]
46. Kaciulis, S.; Mezzi, A.; Calvani, P.; Trucchi, D.M. Electron spectroscopy of the main allotropes of carbon. *Surf. Interface Anal.* **2014**, *46*, 966–969. [CrossRef]
47. Nobili, L.; Magnin, L.; Bernasconi, R.; Livolsi, F.; Pedrazzetti, L.; Lucotti, A.; Balijepalli, S.K.; Mezzi, A.; Kaciulis, S.; Montanari, R. Investigation of graphene layers on electrodeposited polycrystalline metals. *Surf. Interface Anal.* **2016**, *48*, 456–460. [CrossRef]
48. Bolli, E.; Fava, A.; Ferro, P.; Kaciulis, S.; Mezzi, A.; Montanari, R.; Varone, A. Cr segregation and impact fracture in a martensitic stainless steel. *Coatings* **2020**, *10*, 843. [CrossRef]

2

A Microfluidic-Based Fabry-Pérot Gas Sensor

Jin Tao [1], Qiankun Zhang [1], Yunfeng Xiao [2], Xiaoying Li [3], Pei Yao [4], Wei Pang [1], Hao Zhang [1], Xuexin Duan [1], Daihua Zhang [1] and Jing Liu [1,*]

[1] State Key Laboratory of Precision Measurement Technology and Instruments, School of Precision Instruments and Opto-Electronics Engineering, Tianjin University, Tianjin 300072, China; taojin@tju.edu.cn (J.T.); zqkty@tju.edu.cn (Q.Z.); weipang@tju.edu.cn (W.P.); haozhang@tju.edu.cn (H.Z.); xduan@tju.edu.cn (X.D.); dhzhang@tju.edu.cn (D.Z.)

[2] State Key Laboratory for Mesoscopic Physics, School of Physics, Peking University, Beijing 100044, China; yfxiao@pku.edu.cn

[3] Key Laboratory of Optoelectronics Information Technology, School of Precision Instrument and Opto-Electronics Engineering, Tianjin University, Tianjin 300072, China; xiaoyingli@tju.edu.cn

[4] School of Materials Science and Engineering, Tianjin University, Tianjin 300072, China; pyao@tju.edu.cn

* Correspondence: jingliu_1112@tju.edu.cn

Academic Editors: Manabu Tokeshi and Kiichi Sato

Abstract: We developed a micro-gas detector based on a Fabry-Pérot (FP) cavity embedded in a microfluidic channel. The detector was fabricated in two steps: a silicon substrate was bonded to a glass slide curved with a micro-groove, forming a microfluidic FP cavity; then an optical fiber was inserted through a hole drilled at the center of the groove into the microfluidic FP cavity, forming an FP cavity. The light is partially reflected at the optical fiber endface and the silicon surface, respectively, generating an interference spectrum. The detection is implemented by monitoring the interference spectrum shift caused by the refractive index change of the FP cavity when a gas analyte passes through. This detection mechanism (1) enables detecting a wide range of analytes, including both organic and inorganic (inertia) gases, significantly enhancing its versatility; (2) does not disturb any gas flow so that it can collaborate with other detectors to improve sensing performances; and (3) ensures a fast sensing response for potential applications in gas chromatography systems. In the experiments, we used various gases to demonstrate the sensing capability of the detector and observed drastically different sensor responses. The estimated sensitivity of the detector is 812.5 nm/refractive index unit (RIU) with a detection limit of 1.2×10^{-6} RIU assuming a 1 pm minimum resolvable wavelength shift.

Keywords: micro gas sensor; micro Fabry-Pérot cavity; optical fiber; microfluidic channel; MEMS

---

1. Introduction

Gas detectors attract a lot of research interest due to their wide applications in the areas of environmental monitoring [1,2], homeland security [3], anti-terrorism [4], industrial quality control [5–7], *etc.* So far, various stand-alone gas detectors, such as surface acoustic wave detectors [8], chemiresistor detectors [9–15], grating-based optical gas detectors [16], surface Plasmon resonance gas detectors [17,18] and opto-thermal gas detectors [19], have been successfully developed and made commercially available. However, most of them either respond to very few analytes, significantly limiting their application fields [20,21], or respond similarly to a great amount of analytes, so that one cannot be differentiated from others, which can hardly satisfy the increasing demands of applications requiring the identification of a wide range of analytes in a complex mixture.

One of the most promising solutions to this problem is to combine gas detector(s) with a gas chromatography system (GC), to separate gas mixtures by their different velocities when traveling through the separation column [22]. Thus, a detector can be installed at the elution end of the separation column to detect individual analytes sequentially as they elute out in succession. The detector that can be used in a GC usually possesses several features: (1) it can detect a reasonable amount of analytes; (2) it has a flow-through structure and can be easily connected to the GC; and (3) it has a fast sensing response so that it can complete the detection of one analyte before another one elutes out (it takes several seconds to minutes for analytes to elute out from the GC system). Most of the stand-alone detectors are not readily applicable in GC systems because they do not have one or more of the aforementioned feature(s).

The traditional GC detectors are generally divided into destructive and non-destructive detectors. The destructive detectors, such as mass spectrometry and the flame ionization detector, usually detect the current change caused by the ionization of analytes [23,24], and thus the analytes are destroyed after detection. This feature is not desirable for the applications that require multiple detectors to work in concert to acquire complimentary information of the analyte for improving system performance, such as to promote the analyte identification rate in a multi-dimensional GC [22,25,26] and to automate system control in smart GCs [27,28]. In contrast, non-destructive detectors do not destroy analytes during the detection and therefore are much more flexible and applicable than destructive detectors in the system design and integration. One of the most widely used non-destructive detectors in GC is the thermal conductivity detector (TCD), which identifies the analyte by comparing the heat loss caused by passing through the analyte with the loss caused by passing through the carrier gas per unit time [29]. Consequently, it not only keeps the analyte intact during the detection, but also is a universal detector which responds to most of the analytes except the ones that have the same heat capacity as the carrier gas [30]. This capability is very useful for quick sample identification and pre-detection in a multi-detector GC system. Nevertheless, traditional TCD usually has large dead volume, leading to broadened analyte peaks and large sample consumption, and has to work in an elevated temperature, resulting in high power consumption and potential hazards for detecting flammable and explosive gases, which is not desirable for being integrated in miniaturized GC systems designed to satisfy the dramatically increased quick on-site detection demands in recent years [31].

Recently, a lot of microstructured gas detectors have been developed to be integrated with micro-GC and/or microfluidic systems for on-site quick gas detection [32–35]. One of the promising detection schemes is based on the Fabry-Pérot (FP) cavity, which is also widely used in the fields of biosensing, temperature, strain, and humidity sensing because of its simple fabrication and measurement setup. There are mainly two ways to form the FP cavity: using a polymer layer or using the microfluidic channel itself. The polymer layer, either deposited at the end of an optical fiber [33,36] or the bottom of the microfluidic channel [34,37], interacts with the gas analyte and results in the change of its refractive index (RI) or/and thickness, which causes the shift of the interference spectrum. The usage of a polymer improves the sensitivity and selectivity of the sensors; however, it limits the detectable gases to being the ones that have interactions with the polymer, greatly restricting its applications. One the other hand, the microfluidic channel can also form an FP cavity with its inner surfaces as the reflective surfaces. When the analyte is traveling through the microfluidic channel, the RI of the FP cavity changes, leading to the shift of the interference spectrum. This configuration has been used in biosensing [38–40], pressure sensing [41,42], and temperature sensing [43], but it has not yet, to the best of our knowledge, been used in universal gas sensing.

In this paper, we demonstrated the possibility of the microfluidic-based FP sensor for a non-destructive and universal gas detector which has the potential to be further integrated with miniaturized/portable GC systems. The detector, as shown in Figure 1a, is fabricated by inserting a single-mode optical fiber into a hole drilled at the center of a micro-groove curved on a glass slide which is then bonded to a silicon substrate, forming a microfluidic channel. The light coupled into the optical fiber partially reflects at the optical fiber endface and the silicon surface, respectively, generating an interference spectrum.

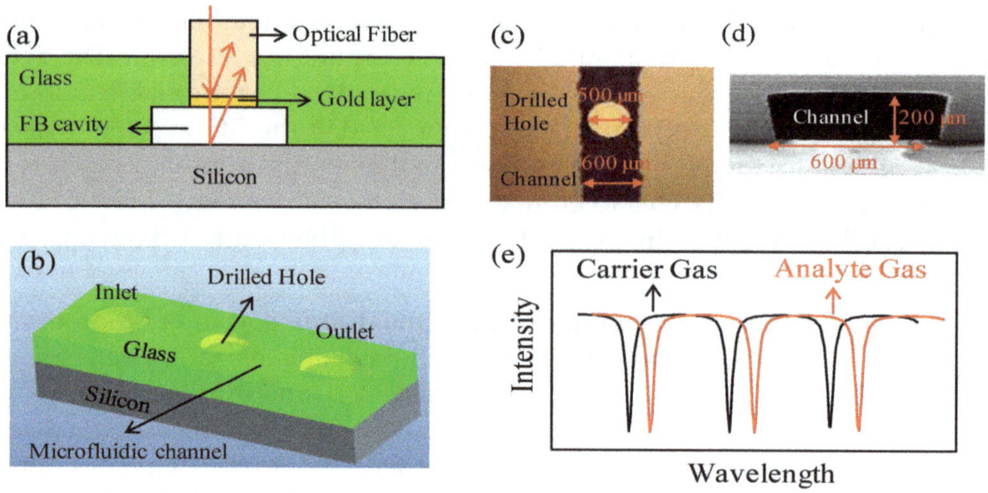

Figure 1. (**a**) Schematic diagram of the cross-section of the microfluidic-based FP-gas detector. Red arrows represent the light beam which was coupled into the optical fiber and reflected at the gold layer and silicon surface, respectively. (**b**) Three-dimensional diagram of the microfluidic-based FP gas detector. (**c**) Optical microscopy image of the top view of the microfluidic channel. (**d**) SEM image of the sectional view of the microfluidic channel. (**e**) Exemplary interference spectrum shift caused by the analyte gas.

When an analyte travels through, the RI of the FP cavity changes. The shift of the resonant wavelength (interference spectrum) is linearly related to the change of the RI of the FP cavity and hence the passage of the analyte. Therefore, by monitoring the interference spectrum shift, the kinetic information of the analyte can be obtained.

There are several unique advantages of the proposed detector. First of all, the detector detects the RI change at the detection point in the microfluidic channel, and, therefore, it can detect an analyte that presents an RI difference from the carrier gas larger than its detection limit, validating its capability of detecting most analytes including organic, inorganic and even inertia gases. Second, since it detects the RI change of the FP cavity, whose two reflectance surfaces are both flush with the inner surface of the microfluidic channel, the detector does not destroy or induce any disturbance to the gas flow, which has the potential to be used in combination with other detectors in a series to improve the detection performance. Third, the detector is embedded in a microfluidic channel, forming a flow-through structure to minimize its dead volume, and at the same time does not involve any analyte absorption/desorption processes, significantly reducing the peak-broadening and speeding up the detection response compared to TCD and other detectors. This merit is extremely important for miniaturized GC to increase its detection efficiency (the number of analytes detected per unit time), because sharp peaks enable it to resolve more analytes per unit time than conventional GC and hence it may resolve as many analytes as a conventional one does in a much shorter time. Fourth, the flow-through structure of the detector allows it to be readily connected to the fluidic channel of the GC. Fifth, it does not need any heating elements or additional gas supply for the proposed detector to be fully functional, which favors the whole system's miniaturization and portability. Last, the fabrication process is compatible with the micro-fabrication process, further lowering the manufacturing cost and benefiting the mass production.

We carried out the initial calibration and characterizations of the proposed gas detector to demonstrate its detection capability. In our experiments, we coated gold layers of various thicknesses at the endfaces of different optical fibers to increase the reflectivity at the optical fiber-air interface, so that an optimal interference spectrum can be obtained. Then, the performance of the detector was tested by various analytes with different physical and chemical properties, which, as expected, responded linearly to the amount of the RI change caused by its exposure to the analyte with a sensitivity of

812 nm/RIU and an estimated detection limit of 1.2×10^{-6} RIU assuming that the minimum resolvable wavelength shift was 1 pm, corresponding to the detection limit of 2800 ppm for C_2H_4.

2. Experimental

2.1. Detector Fabrication

The two- and three-dimensional schematic structures of the detector are shown in Figure 1a,b, respectively, the two reflectance surfaces of which were the endface of an optical fiber and the surface of the silicon substrate, respectively. The optical fiber was inserted into a hole (see Figure 1c) drilled at the center of a micro-groove etched on a glass slide which was then anodic-bonded with a silicon substrate, forming a sealed micro-fluidic channel (its cross-sectional dimension was 600 μm wide by 200 μm deep, as presented in Figure 1d). The optical fiber endface was aligned with the inner surface of the channel to avoid any potential disturbance to the fluidic flow. The hole was then sealed by silicone rubber to prevent any gas leakage. The endface of the optical fiber was coated with gold by the physical vapor deposition method to increase the reflectivity. When an analyte gas was pumped through, the RI of the FP cavity changed, resulting in the interference spectrum shift as shown in Figure 1e.

2.2. Experimental Setup

The whole test system was composed of the optical measurement part and gas delivery part, as shown in Figure 2. The optical measurement part consisted of a laser source, an optical circulator and a photon detector. The laser was scanned from 1510 to 1590 nm at a frequency of 2 Hz with a spectral resolution of 1 pm. The laser output was coupled into Port 1 of the circulator and delivered into the sensing optical fiber through Port 2. The reflected light from the sensing probe was coupled back to Port 2 and delivered into Port 3 which was connected to a photo detector. The light intensity was monitored by a homemade LabVIEW program at a recording rate of 20 kHz. The gas analytes were prepared individually by drawing the pure analyte into a gas-tight syringe until it reached the pressure equilibrium. The outlet of the syringe was connected to the inlet of the microfluidic channel through a capillary column. A syringe pump was used to pump the gas analyte from the syringe into the microfluidic channel at a flow rate of 10 μL/min.

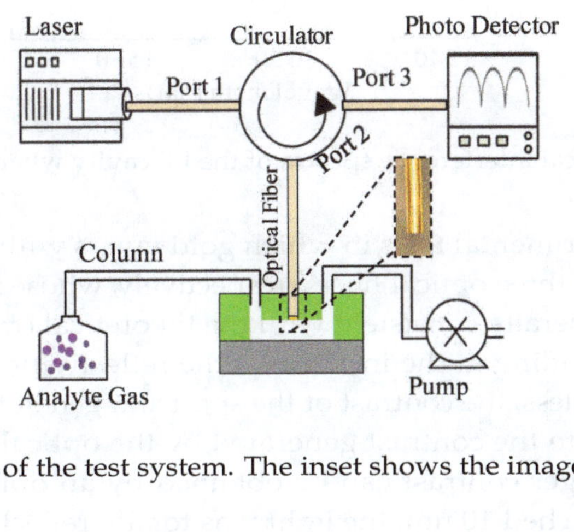

Figure 2. Schematic of the test system. The inset shows the image of the optical fiber.

3. Results and Discussion

The microfluidic-based FP gas detector monitors the interference spectrum shift of the microfluidic FP cavity caused by its RI change when the gas analyte passes through. This section first investigated the effect of the gold coating on the quality of the interference spectrum. Then, various gas analytes, which include organic gases such as CH_4, C_2H_4 and C_3H_6O, inorganic gases such as NH_3, N_2O, and

CO_2, and the inertia gas of He, were used to characterize the detector performance and demonstrate its versatility. Such a wide range of detectable gases can be hardly covered by a single detector.

3.1. Gold Coating Calibration

The reflectivity of the two surfaces of the FP cavity, the optical fiber endface and the silicon surface, decide the quality of the interference spectrum and thus the detection limit of the proposed detector. Here, we used gold coating at the optical fiber endface as a model method to increase its reflectivity while keeping the silicon surface uncoated due to the necessity to simplify the fabrication process of the microfluidic channel. The theoretical calculation was carried out to analyze the effect of the reflectivity of the optical fiber endface on the interference spectrum when the reflectivity of the silicon surface was kept at a constant value of 0.3.

The normalized reflectance intensity R of the FP detector can be described by the following equation:

$$R = \frac{I_r}{I_e} = 1 - \frac{t_1^2 t_2^2}{(1 - r_1 r_2)^2 + 4 r_1 r_2 \sin^2(kd)} \quad (1)$$

where I_r and I_e are the reflectance and input light intensity, respectively; t_1/t_2 and r_1/r_2 are the transitivity and reflectivity of the optical fiber endface and silicon surface, respectively; d is the FP cavity length; and k is the wave vector. Based on Equation (1), the normalized reflected interference spectra are plotted in Figure 3 to visualize the effects of r_1 on the sharpness and contrast of the interference (the value of r_2 is set to be 0.3). The interference spectra correspond to five values of r_1 of 0.3, 0.5, 0.8, 0.96 and 1. As expected, the sharpness of the resonant peaks increases when r_1 increases, while the contrast reaches the maximum value when r_1 equals 0.5.

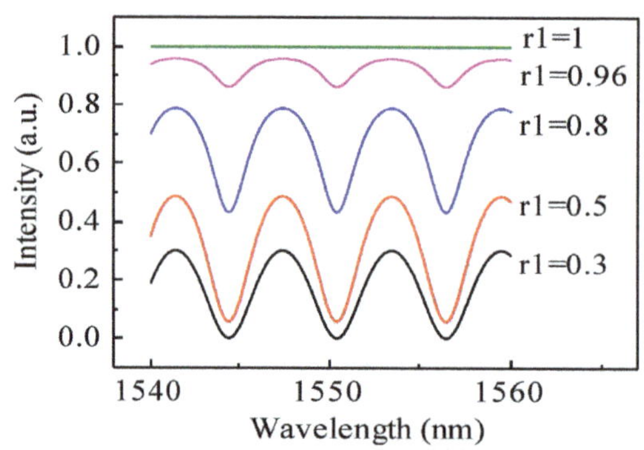

Figure 3. Theoretical interference spectra of the FP cavity when r_2 is set to be 0.3.

We then carried out experimental tests in which gold layers with thicknesses of 5, 8 and 10 nm were coated on the endfaces of three optical fibers, respectively, whose resultant spectra are presented in Figure 4. The trend was generally consistent with the theoretical observation: when the thickness of the gold layer increased, leading to the increase of the reflectivity, the sharpness of the resonant peaks also increased. Nevertheless, the contrast of the spectrum generated by the optical fiber without any gold coating was similar to the contrast generated by the optical fiber with 5 nm gold coating, which may suggest that a larger contrast can be obtained by an optical fiber with a thinner gold layer. When the gold layer reached 10 nm, the light was totally reflected at the optical fiber endface. Since the 5 nm gold coating was the thinnest coating that could be accurately deposited in our lab, and it generated an interference spectrum with reasonably sharp resonant peaks and large contrast at the same time, the optical fiber endface was deposited with 5-nm-thick gold coating in the following experiments. In future development, other methods will be used to increase the reflectivity of both reflectance surfaces to greatly improve the quality of the FP cavity and also that of the detection limit.

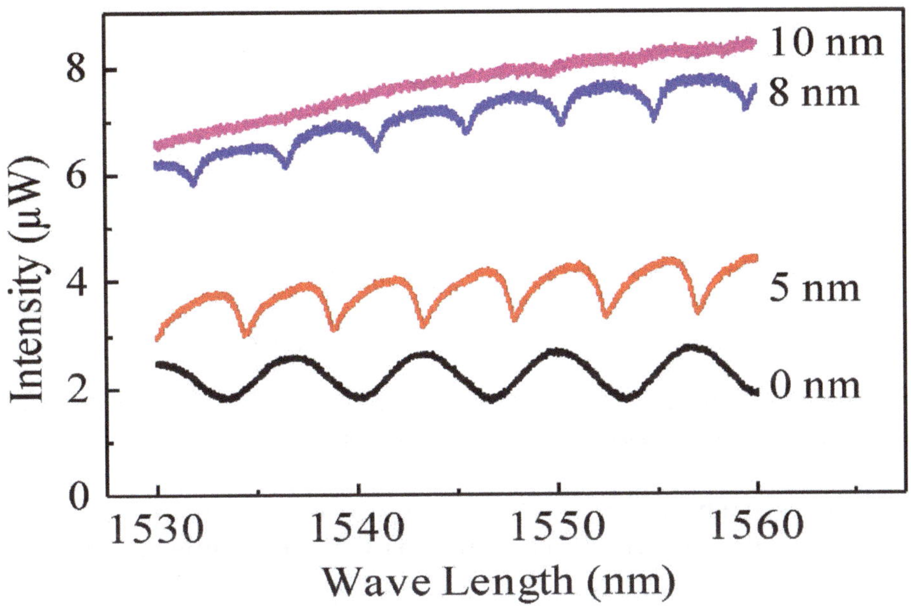

Figure 4. Experimental interference spectra of the FP cavities with various thicknesses of gold coating on the optical fiber enfaces.

3.2. Real-Time Response

We calibrated the responses of the detector to various pure gas analytes which are presented in Figure 5. At the beginning, the ambient air was pumped in the microfluidic channel as the carrier gas (RI = 1.000292) to establish the baseline of the sensor, after which pure gas analyte was injected in. Since the RI of the gas analyte was different from the carrier gas, the RI of the FP cavity changed when the gas analyte passed through, resulting in the shift of the interference spectrum. When the interference shift reached the maximum equilibrium, the carrier gas was switched back into the microfluidic channel to purge out the gas analyte. Consequently, the signal of the detector returned back to the baseline. Each gas analyte was tested by the aforementioned procedures multiple times to demonstrate the detector's repeatability and reliability. Figure 5a shows the real-time responses of our detector to three types of gas analytes: C_2H_4, CO_2 and CH_4, whose RIs are 1.0007198, 1.000449 and 1.000444, respectively. Since it generated the biggest RI change of 4.28×10^{-4} compared to the other two analytes, C_2H_4 had the largest interference shift of around 0.38 nm. On the other hand, CO_2 and CH_4 caused a RI change of 1.57×10^{-4} and 1.52×10^{-4}, respectively, and thus the detector had an interference shift of 0.15 nm and 0.14 nm, respectively. The detector had sharp "on" and "off" response signals to all the analytes (around 0.5 to 1 s, limited by the sampling rate), which is because it has minimum dead volume and does not involve any analyte absorption/desorption processes. The quick response is very important for the micro-GC, because sharp peaks enable it to resolve more analyte peaks per unit time than broad peaks do, thus improving its analysis efficiency. Additionally, the wavelength shifts of the detector upon its exposure to three more analytes of NH_3, N_2O, and C_3H_6O were also recorded and are presented in Figure 5b, which shows the relationship between the interference shifts of the detector with both the absolute RI of the gas analyte and the RI difference of the gas analyte from the carrier gas. As expected, the wavelength shift is linearly proportional to both the absolute RI and the RI difference from air of the analyte gas, from which the sensitivity of the detector is estimated to be 812.5 nm/RIU. Since the wavelength stability of the laser source is around 1 pm, the detection limit of the detector is estimated to be a change of 1.2×10^{-6} RIU, which is similar to the sensors reported in [17,44,45]. Although other undesirable noises, such as the syringe pump, caused vibration, the static noise (2 pm) and the temperature fluctuation degraded it to around 10 pm, and the noise level can be minimized to 1 pm by controlling the system parameters well.

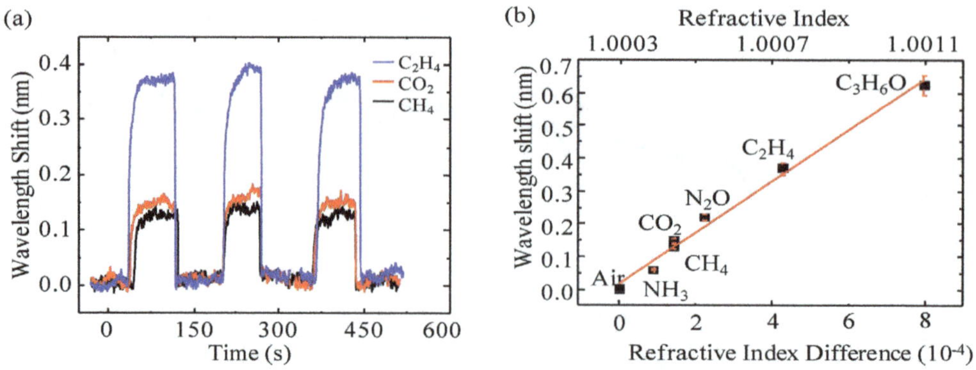

Figure 5. (**a**) Real-time responses of the sensor to C_2H_4, CO_2 and CH_4. (**b**) The wavelength shift corresponding to the RI change caused by exposure to various analytes.

The detector was also calibrated by analytes with concentrations ranging from 100% to 5% by mixing a single analyte with the carrier gas. Figure 6a is the real-time response of the detector to CH_4 and helium (He) with the concentrations of 100%, 50%, 25% and 5%. The absolute values of the interference spectrum shifts of the detector to both analytes declined gradually as the concentrations of both gas analytes dropped from 100% to 5%, and the RI values approached that of the carrier gas. Additionally, the interference spectrum of the detector shifted to a longer wavelength when it was exposed to CH_4, while it shifted to a shorter wavelength when it was exposed to He. The phenomenon can be explained by Equation (1): when the RI change, Δn, is positive/negative, the interference spectrum shift is positive/negative (shift to longer/shorter wavelength). Consequently, CH_4/He, whose RI is larger/smaller than the RI of the carrier gas (air), caused the interference shift to a longer/smaller wavelength. This feature can be used to compare the RI value of the analyte with the RI value of the carrier gas, which may be an important parameter for identifying an unknown gas. In Figure 6b, the values of the absolute interference spectrum shift at equilibrium for different concentrations of CO_2, CH_4, C_2H_4 and He are depicted, in which the absolute value of the wavelength shift increases linearly as the concentration of the analyte increases. From this figure, the sensitivity (detection limit) of the detector in terms of concentration is estimated to be 3.5×10^{-4} pm/ppm (2800 ppm) and 10^{-4} pm/ppm (10,000 ppm) for C_2H_4 and CO_2, respectively, which has the maximum (C_2H_4) and minimum (CO_2) RI difference from the carrier gas, respectively.

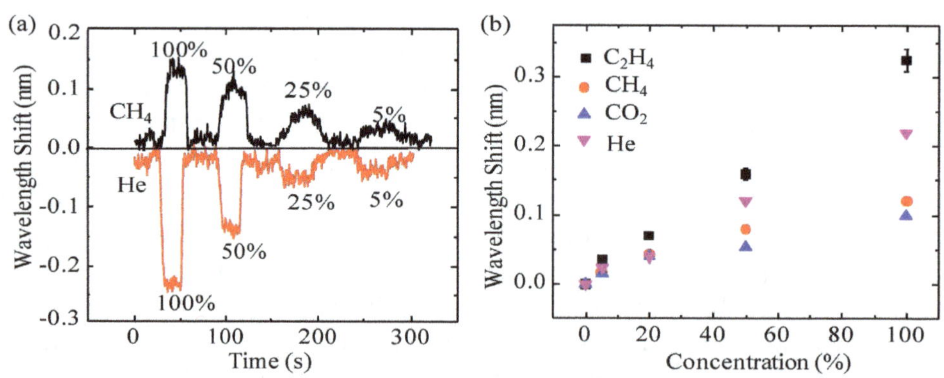

Figure 6. (**a**) Real-time responses of the sensor to CH_4 and He with concentrations of 100%, 50%, 25% and 5%. (**b**) Concentration-dependent wavelength shift upon exposure to C_2H_4, CH_4, CO_2 and He with concentrations of 100%, 50%, 25% and 5%.

4. Materials and Methods

All of the analyte gases used in the experiment were purchased from Best Gas (Tianjin, China) and had purity greater than 99.9%. Pyrex 7740 glass wafer was used for anode bonding with the silicon wafer by the wafer bonder (SB6/8, SUSS MicroTec, Garching, Germany). A fiber sensor system (SM125, Micron Optics Inc., Hackettstown, NJ, USA) was used as the laser source. Single-mode fibers (SMF-28) were purchased from Corning (New York, NY, USA). Universal quick seal column connectors (Part No. 23627) were purchased from Sigma (St. Louis, MO, USA). The syringe pump (Model NO: NE1000) was purchased from New Era Pump System (Farmingdale, NY, USA) The capillary column (Model NO: 160-2255 DEAC1) was purchased from Agilent Technologies (Santa Rosa, CA, USA). The silicone rubber used to seal the column and the FP cavity was purchased from NanDa Inc., Tianjin, China.

5. Conclusions

We developed a non-destructive and universal microfluidic-based FP gas detector. The detector was tested by various analytes with diversified chemical and physical properties with various concentrations, and the results show that it had a sensitivity of 812.5 nm/RIU and a detection limit of 1.2×10^{-6} RIU, assuming the minimum resolvable wavelength shift was 1 pm which is comparable to other optical sensors [36,38,40–42]. The detector senses the RI change in the microfluidic channel where the detector is embedded when the gas analyte passes through. Therefore, it can detect any type of analyte that has a different RI from that of the carrier gas, significantly improving its versatility, which is highly valuated by the quick on-site sample identification applications. The non-destructive detection mechanism and flow-through structure allow the detector to be installed upstream of other detectors to provide complementary information. In addition, the detector also has a fast response due to its minimized dead volume and the disuse of any gas-absorptive materials, which suggests its potential for being used in GC systems. The future work will be focused on promoting the performance of the detector by increasing the quality of the micro-FP cavity to improve the interference spectrum resolution and thus the detection limit.

Acknowledgments: The work is supported by the NSFC (21405109) and the Seed Foundation of Tianjin University (1405), China.

Author Contributions: Jin Tao and Qiankun Zhang designed and performed the experiments. Jing Liu analyzed the data and wrote the paper. Yunfeng Xiao, Xiaoyin Li and Pei Yao provided advice on experiment design and data analysis. Wei Pang, Hao Zhang, Xuexin Duan, and Daihua Zhang attended the discussion of the data analysis. All authors have read and approved the final manuscript.

References

1. Fine, G.F.; Cavanagh, L.M.; Afonja, A.; Binions, R. Metal oxide semi-conductor gas sensors in environmental monitoring. *Sensors* **2010**, *10*, 5469–5502. [CrossRef] [PubMed]
2. Martinelli, G.; Carotta, M.C.; Ferroni, M.; Sadaoka, Y.; Traversa, E. Screen-printed perovskite-type thick films as gas sensors for environmental monitoring. *Sens. Actuators B Chem.* **1999**, *55*, 99–110. [CrossRef]
3. Moore, D.S. Recent advances in trace explosives detection instrumentation. *Sens. Imaging* **2007**, *8*, 9–38. [CrossRef]
4. Wang, J. Electrochemical Sensing of Explosives. *Electroanalysis* **2007**, *19*, 415–423. [CrossRef]
5. Bourrounet, B.; Talou, T.; Gaset, A. Application of a multi-gas-sensor device in the meat industry for boar-taint detection. *Sens. Actuators B Chem.* **1995**, *27*, 250–254. [CrossRef]
6. Schweizer-Berberich, P.M.; Vaihinger, S.; Göpel, W. Characterisation of food freshness with sensor arrays. *Sens. Actuators B Chem.* **1994**, *18*, 282–290. [CrossRef]
7. Ólafsdóttir, G.; Martinsdóttir, E.; Oehlenschläger, J.; Dalgaard, P.; Jensen, B.; Undeland, I.; Mackie, I.M.; Henehan, G.; Nielsen, J.; Nilsen, H. Methods to evaluate fish freshness in research and industry. *Trends Food Sci. Technol.* **1997**, *8*, 258–265. [CrossRef]
8. Ricco, A.J.; Martin, S.J.; Zipperian, T.E. Surface acoustic wave gas sensor based on film conductivity changes. *Sens. Actuators B Chem.* **1985**, *8*, 319–333. [CrossRef]

9. Abraham, J.K.; Philip, B.; Witchurch, A.; Varadan, V.K.; Reddy, C.C. A compact wireless gas sensor using a carbon nanotube/PMMA thin film chemiresistor. *Smart Mater. Struct.* **2004**, *13*, 1045–1049. [CrossRef]
10. Paul, R.K.; Badhulika, S.; Saucedo, N.M.; Mulchandani, A. Graphene nanomesh as highly sensitive chemiresistor gas sensor. *Anal. Chem.* **2012**, *84*, 8171–8178. [CrossRef] [PubMed]
11. Ho, K.C.; Tsou, Y.H. Chemiresistor-type NO gas sensor based on nickel phthalocyanine thin films. *Sens. Actuators B Chem.* **2001**, *77*, 253–259. [CrossRef]
12. Yamazoe, N.; Sakai, G.; Shimanoe, K. Oxide semiconductor gas sensors. *Catal. Surv. Asia* **2003**, *7*, 63–75. [CrossRef]
13. Yamazoe, N. New approaches for improving semiconductor gas sensors. *Sens. Actuators B Chem.* **1991**, *5*, 7–19. [CrossRef]
14. Comini, E.; Faglia, G.; Sberveglieri, G.; Pan, Z.; Wang, Z.L. Stable and highly sensitive gas sensors based on semiconducting oxide nanobelts. *Appl. Phys. Lett.* **2002**, *81*, 1869–1871. [CrossRef]
15. Simon, I.; Bârsan, N.; Bauer, M.; Weimar, U. Micromachined metal oxide gas sensors: Opportunities to improve sensor performance. *Sens. Actuators B Chem.* **2001**, *73*, 1–26. [CrossRef]
16. Zhou, B.; Chen, Z.; Zhang, Y.; Gao, S.; He, S. Active Fiber Gas Sensor for Methane Detecting Based on a Laser Heated Fiber Bragg Grating. *IEEE Photonics Technol. Lett.* **2014**, *26*, 1069–1072. [CrossRef]
17. Bingham, J.M.; Anker, J.N.; Kreno, L.E.; Van Duyne, R.P. Gas Sensing with High-Resolution Localized Surface Plasmon Resonance Spectroscopy. *JACS* **2010**, *132*, 17358–17359. [CrossRef] [PubMed]
18. Sharma, A.K.; Jha, R.; Gupta, B.D. Fiber-optic sensors based on surface plasmon resonance: A comprehensive review. *IEEE Sens. J.* **2007**, *7*, 1118–1129. [CrossRef]
19. Rosengren, L.G. An opto-thermal gas concentration detector. *Infrared Phys.* **1973**, *13*, 173–182. [CrossRef]
20. Morrison, S.R. Selectivity in semiconductor gas sensors. *Sens. Actuators* **1987**, *12*, 425–440. [CrossRef]
21. Coles, G.S.V.; Williams, G.; Smith, B. Selectivity studies on tin oxide-based semiconductor gas sensors. *Sens. Actuators B Chem.* **1991**, *3*, 7–14. [CrossRef]
22. Liu, J.; Sun, Y.; Howard, D.J.; Frye-Mason, G.; Thompson, A.K.; Ja, S.J.; Wang, S.K.; Bai, M.; Taub, H.; Almasri, M.; Fan, X. Fabry-Pérot cavity sensors for multipoint on-column micro gas chromatography detection. *Anal. Chem.* **2010**, *82*, 4370–4375. [CrossRef] [PubMed]
23. Hobbs, P.J.; Misselbrook, T.H.; Pain, B.F. Assessment of Odours from Livestock Wastes by a Photoionization Detector, an Electronic Nose, Olfactometry and Gas Chromatography-Mass Spectrometry. *J. Agric. Eng. Res.* **1995**, *60*, 137–144. [CrossRef]
24. McWilliam, I.G.; Dewar, R.A. Flame ionization detector for gas chromatography. *Nature* **1958**, *181*, 760. [CrossRef]
25. Marriott, P.J.; Chin, S.T.; Maikhunthod, B.; Schmarr, H.G.; Bieri, S. Multidimensional gas chromatography. *Trends Anal. Chem.* **2012**, *34*, 1–20. [CrossRef]
26. Seeley, J.V. Recent advances in flow-controlled multidimensional gas chromatography. *J. Chromatogr. A* **2012**, *1255*, 24–37. [CrossRef] [PubMed]
27. Liu, J.; Seo, J.H.; Li, Y.; Chen, D.; Kurabayashi, K.; Fan, X. Smart multi-channel two-dimensional micro-gas chromatography for rapid workplace hazardous volatile organic compounds measurement. *Lab Chip* **2013**, *13*, 818–825. [CrossRef] [PubMed]
28. Chen, D.; Seo, J.H.; Liu, J.; Kurabayashi, K.; Fan, X. Smart Three-Dimensional Gas Chromatography. *Anal. Chem.* **2013**, *85*, 6871–6875. [CrossRef] [PubMed]
29. Cruz, D.; Chang, J.P.; Showalter, S.K.; Gelbard, F.; Manginell, R.P.; Blain, M.G. Microfabricated thermal conductivity detector for the micro-ChemLab™. *Sens. Actuators B Chem.* **2007**, *121*, 414–422. [CrossRef]
30. Simon, I.; Arndt, M. Thermal and gas-sensing properties of a micromachined thermal conductivity sensor for the detection of hydrogen in automotive applications. *Sens. Actuators A Phys.* **2002**, *97*, 104–108. [CrossRef]
31. Kuo, J.T.; Yu, L.; Meng, E. Micromachined thermal flow sensors—A review. *Micromachines* **2012**, *3*, 550–573. [CrossRef]
32. Zhong, Q.; Steinecker, W.H.; Zellers, E.T. Characterization of a high-performance portable GC with a chemiresistor array detector. *Analyst* **2009**, *134*, 283–293. [CrossRef] [PubMed]
33. Liu, J.; Sun, Y.; Fan, X. Highly versatile fiber-based optical Fabry-Pérot gas sensor. *Opt. Express* **2009**, *17*, 2731–2738. [CrossRef] [PubMed]
34. Reddy, K.; Guo, Y.; Liu, J.; Lee, W.; Oo, M.K.K.; Fan, X. On-chip Fabry-Pérot interferometric sensors for micro-gas chromatography detection. *Sens. Actuators B Chem.* **2011**, *159*, 60–65. [CrossRef]

35. Hossein-Babaei, F.; Paknahad, M.; Ghafarinia, V. A miniature gas analyzer made by integrating a chemoresistor with a microchannel. *Lab Chip* **2012**, *12*, 1874–1880. [CrossRef] [PubMed]
36. Gao, R.; Jiang, Y.; Ding, W.; Wang, Z.; Liu, D. Filmed extrinsic Fabry-Perot interferometric sensors for the measurement of arbitrary refractive index of liquid. *Sens. Actuators B Chem.* **2013**, *177*, 924–928. [CrossRef]
37. Reddy, K.; Guo, Y.; Liu, J.; Lee, W.; Oo, M.K.K.; Fan, X. Rapid, sensitive, and multiplexed on-chip optical sensors for micro-gas chromatography. *Lab Chip* **2012**, *12*, 901–905. [CrossRef] [PubMed]
38. Wei, T.; Han, Y.; Li, Y.; Tsai, H.L.; Xiao, H. Temperature-insensitive miniaturized fiber inline Fabry-Perot interferometer for highly sensitive refractive index measurement. *Opt. Express* **2008**, *16*, 5764–5769. [CrossRef] [PubMed]
39. Lin, C.H.; Jiang, L.; Xiao, H.; Chai, Y.H.; Chen, S.J.; Tsai, H.L. Fabry-Perot interferometer embedded in a glass chip fabricated by femtosecond laser. *Opt. Lett.* **2009**, *34*, 2408–2410. [CrossRef] [PubMed]
40. Tian, Y.; Wang, W.; Wu, N.; Zou, X.; Guthy, C.; Wang, X. A miniature fiber optic refractive index sensor built in a MEMS-based microchannel. *Sensors* **2011**, *11*, 1078–1087. [CrossRef] [PubMed]
41. Xiao, G.Z.; Adnet, A.; Zhang, Z.; Sun, F.G.; Grover, C.P. Monitoring changes in the refractive index of gases by means of a fiber optic Fabry-Perot interferometer sensor. *Sens. Actuators A Phys.* **2005**, *118*, 177–182. [CrossRef]
42. Duan, D.W.; Rao, Y.J.; Zhu, T. High sensitivity gas refractometer based on all-fiber open-cavity Fabry-Perot interferometer formed by large lateral offset splicing. *JOSA B* **2012**, *29*, 912–915. [CrossRef]
43. Kou, J.L.; Feng, J.; Ye, L.; Xu, F.; Lu, Y.Q. Miniaturized fiber taper reflective interferometer for high temperature measurement. *Opt. Express* **2010**, *18*, 14245–14250. [CrossRef] [PubMed]
44. Maharana, P.K.; Jha, R.; Padhy, P. On the electric field enhancement and performance of SPR gas sensor based on graphene for visible and near infrared. *Sens. Actuators B Chem.* **2015**, *207*, 117–122. [CrossRef]
45. Goyal, A.K.; Pal, S. Design and simulation of high-sensitive gas sensor using a ring-shaped photonic crystal waveguide. *Phys. Scr.* **2015**, *90*, 025503. [CrossRef]

3

All Silicon Micro-GC Column Temperature Programming using Axial Heating

Milad Navaei [1], Alireza Mahdavifar [1], Jean-Marie D. Dimandja [2], Gary McMurray [3] and Peter J. Hesketh [1,*]

[1] School of Mechanical Engineering, Georgia Institute of Technology, Atlanta, GA 30324, USA; mnavaei3@gmail.com (M.N.); amahdavifar@gmail.com (A.M.)
[2] Department of Chemistry, Spelman College, Atlanta, GA 30314, USA; jdimandja@spelman.edu
[3] Food Processing Center, Georgia Tech Research Institute (GTRI), Atlanta, GA 30318, USA; gary.mcmurray@gtri.gatech.edu
[*] Author to whom correspondence should be addressed; peter.hesketh@me.gatech.edu

Academic Editor: Manabu Tokeshi

Abstract: In this work we present a high performance micro gas chromatograph column with a novel two dimensional axial heating technique for faster and more precise temperature programming, resulting in an improved separation performance. Three different axial resistive heater designs were simulated theoretically on a 3.0 m × 300 μm × 50 μm column for the highest temperature gradient on a 22 by 22 μm column. The best design was then micro-fabricated and evaluated experimentally. The simulation results showed that simultaneous temperature gradients in time and distance along the column are possible by geometric optimization of the heater when using forced convection. The gradients along the column continuously refocused eluting bands, offsetting part of the chromatographic band spreading. The utility of this method was further investigated for a test mixture of three hydrocarbons (hexane, octane, and decane).

Keywords: gas chromatography; MEMS; joule heating; thermal gradient

---

1. Introduction

There have been many efforts to miniaturize the gas chromatography column system since its introduction by Terry [1] and subsequent efforts by Reston and Kolesar [2] in 1990. Sandia National Lab was the first to integrate a gas chromatography (GC) column, preconcentrator, and chemical sensor arrays into a hybrid system for fast detection of specific analysis commonly present in chemical warfare. Chia-Juang Lu [3] developed the first-generation hybrid MEMS gas chromatograph system, which uses air as the carrier gas and an anodic bonding technique for sealing the silicon etched column to a Pyrex glass. In this system the temperature is regulated using a Kapton embedded resistive wire, with the heat generated by the wire conducted to the bottom of the column. The limitations of this method include the temperature's non-uniformity and the process of heating the column is slow. Significant progress has been made by different research groups [4–8] to miniaturize the GC system to a portable low power, low cost GC system capable of separating and detecting all the volatile organic compounds (VOCs) in a short time; however, band broadening and slow temperature programming are shortcomings of these instruments.

Temperature is the most prominent variable that has significant impact on separation performance, sorbent selectivity and peak spreading of MEMS GC system. Temperature changes the average kinetic energy rate of diffusion and interaction of compounds in the stationary phase. Two methods used commonly for controlling the temperature on a chip are the isothermal and temperature-programmed GC methods. The isothermal GC method provides a higher resolution than the temperature-programmed method, at the cost of a longer processing time and a relatively narrow boiling temperature range. For both methods, the column maintains a constant temperature at any given time. In 1951, Zhukhoviskii [9] introduced the axial temperature gradient, where the temperature is not only varying with time but also in location along the length of the column. Axial temperature-programmed in certain conditions demonstrated an effective method for improving the band broadening and band coelution; however, this method was not adopted for the standard 30 meter column due to the complexity of controlling the temperature at different locations along the column. Phillips [10,11] demonstrated the use of an axial temperature gradient by utilizing direct resistive heating on a short standard column capillary. However, even though Philips and Zhukhoviskii improved the band broadening, they faced many challenges coating the GC column with different thicknesses and generating a temperature gradient. Recently, Zhao [12] demonstrated implementation of this technique on a short micro pack column capillary for separation of a complex mixture of saturated hydrocarbons in the range between C_1 and C_7.

The advancements in micro-fabrication processing and the rise of MEMS GC systems have made it possible to implement an axial temperature gradient which is impractical for standard capillary GC. Axial heating is a good method to integrate into a MEMS GC system mainly due to the small thermal mass, low power consumption, portability and short analysis time of the columns. Despite the tremendous improvement in separation efficiency, the new GC MEMS column still has not resolved issues such as experimental time, peak coelution, and power consumption, especially for separating a complex mixture. Developing a low powered, faster heating time would be advantageous in shortening the separation time and improving the resolution.

Resistive heaters are the standard methods utilized for temperature programming of MEMS columns. As outlined previously, there are two modes: isothermal GC [3] and temperature-programmed [13]. Resistive heater integration on silicon has been demonstrated by many research groups for temperature programming and isothermal analysis; however, the amount of time and power required to ramp and sustain the column at a desired temperature is too large for a portable system. Therefore, fabrication of an all silicon column with a small thermal mass and a heater capable of operating in the temperature ranges of 30–130 °C in less than a minute is vital to design a fast, low-power, temperature programmed, and portable GC systems.

2. Experimental Section

2.1. Computational Modeling

Precise temperature control of a GC column is the most important variable affecting separation performance and peak spreading. In conventional GC systems, for precise temperature control and fast ramping, columns are temperature controlled using a convection oven. However, the time and power required to rapidly increase and sustain the oven's temperature for each temperature set-point is critical. As a result, each sample may take several hours to process. In order to reduce the wait time and minimize the power consumption, we designed a novel heating technique by controlling the temperature gradient throughout the length of the column and provide a rapid heating and cooling response time. The heating elements operate in two modes; (1) isothermal mode (2) simultaneous temperature gradients along the column mode. It was explicitly shown that the temperature gradient along the capillary column improves the band spreading and compression, which ultimately reduces band broadening.

On the reverse side of the column, an axial negative temperature gradient is generated by applying a fixed voltage across the heating element. The temperature is hotter in the center (inlet of the column) than the perimeter (outlet) of the micro-GC, using a linear temperature profile. As the sample passes through different temperature zones, the diffusion rate changes so that the front of the separation peak moves slower relative to the trailing edge of the peak, thereby improving the resolution of the compounds passing through the column. As a result for a given number of theoretic plates, a shorter column length can be used. Typically, shortening the length of a GC column reduces the analysis time at the expense of resolution; however, thermal refocusing of the eluting band will improve the resolution of the shorter columns. The gradient along the column continuously refocuses the eluting bands, offsetting part of the chromatographic band spreading and consequently sharpens the peaks as they move down the column. The more volatile compounds move faster at a higher temperature and will focus at a lower temperature, exiting the column quicker than the non-volatile compounds. The band broadening of compounds as they move down the length of the column could be explained by the following equation [11]:

$$\sigma = \sqrt{HR(X)} \tag{1}$$

where σ is the band standard deviation, *R*(*X*) is the distance as a function of radius, and *H* is the column efficiency. The assumption is that the efficiency of a column is independent of time and the position along the column for a MEMS GC, mainly due to the short length of the column and a small pressure drop. The variance of a band increases in direct proportion to the distance moved down the column and can be explained with the following equation:

$$\left(\frac{\mathrm{d}\sigma^2}{\mathrm{d}t}\right) = \frac{H\overline{u}}{(k+1)} \tag{2}$$

where *k* is the capacity factor and *u* is the linear velocity.

$$L = \int_a^b r^2 + \left(\frac{\mathrm{d}r}{\mathrm{d}\theta}\right)^2 \mathrm{d}\theta \tag{3}$$

On the top of the column, four resistance-heating elements are integrated for side-by-side comparison of the isothermal analysis with the radial heating analysis.

The main challenge facing the implementation of the axial heating design is the ability to control the temperature profile along the length of the column. To achieve this, three heating element designs were investigated to form a hot spot. The heaters were made of a material with high thermal conductivity, so that the applied voltage to the heater dissipates in a relatively short time. Even though the most widely used material for micro-fabrication of resistance heaters on silicon is gold, platinum was chosen due to the material's resistivity. Figure 1 shows the heater design and Table 1 summarizes the dimensions of the designed heater.

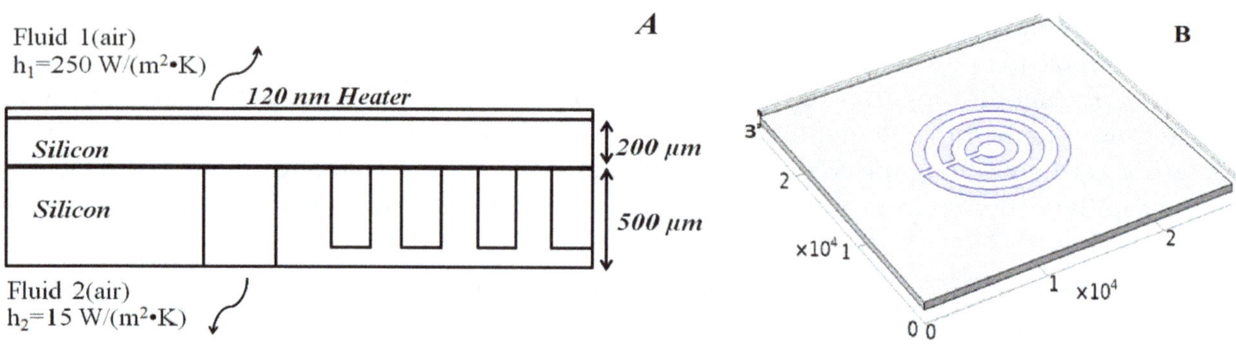

Figure 1. (**A**) Side view of GC column with heat losses; (**B**) layout the heater.

Table 1. The dimensions and resistance of the designed heater.

Heater		Resistance(Ω)	
Length (mm)	**Thickness (nm)**	**Platinum Layer**	**Gold Layer**
4.71	125	7.84	1.52
11.00	125	18.29	3.54
17.28	125	28.75	5.55

In order to investigate the simultaneous temperature gradients induced by the heater on silicon, a new heater was designed and simulated using COMSOL Multiphysics® (COMSOL Group, Stockholm, Sweden). The 3D simulation was performed by coupling the electric current and heat transfer, and applying the numerical methods described in our previous thermal simulations [14,15]. For simplicity, the following assumptions were made: (1) the effect of the insulating SiO_2 layer was insignificant to the overall heat transfer simulation; (2) the convection heat transfer from the edges of the silicon to the surroundings were assumed to be small; (3) the heat losses due to the traverse of helium gas through the column was compensated by increasing the natural convention coefficient on the reverse side of the column; and (4) the heat losses due to low temperatures were small compared to the combined heat loss, and were neglected.

The resistance heating layer was the active surface and simulated with *the shell conductive AC/DC layer* which is governed by:

$$\nabla_t.(-d\sigma\nabla_t V) = 0 \tag{4}$$

where d is the thickness of platinum thin film (125 nm), sigma is the electrical conductivity (S/m), V is the electric potential (V), and ∇_t denoted the gradient operation in the tangential directions. The electrical conductivity of the heater was adjusted by a temperature dependent equation:

$$\sigma = \frac{1}{\rho_0[1 + \alpha(T - T_0)]} \tag{5}$$

where ρ_0 is the resistivity at temperature T_0 ($\rho_0 = 2.41\times10^{-7}$ Ω·m), α is the thermal coefficient of resistance of Platinum ($\alpha = 2.43\times 10^{-3}$ K^{-1}) [16].

The heat generated by the joule heating was coupled with the heat transfer module. The *highly conductive layer* feature of the heat transfer interface was used for joule heating of the thin layer. The power per unit area (W/m^2) generated inside the thin conductive layer is governed by:

$$q_{prod} = dQ_{DC} \tag{6}$$

where Q_{DC} is the power density. At the steady-state, the heat generated by the joule heating is conducted to the silicon column and some dissipates by forced and natural convection from the top and bottom of the silicon piece to the surrounding air.

2.2. Device Fabrication and Characterization

2.2.1. Chip Fabrication

The process for fabrication an all-silicon GC columns was described in our previous work [17,18]. To investigate the axial temperature gradient, three heater designs were investigated. To fabricate the integrated heaters and a temperature sensor on a previously micro-fabricated column, ShipleyMICROPOSIT® S1318 photoresist (Dow, Midland, MI, USA) was spin-coated on the column at 1000 rpm for 40 s. The column was then soft-baked at 160 °C for 60 s and was then exposed using MA-6 mask aligner (SÜSS MicroTec AG, Garching, Germany). The column was then immersed in RA6 developer for 15 s and was post–baked at 90 °C for 3 min. Denton Explorer was used to deposit 10 nm of Titanium as an adhesion layer subsequently 120 nm of platinum was deposited. The column was then immersed in acetone for several minutes and then was rinsed with deionized (DI) water. Figure 2 shows the fabrication process flow of the heaters.

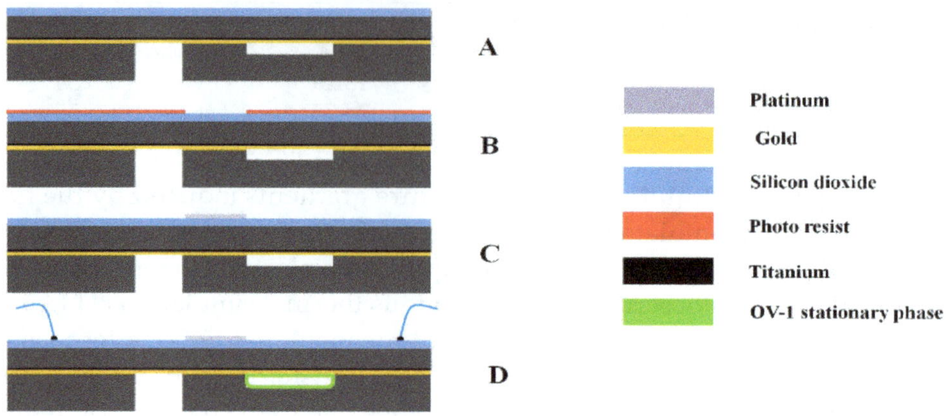

Figure 2. (**A**) 200 nm oxide grown thermally; (**B**) the top-side is coated with photoresists; (**C**) 115 nm of Ti/Pt deposited using E-beam evaporator; (**D**) wire bonded the heaters to the package.

2.2.2. Stage Assembly

A flexible stage was designed to reduce the conductive loss and improve the convective loss by exposing one side to a natural convention and the other side to a forced convection, controlled by an axial fan. Two probes were used to apply a constant power to the heater while another two probes concurrently measured the temperature of the sensor at different locations on the chip. A 6 digit Agilent meter was employed for measuring the resistance change of the platinum sensors. The schematic of the experimental apparatus and the location of the fan relative to the column are presented in Figure 3.

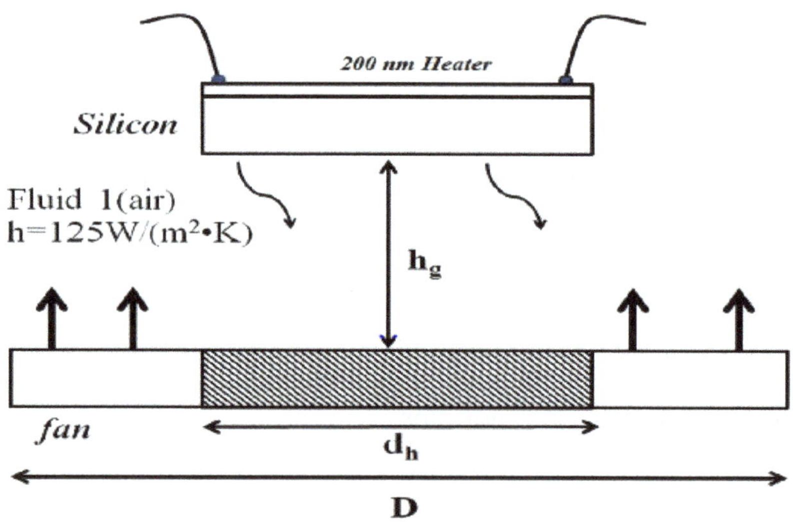

Figure 3. Experimental setup.

3. Results and Discussions

3.1. Computational Modeling

The simulation method was carried out through a transient response, using the room temperature as the initial condition. Figure 4 presents the structure of the column and the calculated temperature field when 6 W power was applied and the force convection heat transfer coefficient was at 150 W/m^2K.

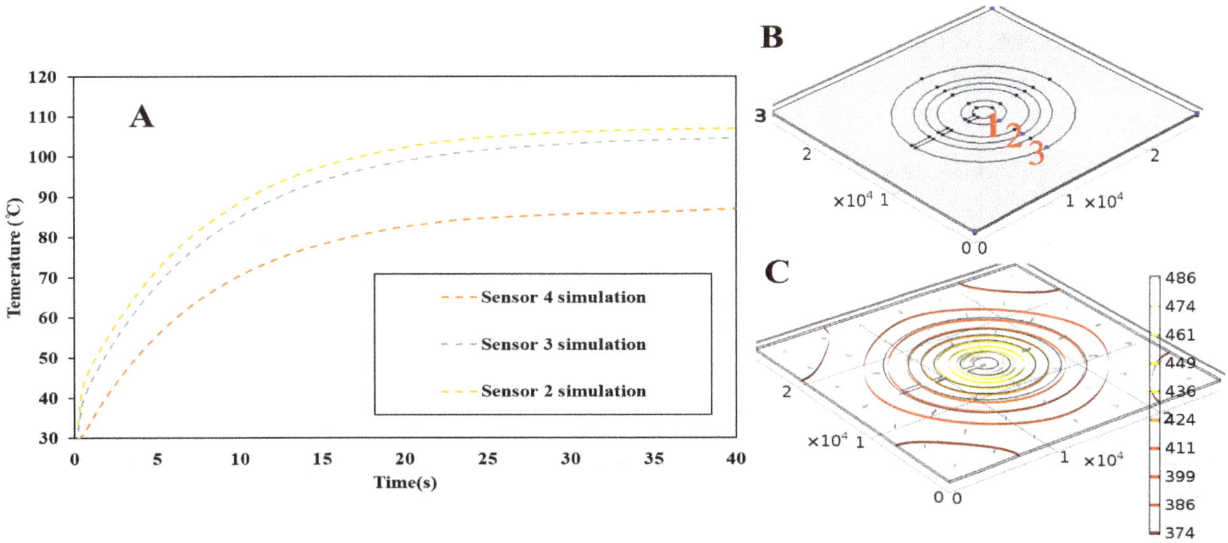

Figure 4. (**A**) Temperature distribution with 6 W of electrical power dissipated in the heating element on the GC silicon column and surrounding air at ambient temperature of 20 °C and (**B**) sensors location; (**C**) temperature profile.

After validation of the model, three heater and temperature sensor designs were investigated. Figure 5 shows the three different heater size and temperature sensor designs. Table 2 provides the dimensions and calculated resistance value of the heaters for each design. The simulation method depicted in Figure 6 shows design b as the best heater design, which provides the largest temperature gradient on the chip.

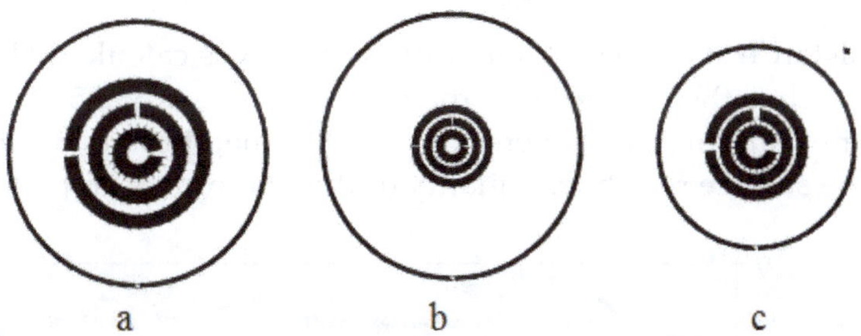

Figure 5. Mask designs of the radial heater and the temperature sensor: (**a**) Design a; (**b**) Design b; (**c**) Design c.

Table 2. The dimensions of the three heater designs and the theoretically calculated the platinum resistance.

Design	Radius 1	Radius 2	Length (mm)	Track Resistance (Ω)
Design a	0.001	0.00175	6.28	6.97
	0.00225	0.003	14.14	15.68
	0.0035	0.00425	21.99	24.40
Design b	0.00075	0.00125	4.71	7.84
	0.00175	0.00225	10.99	18.30
	0.00275	0.00325	17.28	28.75
Design c	0.001	0.002	6.28	5.22
	0.003	0.004	18.85	15.68
	0.005	0.006	31.42	26.14

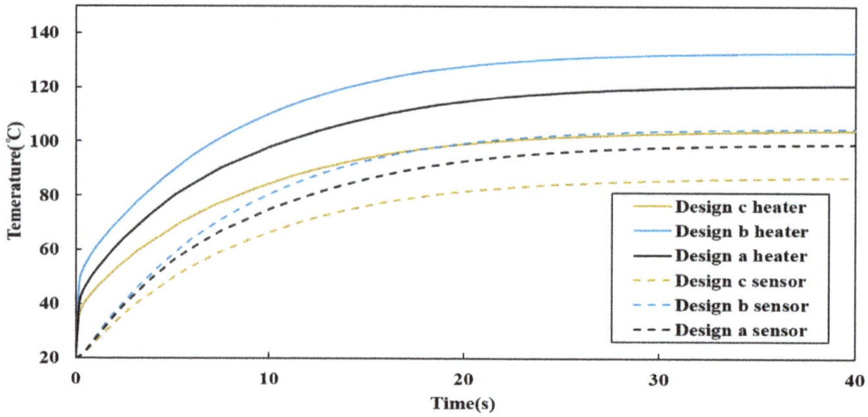

Figure 6. Simulated transient response of temperature over time after applying 6 W of electrical dissipation to the heater, for the three designs.

3.2. Heater Characterization

To validate the simulation data, the electrical properties of a thin film of platinum were characterized. The experiment was carried out by placing the column on a temperature controlled probing station while the change in resistance of the sensor was measured using a digital multimeter. The measurement was performed in a temperature range from 30 to 80 °C and a thermocouple (with accuracy of 0.1 °C) was used to measure the temperature. The Temperature Coefficient of Resistance (TCR) and the base line resistance of the thin film of platinum were calculated by plotting (Figure 7) the measured resistance of the sensor against temperature of the thermocouple and fitting it to the following equation:

$$R = R_0\alpha(T - T_0) + R_0 \tag{7}$$

where R_0 is the base resistance at the base temperature To, α was calculated by dividing the slope of the linear fit with R_0, which is the intercept on the y axis. Hence, $\alpha = 2.54 \times 10^{-3}$ and $R_0 = 18.16\ \Omega$. The TCR values of the thin film platinum were 65% smaller than bulk platinum due to impurities in the platinum and also the surface roughness quality of the platinum film [16].

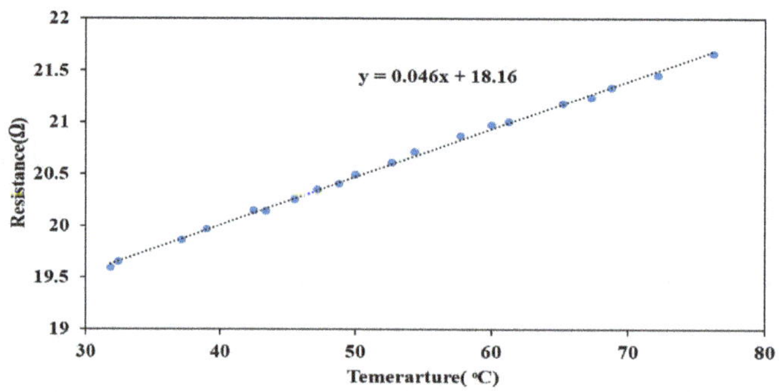

Figure 7. Temperature calibration data for the heater where each data point was obtained by subjecting the column to a uniform temperature on a heated plate.

3.3. Coefficient of Convection Characterization

The convective heat transfer coefficient is critical factor affecting the temperature gradient generated in the column. At a higher convective coefficient a larger temperature gradient is produced at the cost of increasing the power consumption. To obtain a value for the coefficient of convective heat transfer, a model of the column heating geometry was simulated with different heat transfer coefficients of convection, and compared to the experimental data collected with the fan operating at

12 V. Figure 8 shows the convective heat transfer coefficient produced by the fan and was measured to be 125 W/m^2 based on the distance from the column, which agrees with previously reported data [19].

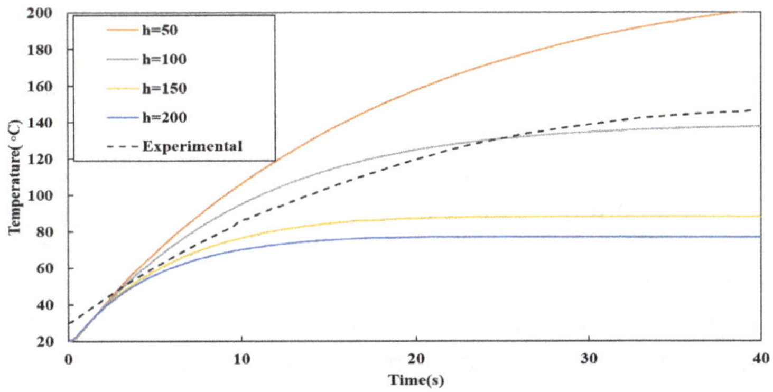

Figure 8. Plot of simulated average temperature transient response of the resistor film with 2 W electrical power applied to the heater compared to experimentally measured temperature on the micro-GC column with convection heat transfer coefficient in air as a parameter.

3.4. Experimental Setup

To test the effectiveness of the heater, the heat transfer performance on the chip was quantified by measuring the resistance change of the platinum sensors at different locations and using temperature profiles using infrared thermography. Here, the 700 µm thick column is heated by resistive heating and cooled by an axial flow fan. A fixed current of 0.6 A is applied to the center of the platinum heater, resulting in heating of the column by joule resistive heating. A 12 V DC fan was employed by placing it at a 20 mm distance from the chip to generate sufficient convective heat transfer in order to cool the column. A thermal image of the GC column was captured using a Ti27 infrared digital camera (Fluke, Everett, WA, USA) with an InSb detector. Furthermore, the temperature of the column at different locations was measured using temperature sensors embedded on the column. A calibration of the temperature sensors were carried out prior to the tests to ensure an accurate temperature measurement.

Figure 9 shows the transient response of the heater in response to an applied current of 0.6 A. Here the temperature is measured at three different locations on the column. The total power applied to the chip was calculated to be 10 W, which is sufficient to generate a temperature gradient of 40 °C in less than 40 s.

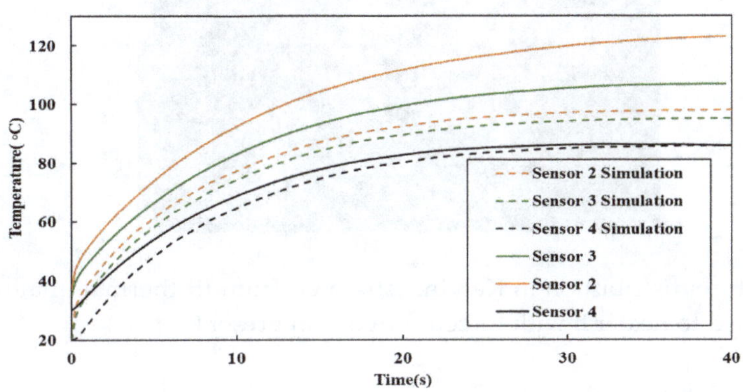

Figure 9. Transient response of heater design b with a 0.5 A applied constant current.

A transient simulation of the heater temperature was performed and the results were compared with the experimental data collected for the 26 by 26 mm column. The simulation data was modified based on the TCR and thermal resistivity of the platinum heater. The simulation response shows a strong agreement between the measured temperatures and the simulation data.

There was a small discrepancy between the experimental and simulation data as a result of the impinging airflow produced by the fan, which results in variation in the local heat transfer coefficient across the column.

The fan to column distance was fixed at 20 mm [19], and the fan was blowing air axially and perpendicular to the surface of the column. The applied power results in the joule heating of the silicon, and the temperature was measured at three different locations on the chip. Figure 10 shows the transient response of the GC column with an embedded heater at two different power levels (I^2R_0), where I is the high current value and R_0 is the base resistance. As expected, the transient heating time constant increases with power and provided a larger temperature gradient from the center to the edge of the column. We expected significant improvement in column performance when the gradient produced results in a temperature different of more than 40 degrees.

Temperature distribution across the silicon column was further investigated using an infrared camera, to confirm the temperature gradients produced at different levels of heat transfer by force convection. Silicon was assumed to diffuse and gray with an emissivity of 0.65 for temperature estimates. Figure 11 shows the IR image of the temperature distribution on the column.

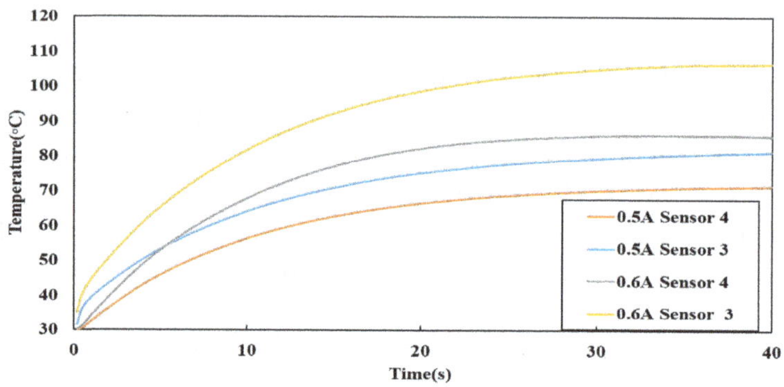

Figure 10. Transient response of the heater at two different power levels.

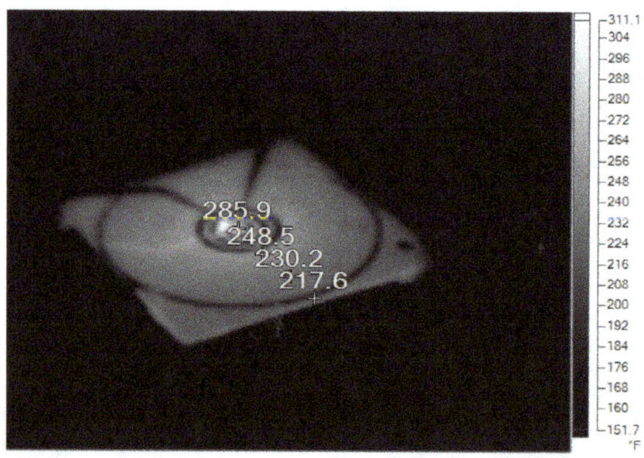

Figure 11. Temperature distribution in Kelvin, estimated from IR thermal imaging with an applied voltage of 13 V applied to heater b with forced convection present.

3.5. Column Performance Evaluation

To evaluate the focusing of the peaks band, the 3 meter GC column was operated in isothermal, temperature gradient (TG), and temperature gradient programmed mode. A simple mixture of hydrocarbons (hexane, octane, and decane) was prepared to test the device over an appreciable range of volatilities.

The column was connected to an Agilent 6890 GC-FID (Agilent Technologies, Santa Clara, CA, USA) and the chromatograms of the three compounds were obtained by injecting 1 μL of the mixture sample (split ratio of 400:1, Pressure 16 psi, and 60 °C center temperature. Figure 12A displays the chromatogram obtained for the isothermal run. The peak widths and shapes of hexane (C_6) and octane (C_8) are acceptable for the length of the chromatographic column, however the peak shape and peak asymmetry of decane (C_{10}) are in clear need of improvement. These runs were performed in triplicate, and the results are consistent within acceptable ranges of statistical variations typically observed in conventional chromatography (5% coefficient of variation or less for retention times, peak area, and peak asymmetry).

Figure 12B shows a chromatogram obtained for a run in which the device was operated isothermally at 60 °C in the thermal gradient mode. The application of the axial thermal gradient is observed to reduce the residence time of the analytes through the system, which results in a reduction in the retention times for all 3 compounds. But more importantly, the peak widths of the analytes are reduced as well, due to the reduction of the peak broadening that typically occurs as the compounds are travelling through the column. Table 3 shows that the peak focusing is significant (from 10%–20% for the compounds in the test sample) and compound specific. It is also observed that for the decane peak the severe tailing that was occurring at isothermal conditions (Figure 12A) has been corrected, and the peak asymmetry has improved substantially. These series of runs were also run in triplicate, and chromatograms were consistent within acceptable statistical ranges. The chromatogram shown in Figure 12B, thus, demonstrates the utility of the negative thermal gradient in the improvement of chromatographic resolution.

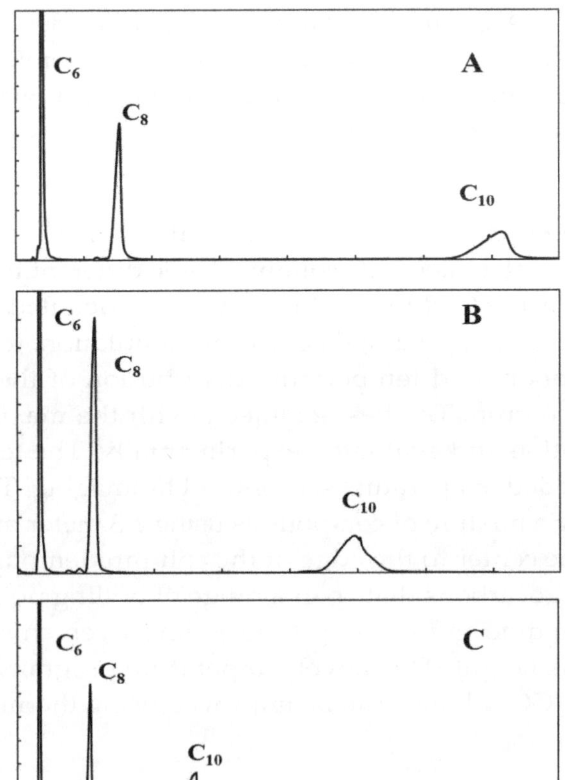

Figure 12. Chromatograph achieved using the 3 meter column at pressure 16 psi and inlet temperature of 240 °C with OV-1 stationary phase using (**A**) 60 °C isothermal temperature (**B**) temperature gradient with center temperature of 60 °C (**C**) Temperature gradient programmed at 250 °C/min from 60 to 110 °C.

Table 3. Peak focusing calculated for Temperature gradient GC (TGGC) and Temperature programmed gradient GC (TPGGC).

Peak	TGGC	TPGGC
C_6	11.19%	19.15%
C_8	20.60%	67.94%
C_{10}	12.41%	69.03%

Figure 12C shows a chromatogram obtained for a run in which the device was operated in a multidimensional temperature programming mode, meaning that there is a temperature program in time (ranging between 60 and 110 °C at 4.2 °C/s) that is superimposed on the temperature program in distance (the negative thermal gradient that was applied in Figure 12B). The combined effect of these two temperature program regimes is a further reduction in retention times that is accompanied with additional peak focusing. Once again, these improvements are compound specific, but significant enhancement is observed for all compounds in the mixture. For a late eluting compound like decane, the impact of both temperature programs is particularly salient. The fast temperature program in time is responsible for the sharp reduction in retention time, and the temperature programming in distance is responsible for the reduction in the peak width, which enhances the peak height due to the fact that the peak area is unchanged. This combined effect is quite advantageous for separations on columns whose length is markedly shorter than those used in conventional chromatography, because it provides a tool to enhance the chromatographic resolution. Overall, the decane peak's retention time has been reduced by over a factor of 2, and its peak height has been enhanced by over a factor of 2 between the conditions used in Figure 12A,C. Similar optimization of bi-dimensional temperature programs were tested for compound mixtures containing a wider variety of chemical functional groups (such as alcohols, ketones, and fatty acid methyl esters), and similar observations were obtained.

4. Conclusions/Outlook

In this work, we investigated the effects of temperature gradients as a function of time and position, $T(t,x)$, along the length of a micro-GC column. Three different heater designs were evaluated through numerical modeling using COMSOL Multiphysics. A detailed geometry of the GC column with thin film resistive heaters was used as a basis for the simulation, to investigate the heating and cooling rate, power consumption, and temperature distribution of the new axial heating method for the all silicon micro-GC column. The best geometry with the maximum temperature gradient was selected for microfabrication and evaluated experimentally. The temperature distribution was evaluated using several embedded temperature sensors and IR imaging. The effectiveness of the heater was investigated by separating a mixture of compounds using a 3 meter micro-GC column. The results with a 30 °C gradient from the center to the edge of the column demonstrate enhanced separations for a test mixture of three hydrocarbons that span a range of boiling points of over 100 °C. The dual temperature programming (in time and space) produced improvements in peak focusing that result in significant increases in peak height. This novel temperature programming method demonstrates that the performance of micro-GC columns can be improved using thermal programing techniques to increase analytical performance.

Acknowledgments: The authors would like to thank the staff of Nanotechnology Research Center and the Agricultural Technology Research Program (ATRP) at Georgia Tech Research Institute for financial support.

Author Contributions: Milad Navaei designed and conducted laboratory experiments, analyzed data and wrote the paper. Alireza Mahdavifar helped with the simulation analysis. Jean-Marie D. Dimandja helped with the chromatography experiments. Peter Hesketh provided assitance with technical and cleanroom process. Gary McMurray helped with directing the project.

References

1. Terry, S.C.; Jerman, J.H.; Angell, J.B. A gas chromatographic air analyzer fabricated on a silicon wafer. *IEEE Trans. Electron Devices* **1979**, *26*, 1880–1886. [CrossRef]
2. Reston, R.R.; Kolesar, E.S. Silicon-micromachined gas chromatography system used to separate and detect ammonia and nitrogen dioxide. I. Design, fabrication, and integration of the gas chromatography system. *J. Microelectromech. Syst.* **1994**, *3*, 134–146. [CrossRef]
3. Lu, C.-J.; Steinecker, W.H.; Tian, W.-C.; Oborny, M.C.; Nichols, J.M.; Agah, M.; Potkay, J.A.; Chan, H.K.L.; Driscoll, J.; Sacks, R.D.; *et al.* First-generation hybrid MEMS gas chromatograph. *Lab Chip* **2005**, *5*, 1123–1131. [CrossRef] [PubMed]
4. Yamanaka, T.; Matsumoto, R.; Nakamoto, T. Study of recording apple flavor using odor recorder with five components. *Sens. Actuators B Chem.* **2003**, *89*, 112–119. [CrossRef]
5. Albert, K.J.; Walt, D.R.; Gill, D.S.; Pearce, T.C. Optical multibead arrays for simple and complex odor discrimination. *Anal. Chem.* **2001**, *73*, 2501–2508. [CrossRef] [PubMed]
6. Kim, S.K.; Chang, H.; Zellers, E.T. Microfabricated gas chromatograph for the selective determination of trichloroethylene vapor at sub-parts-per-billion concentrations in complex mixtures. *Anal. Chem.* **2011**, *83*, 7198–7206. [CrossRef] [PubMed]
7. Zampolli, S.; Elmi, I.; Mancarella, F.; Betti, P.; Dalcanale, E.; Cardinali, G.C.; Severi, M. Real-time monitoring of sub-ppb concentrations of aromatic volatiles with a MEMS-enabled miniaturized gas-chromatograph. *Sens. Actuators B Chem.* **2009**, *141*, 322–328. [CrossRef]
8. Garg, A.; Akbar, M.; Vejerano, E.; Narayanan, S.; Nazhandali, L.; Marr, L.C.; Agah, M. Zebra G.C: A mini gas chromatography system for trace-level determination of hazardous air pollutants. *Sens. Actuators B Chem.* **2015**, *212*, 145–154. [CrossRef]
9. Zhukhovitskii, A.A.; Zolotareva, O.V.; Sokolov, V.A.; Turkel'taub, N.M. New method of chromatographic analysis. *Dokl. Akad. Nauk SSSR* **1951**, *77*, 435–438.
10. Jain, V.; Phillips, J.B. High-speed gas chromatography using simultaneous temperature gradients in both time and distance along narrow-bore capillary columns. *J. Chromatogr. Sci.* **1995**, *33*, 601–605. [CrossRef]
11. Phillips, J.B.; Jain, V. On-column temperature programming in gas chromatography using temperature gradients along the capillary column. *J. Chromatogr. Sci.* **1995**, *33*, 541–550. [CrossRef]
12. Zhao, H.; Yu, L.; Zhang, J.; Guan, Y. Characteristics of TGPGC on short micro packed capillary column. *Anal. Sci.* **2002**, *18*, 93–95. [CrossRef] [PubMed]
13. Agah, M.; Potkay, J.A.; Lambertus, G.; Sacks, R.; Wise, K.D. High-performance temperature-programmed microfabricated gas chromatography columns. *J. Microelectromechan. Syst.* **2005**, *14*, 1039–1050. [CrossRef]
14. Mahdavifar, A.; Aguilar, R.; Peng, Z.; Hesketh, P.J.; Findlay, M.; Stetter, J.R.; Hunter, G.W. Simulation and fabrication of an ultra-low power miniature microbridge thermal conductivity gas sensor. *J. Electrochem. Soc.* **2014**, *161*, B55–B61. [CrossRef]
15. Mahdavifar, A.; Navaei, M.; Aguilar, R.; Hesketh, P.J.; Hunter, G.; Findlay, M.; Stetter, J.R. Transient Thermal Response of Micro TCD for Identification of Gases. In Proceedings of the 224th ECS Meeting, San Francisco, CA, USA, 27 October–1 November 2013. Abstract 2795.
16. Butts, D.A.; Taarea, D. Chapter 19—Electrical properties. In *Smithells Metals Reference Book*, 8th ed.; Gale, W.F., Totemeier, G.C., Eds.; Butterworth-Heinemann: Oxford, UK, 2004.
17. Navaei, M.; Xu, J.; Mahdavifar, A.; Dimandja, J.; McMurray, G.; Hesketh, P. Micro-fabrication of all silicon 3-meter GC column using gold eutectic fusion bonding. *J. Electrochem. Soc.* **2015**, in press.
18. Navaei, M.; Xu, J.; Hesketh, P.; Wallace, R.; McMurray, G. Micro gas chromatography system for detection of volatile organic compounds released by Fungi. In Proceedings of the 224th ECS Meeting, San Francisco, CA, USA, 27 October–1 November 2013.
19. Stafford, J.; Walsh, E.; Egan, V.; Grimes, R. Flat plate heat transfer with impinging axial fan flows. *Int. J. Heat Mass Transf.* **2010**, *53*, 5629–5638. [CrossRef]

4

Balloon Pump with Floating Valves for Portable Liquid Delivery

Yuya Morimoto [1,2], Yumi Mukouyama [1], Shohei Habasaki [1,2] and Shoji Takeuchi [1,2,*]

1 Center for International Research on Integrative Biomedical Systems (CIBiS), Institute of Industrial Science (IIS), The University of Tokyo, 4-6-1 Komaba, Meguro-ku, Tokyo 153-8505, Japan; y-morimo@iis.u-tokyo.ac.jp (Y.M.); yumi.mukouyama@gmail.com (Y.M.); habasaki@iis.u-tokyo.ac.jp (S.H.)

2 Takeuchi biohybrid Innovation Project, Exploratory Research for Advanced Technology (ERATO), Japan Science and Technology (JST), Komaba Open Laboratory (KOL) Room M202, 4-6-1, Komaba, Meguro-ku, Tokyo 153-8904, Japan

* Correspondence: takeuchi@iis.u-tokyo.ac.jp

Academic Editors: Manabu Tokeshi and Kiichi Sato

Abstract: In this paper, we propose a balloon pump with floating valves to control the discharge flow rates of sample solutions. Because the floating valves were made from a photoreactive resin, the shapes of the floating valves could be controlled by employing different exposure patterns without any change in the pump configurations. Owing to the simple preparation process of the pump, we succeeded in changing the discharge flow rates in accordance with the number and length of the floating valves. Because our methods could be used to easily prepare balloon pumps with arbitrary discharge properties, we achieved several microfluidic operations by the integration of the balloon pumps with microfluidic devices. Therefore, we believe that the balloon pump with floating valves will be a useful driving component for portable microfluidic systems.

Keywords: microfluidic device; portable device; optofluidic lithography

1. Introduction

Microfluidic devices have been extensively developed to enable their application in chemical, biological, and biomedical processes because of their rapid processing and high sensitivity by minimizing the sample size [1,2]. Recently, integrated systems comprising microfluidic devices and liquid feed devices have garnered attention for use in point-of-care diagnostic testing and scientific studies without employing expensive and bulky equipment [3,4]. In such systems, the liquid feed devices with driving sources enable the precise and continuous control of samples [5–10]; however, the use of these devices result in low application possibilities for *in situ* use of the systems because the driving sources have low portability and need external components (e.g., electric sources) for their operation.

To solve the abovementioned problem, portable microfluidic systems based on osmotic pressure [11,12], capillary flow [13], negative pressure generated using a vacuumed chamber [14,15], surface tension of droplets [16], and finger power [17–20] have been developed to generate flow in microfluidic channels without the use of external components. Although these microfluidic systems allow *in situ* use as portable devices, they do not provide continuous liquid delivery and closed channels and/or inlets for preventing the evaporation and contamination of liquids. As an improved microfluidic system that can solve the abovementioned issues, Gong *et al.* integrated a balloon pump with a syringe to manipulate samples in a closed system [21]. When samples are infused via the syringe, the balloon pump allows the storage of liquids inside it by inflating the balloon and facilitates the transport of liquids by deflating the balloon. Therefore, the balloon pump provides continuous liquid delivery and ensures the conservation of independent environments, thereby making it an attractive candidate for use as a portable microfluidic system. However, the discharge liquid flow rates cannot be easily regulated in the balloon pump because the flow rates depend on the deflation pressure of the balloon. To change the flow rates, the balloon pump needs to be redesigned to incorporate additional components such as microchannels that work as flow resistors using specialized equipment for microfabrication.

In this study, we develop a balloon pump with floating valves to change the flow rates of liquids (Figure 1). The advantages of the balloon pumps are: (i) fabrication of floating valves without specialized equipment for microfabrication; (ii) changeable discharge flow properties by conditions of the floating valves; and (iii) liquid delivery without any additional manipulation. As a fabrication technique for floating valves, we apply optofluidic lithography to a photoreactive resin in microfluidic channels made from polydimethylsiloxane (PDMS) [22]. By mounting an exposure system to a microscope, we can prepare the floating valves in the balloon pump without use of any microfabrication techniques. Using this method, we can control the length and width of the floating valves by changing the exposure pattern. Furthermore, we can vary the shapes of the floating valves in the PDMS microchannels by changing the oxygen concentration in the balloon pump because the oxygen absorbed in the PDMS channels inhibits the polymerization of the photoreactive resin [23]. By controlling the floating valve dimensions, we can easily prepare flow resistors with arbitrary properties and hence control the properties of liquid discharge. The floating valves are placed in the microchannels and create narrow gaps with the walls of the microchannels. Because the narrow gaps can act as flow resistors and decrease the flow rates [24,25], the balloon pump can change the discharge flow rates according to the dimensions of the floating valves. Consequently, as a demonstration of microfluidic operations using the proposed balloon pump, we present the manipulation of microsized beads in dynamic microarray devices [26] and the formation of laminar flows. Because these microfluidic operations need flow control, the demonstration indicates the potential of the balloon pumps for portable microfluidic systems to easily generate controlled flows without the use of external driving elements.

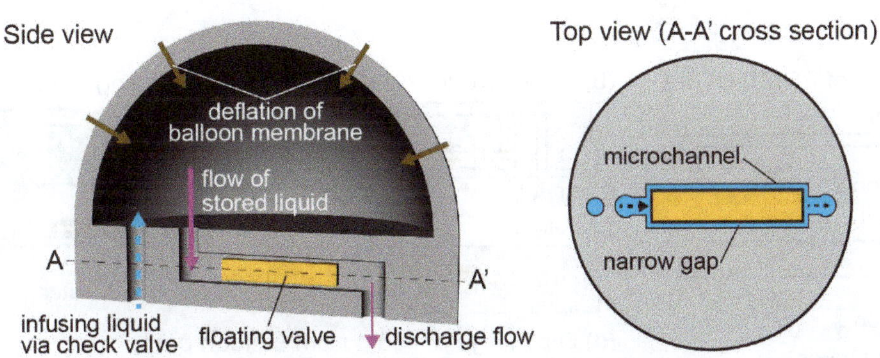

Figure 1. Conceptual illustration of a balloon pump with a floating valve.

2. Experimental

2.1. Materials

For the fabrication of the balloon pump, we used PDMS and a curing agent (Sylgard 184 Silicone Elastomer, Dow Corning Toray Co., Ltd., Tokyo, Japan), a photoreactive acrylate resin (R11, 25–50 μm layers, EnvisionTEC, Dearborn, MI, USA), parylene (parylene-C, Specialty Coating Systems, Inc., Indianapolis, IN, USA), and SU-8 (SU-8 50, MicroChem Corp., Westborough, MA, USA). The materials used for manufacturing the floating valves were polyethylene glycol diacrylate (PEGDA) (Sigma-Aldrich, St. Louis, MO, USA, average Mn = 700) and phenylbis

((2,4,6-trimethylbenzoyl)phosphine oxide) (Sigma-Aldrich). To demonstrate the microfluidic operations of the balloon pump with the floating valves, we used 100 μm beads (PS-Red-Particles, microParticles GmbH, Berlin, Germany), Tween 20 (Kanto Chemical Co., Inc., Tokyo, Japan), and blue ink (Pilot Corp., Tokyo, Japan).

2.2. Device Design and Fabrication

Figure 1 shows a schematic illustration of the balloon pump with the floating valves. The balloon pump works as a pneumatically driven pump because of the inflation and deflation of the balloon membrane. In addition, the floating valves work as flow resistors by the formation of narrow flow paths with the channel walls. By adjusting the width and length of the flow paths, we can control the discharge characteristics of the balloon pump.

To fabricate the balloon pump, we integrated a balloon layer, an intermediate layer, and a microchannel layer, all made from PDMS. We shaped these three layers by molding. The molds for the balloon layer and intermediate layer were made by carrying out stereolighography on a photoreactive acrylate resin using a modeling machine (Perfactory, EnvisionTEC). After the fabrication of the molds, we exposed them to ultraviolet (UV) light for over 60 s using a laser machine (UV-LED, Keyence Corp., Osaka, Japan) to ensure complete curing. Then, we coated them with a 2 μm parylene layer using a chemical vapor deposition machine (Parylene Deposition System 2010, Specialty Coating Systems, Inc.) to avoid direct contacts between PDMS and the surface of the resin mold that cause non-solidification of PDMS. The mold for the microchannel layer was made from SU-8 by standard soft lithography techniques. After spin-coating SU-8 on a silicon wafer to form a layer with a height of 60 μm, we fabricated the SU-8 mold by UV light exposure through a photomask (Clean Surface Technology Co., Ltd., Kanagawa, Japan) designed using a mask exposure machine (D-Light DL-1000, NanoSystem Solutions, Inc., Okinawa, Japan).

We filled the three molds with a PDMS-curing agent mixture in the ratio of 10:1 (*w*/*w*). After peeling the cured PDMS layers from the molds manually without using any organic solvents, we coated a specific surface of the intermediate layer with pre-cured PDMS using a pipette and heated it for 55 min at 60 °C to fabricate a semi-cured PDMS layer. Finally, we bonded all layers by heating them again for 90 min at 75 °C (Figure 2a).

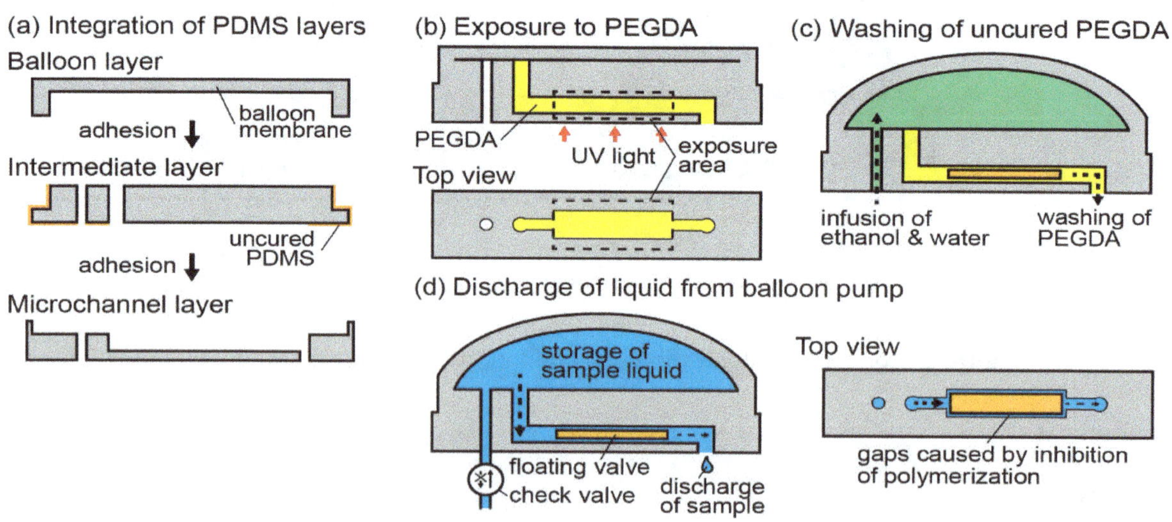

Figure 2. Process flow of the preparation of a balloon pump with a floating valve: (**a**) fabrication of the balloon pump by the integration of polydimethylsiloxane (PDMS) layers; (**b**) solidification of polyethylene glycol diacrylate (PEGDA) using UV light in a microchannel; (**c**) washing of uncured PEGDA using ethanol and water; and (**d**) discharge of the liquid from the balloon pump after the liquid is infused via a check valve.

Next, we constructed the floating valves in the microchannel of the balloon pump. First, we prepared PEGDA with 1% (*w*/*v*) of phenylbis as the photoreactive resin for the floating valves. The photoreactive resin was introduced into the microchannel via an outlet while ensuring that the resin did not touch the balloon membrane. Then, we exposed the resin to UV light in the required shapes of the valves using a microscope (IX71, Olympus Corp., Tokyo, Japan) equipped with a digital micromirror device (DMD) (Figure 2b). Using the DMD, we were able to make the floating valves in arbitrary two-dimensional extruded shapes by changing the exposure patterns, and we were able to fabricate multiple floating valves in one microchannel by changing the exposure area. After the exposure process, we washed the microchannel by infusing ethanol and water via an inlet and aspirating them from an outlet to remove the non-cured resin (Figure 2c). The fabricated valves did not stick to the wall of the microchannel because the oxygen layers on the PDMS surface worked as a PEGDA polymerization inhibitor; therefore, liquids could pass though the gaps between the microchannel wall and floating valve. In the fabrication process of the floating valves, the length of the valve in the flow direction was longer than that of the exposure pattern because of the flow of the photoreactive resin in the microchannel. To make up for the difference between the two lengths, we adjusted the length of the exposure pattern according to a calibration line (Supplementary Figure S1). To increase the widths of the floating valves, we degassed the PDMS devices in a vacuum desiccator (AS ONE Corp., Osaka, Japan), to reduce the thickness of the oxygen layers.

2.3. Setup of the Balloon Pump

We attached a check valve (PU Celsite Port, Toray Medical Co., Ltd., Tokyo, Japan) at the inlet of the balloon pump to prevent backward flow of liquids. Then, we carefully infused liquids into the balloon pump via the check valve using a syringe to avoid trapping air bubbles under the balloon membrane. In case air bubbles were trapped, we infused water again after removal of the previously loaded water with air bubbles. Because of the actions of the check valve and floating valve, the balloon pump stored liquids by the inflation of the balloon membrane. Because liquids could pass through the gaps between the floating valves and channel walls, the pump gradually discharged the liquids by the deflation of the balloon (Figure 2d).

2.4. Evaluation of the Balloon Pump

To evaluate the behavior of the balloon pumps, we checked their external appearance and their discharge characteristics. The external appearance of the balloon pumps was obtained by taking pictures using a digital camera (EOS Kiss X6i, Canon Inc., Tokyo, Japan). For checking the discharge characteristics of the balloon pumps, we stored water in the pump using a syringe and check valve and measured the volume of the discharge water at the outlet of the balloon pump using a pipette.

To check the flow regulation properties of the floating valve in the microchannels, we prepared floating valves in the PDMS microchannels composed of the intermediate layer and microchannel layer; this system was used to control the input flow pressure using a pressure-driven flow pump (MFCS-100, Fluigent, Inc., Villejuif, France). In the experiment, we measured the volume of the discharge water at the outlet of the microchannel using a pipette.

For evaluating the characteristics of the discharge caused by the deflation of the balloon membrane, we prepared devices composed of the balloon layer and intermediate layer to eliminate the effects of the microchannel. We controlled the volume of the stored water using a syringe pump (KDS210, KD Scientific, Holliston, MA, USA) and connected a pressure gauge (GP-M 001, Keyence Corp.) to an outlet of the device to measure the output pressure of the discharge water.

2.5. Microfluidic Operations Using the Balloon Pump

We integrated laminar flow devices and dynamic microarray devices with the outlet of the balloon pump. Both devices were prepared by bonding PDMS channels and glass plates. In the case of the demonstration using laminar flow devices, we connected the balloon pumps containing water and water including 10% (v/v) blue ink separately to two inlets of the laminar flow device. The dynamic microarray devices enabled us to make arrays of microsized beads. For the demonstration using the dynamic microarray devices, we placed microsized beads at an inlet of the dynamic microarray device and then connected the balloon pump containing water with 0.5 wt % Tween 20 to the device. The discharge water from the balloon pump pushed the stored microsized beads to the dynamic microarray devices. In both the experiments, we used a microscope (IX71, Olympus Corp.) for observation.

3. Results and Discussion

3.1. Regulation Properties of the Floating Valve

The floating valves moved freely in the microchannel according to the flow direction since the oxygen layer on the PDMS surface worked as a PEGDA polymerization inhibitor (Figure 3a) (Movie S1). In this state, by designing an appropriate exposure pattern using the DMD, we succeeded in controlling the shape of the floating valve (Figure 3b). In addition, by shrinking the oxygen layer on the PDMS channel surface using a degassed balloon pump, larger floating valves were fabricated in the channels (Figure 3b). These results show that our fabrication method can change the configuration of the balloon pump easily without any change in the fabrication tools such as photomasks and molds.

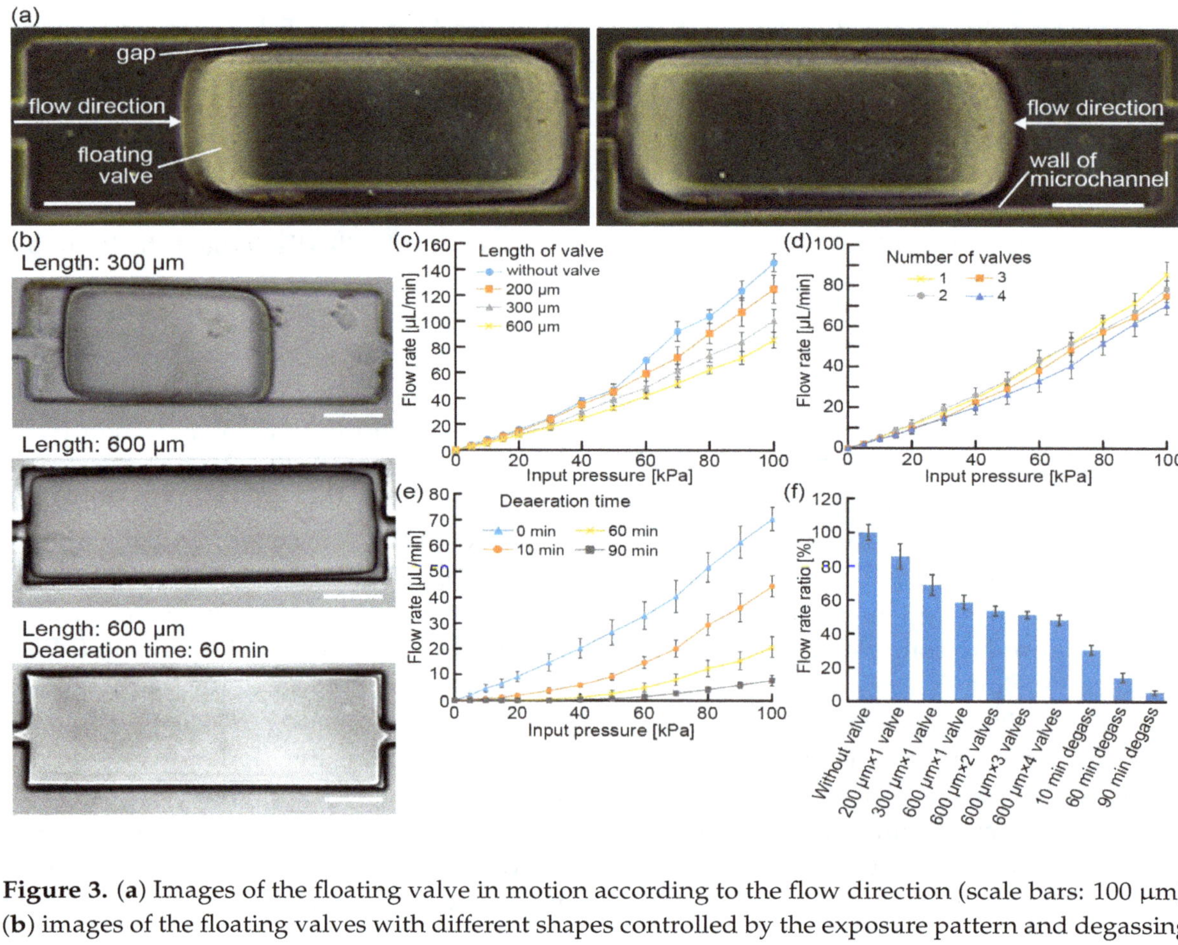

Figure 3. (**a**) Images of the floating valve in motion according to the flow direction (scale bars: 100 μm); (**b**) images of the floating valves with different shapes controlled by the exposure pattern and degassing time (scale bars: 100 μm); (**c**–**e**) changes in the regulation properties of the floating valves with varying input pressure of liquids when the (c) length, (d) number (length: 600 μm), and (e) degassing time of the floating valves were changed; and (**f**) summary of the regulation properties of the floating valves fabricated under different conditions at 100 kPa input pressure.

To evaluate the floating valves, we checked their flow regulation properties in the microchannel against the input flow pressure using the pressure-driven pump. Using the floating valves with various shapes, we could change the discharge flow rates under the same input flow pressure by controlling the length and number of the valves (Figure 3c,d): when the number or length of the valves increased, the discharge flow rates decreased. The results indicate that the formation of long gaps between the valves and channel walls can result in small flow rates because an increase in the length and number of the valves causes the gap to be long. We believe that the gaps caused the flow path to become narrow and regulated the flow speed because of a loss in the flow pressure. In addition, these results showed that slopes of discharge flow rates increased according to increase of the input flow pressures differently from expected principle of Hagen-Poiseuille flow. We think that extension of the PDMS microchannels by applied pressures caused accelerated increase of the discharge flow rates because we observed that applied pressure deformed the microchannels (Movie S1). Furthermore, we managed to decrease the discharge flow rates by changing the degassing time of the PDMS devices (Figure 3e). The use of degassed PDMS devices for a longer period (up to 90 min) caused the floating valves to generate flows with lower flow rates since the gaps between the valves and channel walls became narrower. However, degassing for over 120 min caused clogging of the microchannel because of the dissipation of the oxygen layer on the PDMS surface. From these results, we confirmed that the configuration changes of the floating valves allowed us to easily control the discharge characteristics at the same input flow pressure (Figure 3f). Because various microchannels with an extensive range of lengths were necessary to facilitate the control of the discharge characteristics (Figure S2), the floating valves were appropriate to control the flow rates of the balloon pumps without requiring any design changes of the microchannels.

3.2. Discharge Characteristics of the Balloon Pump

For the evaluation of the balloon pump, we investigated the inflation and deflation properties of the balloon membranes (diameter: 15 mm). First, we checked the maximum amount of stored water by the inflation of the balloon membrane. We prepared balloon membranes with different thicknesses and infused water using a syringe until leakage started. From the results, we confirmed that the balloon pump with a thinner balloon membrane could store more amount of water than that with a thicker membrane (Figure S3a). On the other hand, when we measured the deflation pressure of the balloon membrane using the pressure gauge, we found that thick balloon membranes were necessary to achieve reproducible deflation of the membrane (Figure S3b). These results show that a balloon membrane with an appropriate thickness is required to satisfy the conditions of a good volume of stored liquids as well as reproducible deflation. Therefore, we used a balloon membrane with 0.4 mm thickness in the balloon pumps for further experiments.

Using the balloon pump, we discharged liquids without using any external driving source such as an electrical source. When we infused water into the balloon pump via the check valve, the balloon pump stored water in the balloon membrane without any leakage (Figure 4a). To analyze the properties of the balloon membrane, we clarified the relationship between the volume of the stored water and the flow pressure applied using a pressure-driven pump (Figure 4b). The results indicate that the input flow pressure induced by the deflation of the balloon membrane changes according to the volume of the stored water. Because of the change in the input flow pressure, the discharge flow rates of the balloon pumps varied in response to the volume of the stored water (Figure 4c). In this state, we analyzed the regulation characteristics of the floating valves and found that when the number and length of the floating valves increased, the discharge flow rates of the balloon pumps decreased. This result shows that the floating valve works as a flow regulator in the balloon pumps. However, we observed that the balloon pumps with the floating valves fabricated in degassed PDMS devices did not deliver water because the input flow pressure of the balloon pumps was not sufficiently more than the loss of pressure around the floating valves. Furthermore, the balloon pump enabled the delivery of water over 10 h, and discharged properties of the balloon pump followed basic balloon principles [21] (Figure 4d).

Based on these results, we believe that balloon pumps with floating valves can be used as portable pumps for liquid delivery.

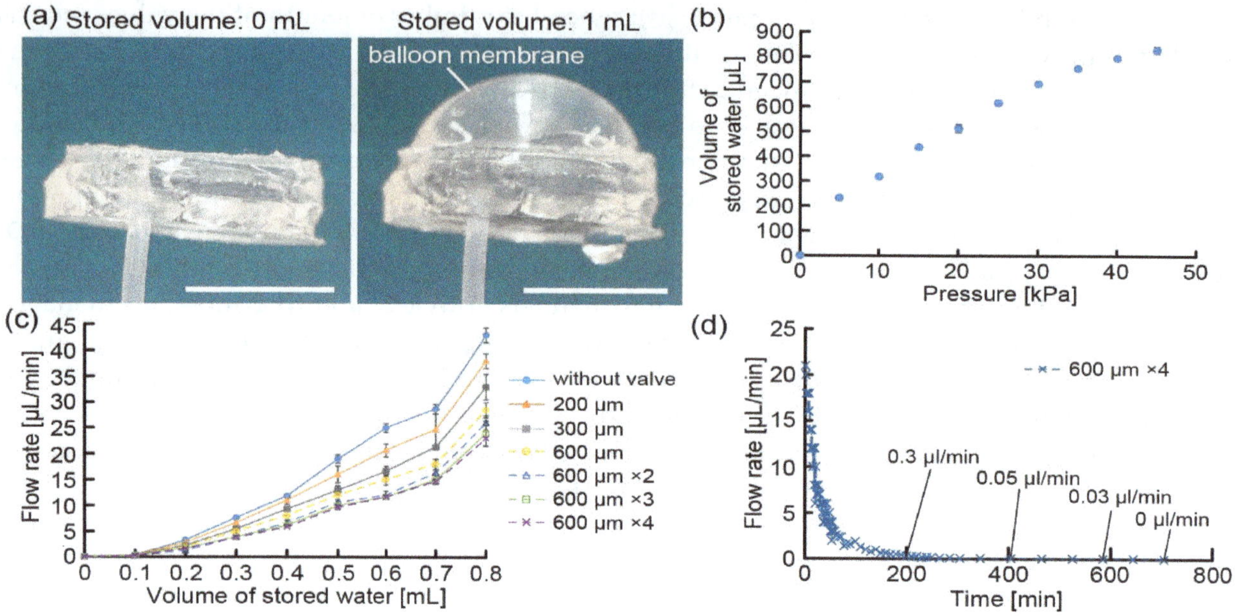

Figure 4. (**a**) Images of the inflation of the balloon membrane when storing water (scale bars: 1 cm); (**b**) relationship between the maximum volume of stored water and the applied input pressure; (**c**,**d**) plots of the discharge flow rates for different floating valves according to the (c) volume of stored water and (d) time after the storage of 0.8 mL of water.

3.3. Microfluidic Operations Using the Balloon Pump

As a demonstration of the microfluidic operations using the balloon pump, we integrated a laminar flow device and dynamic microarray device with the balloon pumps. We succeeded in forming laminar flows in the device by the liquid delivery of individual balloon pumps. In this state, the balloon pumps can facilitate changes in the widths of the laminar flows by altering the combination of the balloon pumps because we can control the discharge flow rates from the balloon pumps according to the type of the floating valves and the volume of the stored liquids (Figure 5a). Because the balloon pumps provide liquid delivery for more than a few hours, they could also achieve the formation of laminar flows for over 2 h (Figure 5b). In this regard, role of diffusion increased after 2 h because flow rates are substantially smaller due to basic balloon principles. Because some microfluidic devices need the formation of laminar flows for several tens of minutes to prepare fiber-shaped samples [27], the balloon pumps can be used in seamless *in situ* sample fabrication.

Furthermore, when microbeads were placed at the inlet of the dynamic microarray device, the discharge water from the balloon pump delivered the microbeads to the device channels. As a result, we achieved the induction of the microbeads into the trapping area in order and the formation of an array of microbeads (Figure 5c). In addition, the speeds of microbead delivery in the channel were controlled by the type of the floating valves in the balloon pumps (Movie S2). Because the dynamic microarray device allows the use of various types of beads such as polymer capsules with microbes and collagen beads with cells by controlling the speed of sample delivery [28,29], we infer that the balloon pump is a useful tool for the *in situ* array formation of various bead-shaped samples in the dynamic microarray devices.

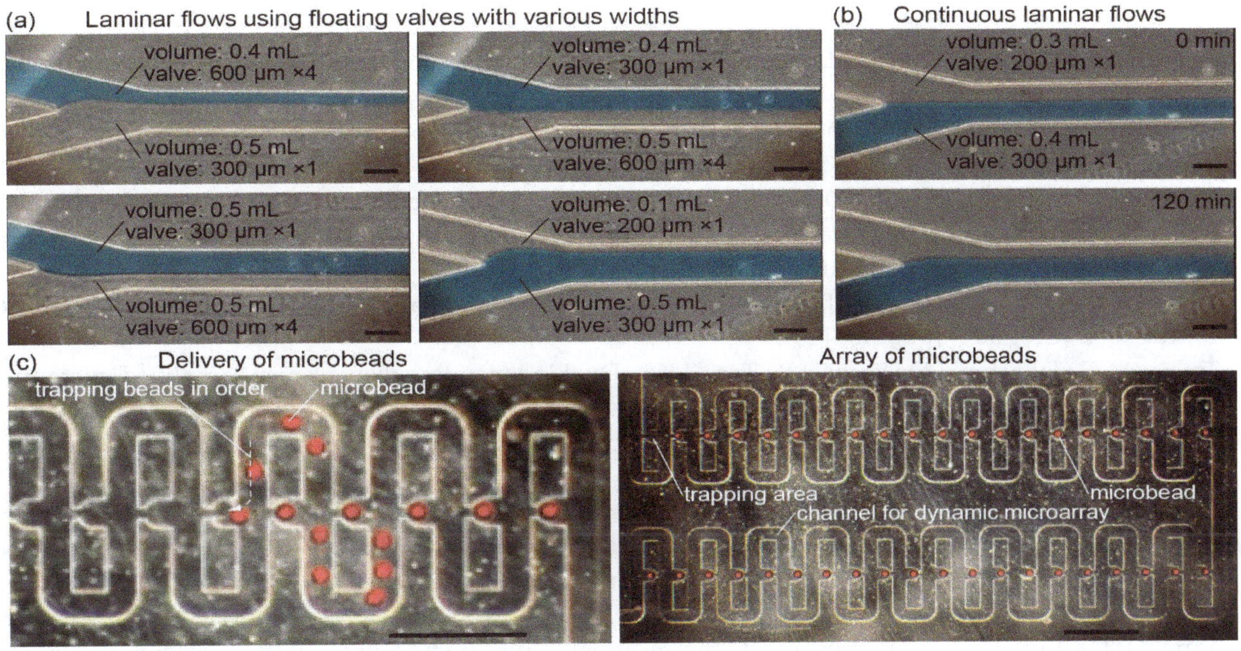

Figure 5. (**a**) Laminar flows formed by using balloon pumps with various floating valves and volumes of stored liquids; (**b**) continuous formation of the laminar flows; and (**c**) formation of an array of microbeads in the dynamic microarray device using the balloon pump with four 600 µm width floating valves and 0.3 mL of stored water. Scale bar: (a,b) 100 µm; and (c) 500 µm.

4. Conclusions

In this study, by combining a balloon membrane, microchannels, and floating valves, we developed balloon pumps for *in situ* liquid delivery without the use of external sources. The advantages of the balloon pumps are as follows: (i) easy preparation of various types of floating valves because of photopolymerization using a DMD system; (ii) changeable discharge flow properties by adjusting the number and width of the floating valves; and (iii) liquid delivery without any additional manipulation. Thus, the balloon pumps can be used for *in situ* microfluidic operations instead of conventional pumps. In addition, the balloon pump provides adjustability of discharge flow properties to users who mount a DMD system to a microscope because the users can change dimensions of the floating valves by exposure condition. Therefore, we believe that the balloon pump with the floating valves will be a useful tool for manipulating liquids and samples in microfluidic devices for point-of-care analyses.

Supplementary Materials
Figure S1: Relationship between the lengths of the exposure area and fabricated floating valves, Figure S2: Plots of the discharge flow rate via microchannels with different lengths at various input pressures, Figure S3: Influence of change of balloon membrane thickness, Movie S1: Motion of the floating valve according to the flow direction. Movie S2: Formation of an array of microbeads in the dynamic microarray device using the balloon pumps under various conditions.

Acknowledgments: The authors thank Maiko Onuki for her technical assistance and Teru Okitsu for his valuable comments. This work was partially supported by Health and Labor Sciences Research Grants from Japan Agency for Medical Research and Development (AMED).

Author Contributions: Yuya Morimoto: Responsible for microfluidic demonstration of the balloon pump, data analyses, and paper composition. Yumi Mukouyama: Responsible for analyzing the properties of the balloon pump. Shohei Habasaki: Responsible for the preparation of the photoreactive resin and systems for the polymerization of the photoreactive resin. Shoji Takeuchi: Responsible for the overall planning and modification of the paper.

References

1. Griffiths, A.D.; Tawfik, D.S. Miniaturising the laboratory in emulsion droplets. *Trends Biotechnol.* **2006**, *24*, 395–402. [CrossRef] [PubMed]
2. Dittrich, P.S.; Manz, A. Lab-on-a-chip: Microfluidics in drug discovery. *Nat. Rev. Drug Discov.* **2006**, *5*, 210–218. [CrossRef] [PubMed]
3. Chin, C.D.; Laksanasopin, T.; Cheung, Y.K.; Steinmiller, D.; Linder, V.; Parsa, H.; Wang, J.; Moore, H.; Rouse, R.; Umviligihozo, G.; *et al.* Microfluidics-based diagnostics of infectious diseases in the developing world. *Nat. Med.* **2011**, *17*, 1015–1019. [CrossRef] [PubMed]
4. Yager, P.; Domingo, G.J.; Gerdes, J. Point-of-care diagnostics for global health. *Annu. Rev. Biomed. Eng.* **2008**, *10*, 107–144. [CrossRef] [PubMed]
5. Tracey, M.C.; Johnston, I.D.; Davis, J.B.; Tan, C.K.L. Dual independent displacement-amplified micropumps with a single actuator. *J. Micromech. Microeng.* **2006**, *16*, 1444–1452. [CrossRef]
6. Cui, Q.; Liu, C.; Zha, X.F. Study on a piezoelectric micropump for the controlled drug delivery system. *Microfluid. Nanofluid.* **2007**, *3*, 377–390. [CrossRef]
7. Pecar, B.; Vrtacnik, D.; Resnik, D.; Mozek, M.; Aljancic, U.; Dolzan, T.; Amon, S.; Krizaj, D. A strip-type microthrottle pump: Modeling, design and fabrication. *Sensors* **2013**, *13*, 3092–3108. [CrossRef] [PubMed]
8. Yang, Y.J.; Liao, H.H. Development and characterization of thermopneumatic peristaltic micropumps. *Microelectron. Eng.* **2009**, *19*, 025003. [CrossRef]
9. Yang, L.J.; Lin, T.Y. A PDMS-based thermo-pneumatic micropump with Parylene inner walls. *J. Microelectron. Eng.* **2011**, *88*, 1894–1897. [CrossRef]
10. Shen, M.; Yamahata, C.; Gijs, M.A.M. A high-performance compact electromagnetic actuator for a PMMA ball-valve micropump. *J. Micromech. Microeng.* **2008**, *18*, 025031. [CrossRef]
11. Su, Y.C.; Lin, L.W.; Pisano, A.P. A water-powered osmotic microactuator. *J. Microelectromech. Syst.* **2002**, *11*, 736–742.
12. Herrlich, S.; Spieth, S.; Messner, S.; Zengerle, R. Osmotic micropumps for drug delivery. *Adv. Drug Deliv. Rev.* **2012**, *64*, 1617–1627. [CrossRef] [PubMed]
13. Martinez, A.W.; Phillips, S.T.; Whitesides, G.M. Three-dimensional microfluidic devices fabricated in layered paper and tape. *Proc. Natl. Acad. Sci. USA* **2008**, *105*, 19606–19611. [CrossRef] [PubMed]
14. Dimov, I.K.; Basabe-Desmonts, L.; Garcia-Cordero, J.L.; Ross, B.M.; Lee, L.P. Stand-alone self-powered integrated microfluidic blood analysis system (SIMBAS). *Lab Chip* **2011**, *11*, 845–850. [CrossRef] [PubMed]
15. Li, G.; Luo, Y.; Chen, Q.; Liao, L.; Zhao, J. A "place n play" modular pump for portable microfluidic applications. *Biomicrofluidics* **2012**, *6*, 014118. [CrossRef] [PubMed]
16. Meyvantsson, I.; Warrick, J.W.; Hayes, S.; Skoien, A.; Beebe, D.J. Automated cell culture in high density tubeless microfluidic device arrays. *Lab Chip* **2008**, *8*, 717–724. [CrossRef] [PubMed]
17. Qiu, X.B.; Thompson, J.A.; Chen, Z.Y.; Liu, C.C.; Chen, D.F.; Ramprasad, S.; Mauk, M.G.; Ongagna, S.; Barber, C.; Abrams, W.R.; *et al.* Finger-actuated, self-contained immunoassay cassettes. *Biomed. Microdevices* **2009**, *11*, 1175–1186. [CrossRef] [PubMed]
18. Li, W.T.; Chen, T.; Chen, Z.; Fei, P.; Yu, Z.; Pang, Y.; Huang, Y. Squeeze-chip: A finger-controlled microfluidic flow network device and its application to biochemical assays. *Lab Chip* **2012**, *12*, 1587–1590. [CrossRef] [PubMed]
19. Chen, A.; Pan, T. Manually operatable on-chip bistable pneumatic microstructures for microfluidic manipulations. *Lab Chip* **2014**, *14*, 3401–3408. [CrossRef] [PubMed]
20. Iwai, K.; Shih, K.C.; Lin, X.; Brubaker, T.A.; Sochol, R.D.; Lin, L. Finger-powered microfluidic systems using multilayer soft lithography and injection molding processes. *Lab Chip* **2014**, *14*, 3790–3799. [CrossRef] [PubMed]
21. Gong, M.M.; MacDonald, B.D.; Nguyen, T.V.; Sinton, D. Hand-powered microfluidics: A membrane pump with a patient-to-chip syringe interface. *Biomicrofluidics* **2012**, *6*, 044102. [CrossRef] [PubMed]
22. Beebe, D.J.; Moore, J.S.; Bauer, J.M.; Yu, Q.; Liu, R.H.; Devadoss, C.; Jo, B.H. Functional hydrogel structures for autonomous flow control inside microfluidic channels. *Nature* **2000**, *404*, 588–590. [CrossRef] [PubMed]
23. Dendukuri, D.; Panda, P.; Haghgooie, R.; Kim, J.M.; Hatton, T.A.; Doyle, P.S. Modeling of oxygen-inhibited free radical photopolymerization in a PDMS microfluidic device. *Macromolecules* **2008**, *41*, 8547–8556. [CrossRef]

24. Kirby, B.J.; Shepodd, T.J.; Hasselbrink, E.F., Jr. Voltage-addressable on/off microvalves for high-pressure microchip separations. *J. Chromatogr. A* **2002**, *979*, 147–154. [CrossRef]
25. Kim, D.; Beebe, D.J. A bi-polymer micro one-way valve. *Sens. Actuator A Phys.* **2007**, *136*, 426–433. [CrossRef]
26. Tan, W.H.; Takeuchi, S. A trap-and-release integrated microfluidic system for dynamic microarray applications. *Proc. Natl. Acad. Sci. USA* **2007**, *104*, 1146–1151. [CrossRef] [PubMed]
27. Onoe, H.; Okitsu, T.; Itou, A.; Kato-Negishi, M.; Gojo, R.; Kiriya, D.; Sato, K.; Miura, S.; Iwanaga, S.; Kuribayashi-Shigetomi, K.; *et al.* Metre-long cell-laden microfibres exhibit tissue morphologies and functions. *Nat. Mater.* **2013**, *12*, 584–590. [CrossRef] [PubMed]
28. Morimoto, Y.; Tan, W.H.; Tsuda, Y.; Takeuchi, S. Monodisperse semi-permeable microcapsules for continuous observation of cells. *Lab Chip* **2009**, *9*, 2217–2223. [CrossRef] [PubMed]
29. Morimoto, Y.; Tanaka, R.; Takeuchi, S. Construction of 3D, Layered Skin, Microsized Tissues by Using Cell Beads for Cellular Function Analysis. *Adv. Healthc. Mater.* **2013**, *2*, 261–265. [CrossRef] [PubMed]

5

Unconventional Electrochemistry in Micro-/Nanofluidic Systems

Sahana Sarkar, Stanley C. S. Lai and Serge G. Lemay *

MESA+ Institute for Nanotechnology, University of Twente, P.O. Box 217, 7500 AE Enschede, The Netherlands; s.sarkar-1@utwente.nl (S.S.); s.c.s.lai@utwente.nl (S.C.S.L.)

* Correspondence: s.g.lemay@utwente.nl

Academic Editors: Manabu Tokeshi and Kiichi Sato

Abstract: Electrochemistry is ideally suited to serve as a detection mechanism in miniaturized analysis systems. A significant hurdle can, however, be the implementation of reliable micrometer-scale reference electrodes. In this tutorial review, we introduce the principal challenges and discuss the approaches that have been employed to build suitable references. We then discuss several alternative strategies aimed at eliminating the reference electrode altogether, in particular two-electrode electrochemical cells, bipolar electrodes and chronopotentiometry.

Keywords: electrochemistry; reference electrode; bipolar electrode; floating electrode; potentiometry

---

1. Introduction

One of the main challenges in creating micro- and nanodevices for chemical analysis is downscaling the measurement system that is ultimately used for readout. Several features of electrochemistry render it a desirable mechanism for transducing chemical information into electrical signals [1–15]: The fabrication of electrodes suitable for electrochemistry is largely compatible with the methods employed for creating micro- and nanofluidic channels, it requires minimal additional (relatively low-cost) equipment, its sensitivity often increases with the downscaling of the electrode dimensions, it directly yields electrical signals without an intermediary transduction step (e.g., light), and it operates at relatively low power. Nonetheless, electrochemical methods can prove challenging to implement in micro- and nanosystems: While the concepts and instrumentation required for such measurements are well developed on the macroscopic scale, subtle, unobvious adjustments and compromises are often necessary upon downscaling. This complexity often goes unrecognized in the design of miniaturized systems, limiting accuracy and performance.

The aim of this review is to introduce the key concepts that influence electrochemical measurements in micro- and nanoscale measurement systems. Our target audience consists of scientists and engineers working on miniaturizing electrochemical measurement systems. We assume that the reader is already familiar with the methods used to fabricate micro-/nanofluidic devices and with basic electrochemical principles [16,17], and concentrate on elucidating some of the key factors that influence electrochemical measurements in miniature systems. We pay particular attention to how the electrostatic potentials of electrodes are established, determined, and controlled - or not, as is often the case. We first discuss reference electrodes, a key component of most macroscopic electrochemical measurement systems. This allows introducing the notation used in the reminder of the article as well as some important concepts that are sometimes misunderstood. We then discuss two classes of systems in which the conventional electrode biasing scheme is abandoned, namely, electrochemical cells without a reference electrode and bipolar electrodes. We end with a brief discussion of potentiometric measurements, in which the potential of an electrode is not controlled but is instead employed for detection. Unless stated otherwise, we assume that the test solution consists of water containing both redox-active analyte molecules as well as a much higher concentration of inert salt ions, the so-called supporting electrolyte. This situation is typical for, e.g., biomedical samples. We concentrate on fluidic devices and exclude individual miniature electrodes used in conjunction with macroscopic measurement cells, conventional electrodes modified with nanomaterials, and electrochemical scanning probe techniques, which are reviewed extensively elsewhere [18–21].

2. Anatomy of an Electrode

Before discussing specific electrochemical systems, we introduce a few key concepts that will recur throughout this review[1]. The interface between a solution (an ionic conductor) and an electrode (an electronic conductor, typically a metal, but also potentially a semiconductor or a macromolecule) can be represented by a capacitor C and a (nonlinear) resistor R in a parallel configuration, as shown in Figure 1. Here, C represents the buildup of charge in the so-called electrical double layer (EDL) that develops at this interface. The EDL consists of electrons (or holes) in the electrode and compensating ions in the solution. These lead to an electric field—and thus an electrostatic potential difference—between the solution and the electrode. The EDL is highly local, for example, extending only on the order of ~1 nm for water at physiological concentrations. The resistor R, on the other hand, represents the transfer of electrons between the electrode and the redox species in solution via electrochemical reactions.

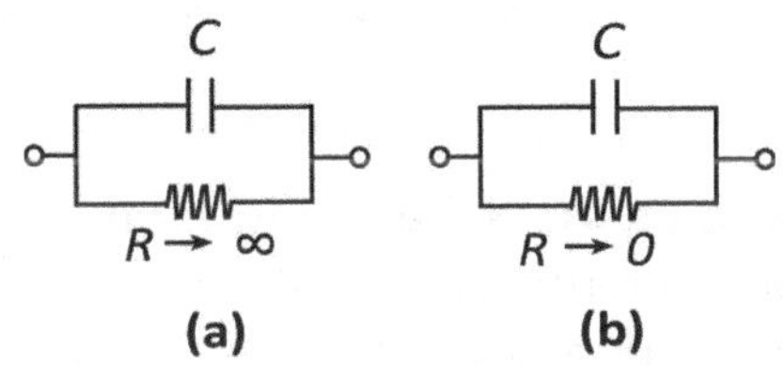

Figure 1. Equivalent circuits for (**a**) a polarizable and (**b**) a non-polarizable interface.

Electrodes can be qualitatively classified as polarizable or non-polarizable. In the case of a polarizable electrode, R is very high and it is therefore possible to alter the potential difference across the interface without injecting significant current into the measurement cell. On the contrary, if R is very low, changing the potential difference across the capacitor requires the application of very large currents, as charge is "leaked" through the interface. This short-circuit-like behavior is referred to as a non-polarizable interface. In practice, no electrode is ever fully polarizable or non-polarizable; whether an electrode represents a good approximation to either depends on the magnitude of the voltages and currents that occur in a particular measurement.

3. Reference Electrodes

In macroscopic systems, electrochemical measurements are typically carried out in a three-electrode configuration [16], as shown schematically in Figure 2a. The working (or indicator) electrode (WE) is the electrode where the analytical measurement takes place: An electrochemical reaction occurs if the potential difference between this electrode and the adjacent solution is such as to favor electron transfer, leading to a current. This electrode is coupled to an electrode of a known, defined potential, called the reference electrode (RE). The (conceptual) circuit diagram of this two-electrode system is depicted in Figure 2b. Importantly, potentials applied to the WE are always with respect to the potential of the RE. Thus, an RE provides a reference point for the potential (similar to the role of ground in electronic circuits). However, it is important to note that the actual electrostatic potential difference between the RE and the solution may not be (and, in practice, rarely is) zero, and one therefore needs to specify the type of RE when stating the voltage of a WE (e.g., "1 V *vs.* Ag/AgCl (3 M KCl)" for a silver/silver chloride reference electrode immersed in a 3 M potassium chloride solution). Similarly, an often overlooked nuance is that applying an *external* potential of 0 V with respect to the RE does not insure that no potential difference exists between the WE and the adjacent solution.

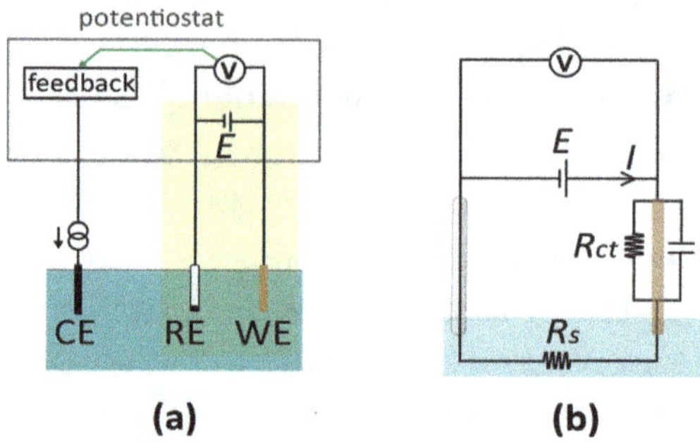

Figure 2. (**a**) Schematic of a conventional electrochemical cell for voltammetric measurement. The cell consists of three electrodes, termed the working (WE), reference (RE), and counter electrode (CE), immersed in the electrolyte solution. A potential, *E*, is applied to the WE with respect to the RE. If the current through the RE would be high enough to cause a potential shift, a CE is introduced to minimize the current through the RE. At low currents, it is instead possible to operate with a two-electrode configuration and eliminate the CE altogether (highlighted in green), simplifying the detection circuitry. (**b**) Equivalent circuit diagram of a two-electrode setup. R_S: solution resistance; R_{ct}: charge-transfer resistance at the WE; *C*: electrical double layer capacitance at the WE. This circuit treats the RE as ideally non-polarizable.

Any electrode system can serve as an RE as long as it approaches ideal non-polarizability, meaning that its interfacial potential remains essentially fixed with the passage of currents [16,22]. The amount of current that can pass depends on the specific RE system and design, but in general non-polarizability breaks down at "high" currents [22], and the reference potential will vary (for a commercial, macroscopic RE, this is typically in the order of μA's). Consequently, the WE potential is not controlled accurately at high currents, as a (undefined and variable) part of the applied potential between the WE and RE, *E*, is dropped at the RE-electrolyte interface. To circumvent this issue, one can introduce a third electrode, the counter (or auxiliary) electrode (CE). In this three-electrode setup, the current from the WE is routed through the CE, which acts as the electron source or sink for the reaction at the WE. The terminal controlling the RE has a high input impedance, rendering the current drawn through the RE negligible, and the RE interfacial potential thus remains constant. The technical implementation for potential control and current measurement in a three-electrode setup employs a potentiostat. Conceptually, this instrument monitors the potential difference between WE and RE, which is used as a feedback signal to control the current passing through the CE so that the actual potential difference matches the desired (applied) potential difference. A detailed description of the workings of a potentiostat can be found in many textbooks on electrochemistry and electrochemical instrumentation [16,23]. As a final note, it should be borne in mind that a CE (and potentiostat by extension) is only required if the current in the system is large, and may be bypassed in miniaturized sensors if currents of the order of a few μA are measured that can be directly passed through a RE without significantly affecting its potential. In our experience, this condition is easily satisfied in most micro- and nanoscale systems. This results in compact simplified electronics, shown by the yellow box in Figure 2a, which essentially consists of a power source and an ammeter connected in series with the two electrodes.

Solution resistance. While in principle the RE only sets the electrostatic potential near its surface, the solutions employed in electrochemical measurements are ionic conductors. As a result, the potential of a solution when no electrical current is flowing through it is uniform throughout its entire bulk volume and is set by the RE. An important exception occurs at the boundaries of the liquid, where EDLs can develop as discussed above. This is particularly relevant near the surface of the WE, where a potential difference is required to drive electrochemical processes. However, if a net current, I, is flowing through the solution, an electric field can develop according to Ohm's law ($E = IR_s$, where R_s is the solution ionic resistance), and part of the applied voltage is dropped in the solution between the RE and WE. These ohmic voltage drops can be minimalized either by reducing the current (e.g., by decreasing the analyte concentration or reducing the size of the electrode) or by minimizing the electrolyte resistance between the RE and WE (e.g., by increasing the conductivity of the electrolyte solution or placing the RE close to the WE to decrease the length of the resistive path). In most electroanalytical measurements, the analyte concentration is much lower than the electrolytic (salt) concentration; therefore, these ohmic voltage drops may reasonably be neglected. However, if an electrolytic solution of low conductivity (usually due to low ionic strength) is used, IR_s may be significant and needs to be taken into account when considering the WE potential ($E_{WE} = E - IR_s$). This can be particularly significant in fluidic devices where confinement of the liquid easily leads to higher values of R_s than is typical in macroscopic experiments.

Requirements. At this point, it is worth discussing the technical requirements of a reference electrode. A RE should have a potential which is stable over time [22] and which is not significantly altered by small perturbations to the system—in particular, the passage of a small current. Some of the main considerations while designing a RE are discussed in depth by Shinwari *et al.* [22]. Commercial REs typically employ a macroscopic piece of metal (providing an "infinite" reservoir of redox species) coated with a sparingly soluble metal salt (such that the interfacial concentration is determined by the solubility product of the salt), immersed in a contained reference solution, and the entire system is connected to the test solution by a salt bridge (to prevent composition changes of the reference solution while minimizing the liquid junction potential) [16,24]. While such electrode systems are straightforward to realize on the macroscale, implementing REs in miniaturized systems requires careful considerations in the downscaling of all these components [22,25].

Miniaturized REs. Several analogues to conventional REs have been demonstrated using microfabrication, and several techniques are available for their manufacture such as thin film deposition [26–30], electroplating [31,32], or screen printing [33,34] of the metal followed by ion exchange reactions or electrochemical coating. The interface to the test solution and reference solution chamber is typically implemented using gels or nanoporous membranes/glass. For example, an Ag/AgCl electrode was replicated by a thin-film deposition of Ag supported over Pt, after which AgCl was formed by oxidizing it in a solution containing chloride ions [31]. In another example, miniaturization of the liquid junction Ag/AgCl was demonstrated by covering a deposited thin film of silver with a layer of polyamide. This layer had a slit at the center where AgCl was grown; the liquid junction was formed with photo-curable hydrophilic polymer [35].

However, the stability of such miniaturized references electrodes is often limited, and typical problems include limited lifetimes, poor reproducibility, and drifting electrode potentials [22,36]. A common cause is the rapid consumption of the electrode material due to its small size. In general, electrode consumption can be divided into an electrochemical (Equation (1)) and a chemical (Equation (2)) pathway.

$$\mathrm{AgCl\,(s)} + \mathrm{e^-} \rightleftharpoons \mathrm{Ag\,(s)} + \mathrm{Cl^-\,(aq)}\ \text{(electrochemical)} \quad (1)$$

$$\mathrm{AgCl\,(s)} + n\mathrm{Cl^-\,(aq)} \rightleftharpoons \mathrm{AgCl_{(n+1)}}^{n-}\,\mathrm{(aq)},\ \text{where}\ 0 < n < 3\ \text{(chemical)} \quad (2)$$

In the electrochemical pathway, the passage of a small current through a miniaturized RE can already be sufficient to induce complete consumption of the electrode material within experimental time scales. For example, a microscopic Ag/AgCl RE of an area of 100 μm^2 (AgCl thickness 100 nm) exposed to a current of only 10 pA would be completely consumed within approximately one hour. The chemical pathway relates to the non-zero solubility of the metal salt, where the dissolved and solid species are only in chemical equilibrium as long as the solution is saturated with the metal salt. If the RE is exposed to a non-saturated solution, or the solution is continuously replenished (such as in flow systems), dissolution of the metal salt will occur. This issue is further exacerbated in the case of

Ag/AgCl electrodes, where there is a non-negligible formation of aqueous $AgCl_{(n+1)}^{n-}$ ion complexes in chloride-containing solutions [37,38]. At physiological electrolyte concentrations, this leads to an equilibrium concentration of dissolved AgCl in the μM range, sufficient to completely dissolve a 100 $\mu m^2 \times$ 100 nm AgCl layer in ~0.1 μL of electrolyte solution.

Another common cause for the limited stability of miniaturized REs is the possible contamination of the reference solution via non-ideal ("leaky") bridging membranes. This issue can be alleviated by eliminating the salt bridge and reference solutions. Such systems are commonly termed quasi- or pseudo-RE. While the terms are often used interchangeably, there is a subtle but important difference between the two. A quasi-RE simply omits the reference solution and immerses the electrode directly into the test solution [28,29,39–45]. A clearly defined redox couple, however, sets the electrode potential, and any fluctuations result from changes in the activity coefficients of this couple. For example, a common Ag/AgCl quasi-RE consists of a silver electrode coated with silver chloride salt and in contact with the chloride-containing test solution; here, the Ag/Ag^+ couple sets the solution potential [30]. On the other hand, a pseudo-RE refers to a large surface area electrode (such as a platinum or silver wire) directly exposed to the solution [42,43,45]. In this case, which redox couple sets the reference potential is undefined, and the reference potential remains reasonably constant by virtue of the large surface area, with even low reactivity being sufficient to take up small currents without significant polarization of the electrode. In both cases, the RE can be calibrated by measuring its potentials relative to a conventional RE. Thus, while miniaturizing REs still present challenges, rational design can provide a microscopic RE which is sufficiently stable given the requirements for a specific measurement.

Finally, it is worthwhile to consider the placement of electrodes in microfabricated systems. In a macroscopic system, the CE is placed far from the WE and RE, such that the substances produced at the CE do not reach the WE surface to interfere with the measurements there. However, in microscopic systems, this might not be possible due to space requirements, and such interference needs to be taken into account in order to avoid undesirable shifts in the reference potential.

4. Systems without a Reference Electrode

Considering the difficulties inherent in implementing miniaturized high-quality reference electrodes, it is natural that considerable effort has been devoted to creating analytical systems in which the role of the reference is minimized or omitted altogether. Doing so comes at a price since in such cases the interfacial potentials that drive electron-transfer reactions at the system's electrodes is no longer explicitly controlled. As a result, no universally applicable alternative to the conventional combination of potentiostat and reference electrode has evolved. Nonetheless, reliable alternatives can be implemented in some particular geometries and/or when sufficient information about the solution to be analyzed is available.

The basic configuration for a reference-free, two-electrode system is sketched in Figure 3. While this represents the simplest case of a system without an RE, the discussion of the solution potential in the following is general, and can be extended to incorporate additional electrode elements. The most important feature of the system of Figure 3 is that the interfacial potential differences at the two electrodes is not controlled separately since only the total potential difference between the two electrodes is accessible experimentally. The potential of the bulk electrolyte phase, E_s, is thus instead free to float to different values. This is in stark contrast with the case where one of the electrodes is an RE; in that case, there is no change in the potential difference at the RE interface, and the potential of the electrolyte is pinned to the RE potential.

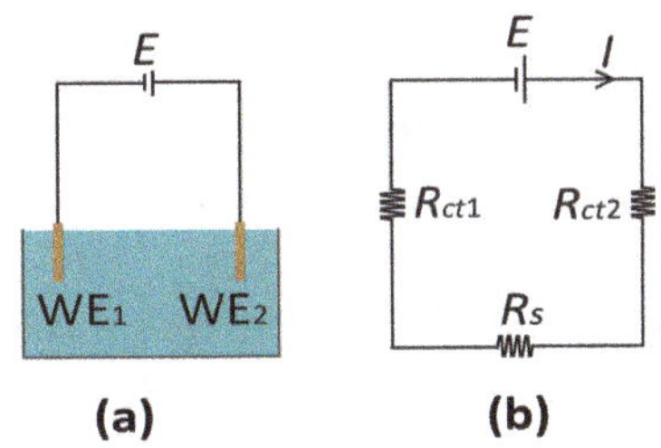

Figure 3. (**a**) Reference-less two-electrode system where E is the applied potential between the two WEs. (**b**) Corresponding equivalent-circuit diagram. R_S: solution resistance; $R_{ct1,2}$: (charge transfer) resistance at the $WE_{1,2}$.

What sets the potential of the solution in the experiment of Figure 3? The passage of a current at one of the electrodes causes charge to be injected in this solution. As discussed above in the context of reference electrodes, this charge accumulates at the boundaries of the bulk phase. For example, an oxidation reaction taking place at an electrode causes the withdrawal of electrons from the solution and the accumulation of positive charge at its boundaries, in turn causing the electrostatic potential of the solution to become more positive. This acts as a negative feedback mechanism, as the shift in solution potential acts to inhibit the electrochemical process that caused it (in our example, the oxidation current decreases by making the solution more positive with respect to the electrode). The solution eventually settles to a stationary steady state at a potential such that no net charge injection takes place, that is, the total current being injected into the solution vanishes:

$$\sum_j I_j = 0, \text{ where } I_j = I_j\left(V_j - E_s\right) \tag{3}$$

here, I_j is the current through the jth electrode, which is a function of its interfacial potential difference $(V_j - E_s)$, V_j is the potential applied to the electrode, and E_s is the solution potential (neglecting ohmic drops for ease of notation) with respect to a common reference point in the circuit such as signal ground. In principle, if the relations between current and interfacial potential at each of the electrodes are known (because, e.g., they can be derived from fundamental electrochemical kinetic theory or they have been experimentally determined), then it is possible to solve for the unique value of E_s that satisfies Equation (3) and to deduce the current through each of the electrodes. This procedure essentially amounts to solving the equivalent circuit shown in Figure 3b, where the electrochemical reactions are represented by (highly nonlinear) resistors R_{ct1} and R_{ct2}, and Equation (3) is the direct application of Kirchhoff's current law.

For the two-electrode system of Figure 3a, Equation (3) reduces to the statement that the solution potential will shift in such a way that the reduction current at the more negative of the two electrodes is equal in magnitude to the oxidation current at the more positive electrode. This scenario was discussed in detail by Xiong and White [46], where it was, for example, shown explicitly that increasing the area of one of the electrodes causes the solution potential to shift closer to that electrode's open-circuit potential because that electrode's effective resistance becomes smaller.

A further consequence of Equation (3) is that parasitic pathways for a current—such as may result from a minor leak—can sometimes have a significant influence in a microsystem without a reference electrode. In conventional electrochemical cells, such a parasitic current can be accommodated by the counter electrode (or the reference for low-current systems) without influencing the signal measured at the working electrode. For a floating solution potential, however, even relatively small uncompensated currents can lead to drift. This was illustrated by Sarkar *et al.* [47], who showed how the (large) redox-cycling current between two electrodes separated by 65 nm can be controlled by the (much smaller) current to an additional electrode located outside the nanofluidic device [47].

5. Bipolar Electrodes

A bipolar electrode (BPE) is a *floating* conductor which facilitates opposing electrochemical reactions (oxidizing and reducing) on spatially separated regions of its surface. Two example systems are shown in Figure 4. Figure 4a represents the (conceptually) simplest case. Here, two electrolyte solutions are physically separated by a BPE, such that the only current path between them is through the BPE. Since it is a good conductor, the entire BPE is essentially at the same potential, while the relative potential of the two electrolyte solutions can be changed independently. Consequently, the local interfacial potential difference of the BPE with the adjacent solution is different at the two ends. If suitable species are present in the two reservoirs, reduction and oxidation processes may occur at the two ends of a BPE, thereby coupling two, otherwise isolated, electrochemical systems.

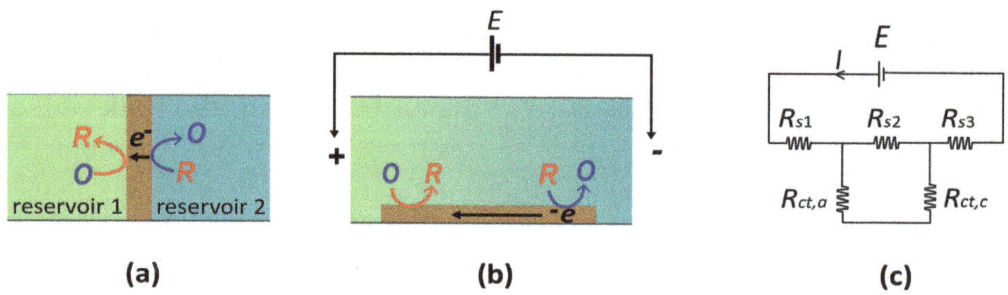

Figure 4. (**a**) Schematic diagram of a bipolar electrode (brown) in contact with two separate reservoirs. (**b**) Alternative concept of a bipolar electrode in which a uniform electric field is applied along a channel filled with electrolytic solution. A band electrode exposed to this solution exhibits bipolarity at its opposing ends (cathodic at left and anodic on right). (**c**) Equivalent circuit for panel (b). E is the potential applied across the solution, R_s is the resistance of the solution, and R_{ct} is the charge transfer resistance across the anodic/cathodic ends of the bipolar electrode (BPE).

Alternatively, a BPE can be located in a single reservoir (Figure 4b). Two additional electrodes are then placed at the ends of the reservoir, and applying a large current between them induces an electric field in the electrolyte due to its finite conductivity (ohmic drop). As shown in Figure 4b, this spatially heterogeneous solution potential leads to a gradient of electrostatic potential differences along the length of the BPE (that is, between the electrode and the solution). If a sufficiently large potential difference between the two ends of the bipolar electrode is induced, it becomes possible to drive an oxidation reaction at one end and a reduction at the opposite end of the same electrode, similarly to the case of Figure 4a.

From a purely conceptual point of view, the scenarios shown in Figures 3a and 4b are very closely related. In each case, one element of the electrochemical circuit—the solution in Figure 3a and the BPEs in Figure 4—is free to adjust its electrostatic potential in response to redox reactions taking place at spatially separated regions. It is therefore unsurprising that the same basic principles apply for determining the potential to which the BPE drifts in response to electrochemistry. In fact, Equation (1) carries over directly to this case, where now E_s represents the potential of the bipolar electrode, and j is an index that runs over the different regions of this electrode (for the case of a continuous gradient as in Figure 4b, the sum becomes an integral over the electrode surface, but the underlying principle remains unchanged).

The defining feature of BPEs is that they are floating electrodes, yet can be induced to facilitate electrochemical reactions of choice at their interface. This is particularly attractive for miniaturized systems, as abolishing the need for contacts to solution (*i.e.*, reference electrodes) simplifies fabrication and instrumentation. Furthermore, it enables an arbitrarily large number of BPEs (such as arrays of BPEs imbedded in insulating matrices [48]) to be driven simultaneously. The use of BPEs in the micro-/nanoscopic domain was pioneered by Bradley *et al.* [49], who demonstrated the use of bipolar electrochemistry to create electrical contacts in microcircuits by employing copper electrodeposition as the cathodic reaction. This work was followed by a dramatic increase in the investigation of bipolar electrochemistry—in particular, by the groups of Kuhn [50–55] and Crooks [56–59]. A recent review by Sequeira *et al.* [60] discusses bipolar electrochemistry and their many varied applications that several contemporary groups are presently exploring. As a particularly striking example, Mallouk, Sen, and colleagues [61–63] demonstrated a locomotion mechanism for bipolar microswimmers based on electrochemical reactions taking place at both ends of the swimmer. Another intriguing variant is to use the bipolar electrode to couple the reaction of a target analyte to a second, separate reaction that produces an optically active species. Using the latter's fluorescent properties allowed for the demonstration of the highly sensitive, fluorescence-mediated detection of species that are not themselves optically active [48,64].

Implicit bipolar behavior. Apart from devices that explicitly exploit bipolar electrochemistry as their mode of operation, this effect has an important consequence for the design and validation of electrochemical detection devices. Any conductor in contact with solution has the potential to act as a bipolar electrode if its potential is not controlled. This is a very different situation from conventional electronic devices, where leaving a particular component unconnected typically means that it can be safely ignored, at best, or a source of stray capacitance, at worse. A well-documented example of a system where bipolar electrochemistry is implicitly utilized is scanning electrochemical microscopy (SECM in the positive feedback mode), where a conducting sample can be left unbiased but then acts as a bipolar electrode [65,66]. Similarly, floating electrodes imbedded in nanochannels were shown to act as "short circuits" to a reference located outside the nanofluidic device [47,67]. Last but not least, it is important to keep in mind that all solvents—especially water—are liable to electrochemical breakdown; if a sufficient potential gradient is applied, any floating metal features in a device can become implicated in reactions involving water, protons, hydroxide, or dissolved oxygen, leading to unintended currents flowing through the system [68,69].

6. Potentiometry

The main theme of this review has been the control of potentials in electrochemical systems. For completeness, we discuss here very briefly potentiometry, the branch of analytical chemistry concerned with the measurement of potentials as a detection mechanism. It is difficult to understate the importance of potentiometry as it forms the basis for many widely used technologies, starting with pH-sensitive electrodes and extending to a wide family of other ion-selective electrodes [70–73].

In its most common form, potentiometry is an equilibrium technique, with the potential of a working electrode being measured with respect to a reference. This makes it particularly sensitive to the choice of RE, which becomes challenging to implement in miniaturized systems given all the complications discussed above. Commonly used for concentration determination, lower detection limits of such techniques can be achieved with downscaling, and extensive work has been carried out in the development of so-called nanopotentiometry. Much of this work has focused on nanostructured thin films interfaced to macroscopic electrodes [70,72]. To what extent these approaches and materials can be adapted in the context of, e.g., lithographically fabricated micro- and nanodevices largely remains an interesting question for future work. Thus, while this is an area where we expect major developments will likely happen in the near future, we do not attempt to discuss specific works at this time.

One variant that may lend itself more readily to integrated miniature systems is so-called chronopotentiometry [12], in which the potential of an electrode is monitored as a function of time using high-impedance readout circuitry. Before equilibrium is established, electrochemical reactions occurring at an electrode cause its potential to shift over time. The rate of change of the potential is proportional to the electrochemical current and inversely proportional to the electrode capacitance; hence, a concentration can in principle be extracted from the time-dependent data. To explicitly illustrate this principle within the nanogaps [47], we show in Figure 5 measurements of the potential of a floating electrode over time as it accumulates charge due to redox cycling in a nanofluidic device (consisting of two electrodes, one of which is floating). The evolution of the potential over time reflects the rate of electrochemical charge transfer, which itself depends on the composition of the solution.

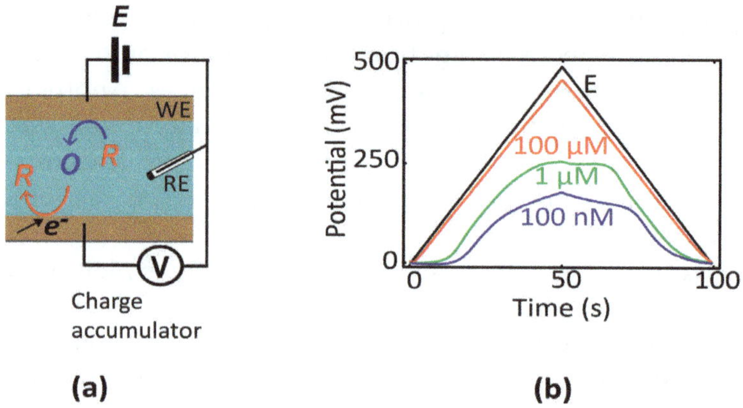

Figure 5. (**a**) Schematic diagram of a two-electrode nanogap system in contact with a solution containing reversible redox species. The bottom (unbiased) electrode accumulates charge over time, and the resulting potential shift is used as readout signal. (**b**) Chronopotentiometric signal *versus* concentration of redox species (100 μM, 10 μM, and 1 nM $Fc(MeOH)_2$ in 0.1 M KCl) in response to a triangular potential wave applied to the top electrode (black line).

Furthermore, since small electrodes normally have lower capacitances, the potentiometric signal is more sensitive in this case, making the method a logical candidate for miniaturization. Based on the additional consideration that the readout of potentials is relatively straightforward to implement in conventional complementary metal-oxide semiconductor (CMOS) electronics [74], Zhu *et al.* [75–77] suggested that chronopotentiometry is particularly well suited for systems in which fluidics and electronics are implemented on a single, highly integrated chip. Whether this type of electrochemical analysis can offer a competitive alternative to existing methods is presently an open question.

7. Summary and Outlook

The emergence of point-of-care diagnostic systems has led to the rapprochement of micro-/nanofluidics and electrochemical sensing methods, and this trend can be expected to strengthen in the coming years. Although electrochemistry is an exhaustive subject and a vast amount of information is available in the literature, it is not always straightforward for new researchers in the area of microsystems to identify the concepts and approaches that are most relevant for building practical miniaturized devices. This is particularly true because some standard ingredients of electrochemical analysis—especially the use of optimized reference electrodes—are surprisingly challenging to scale down. While universally applicable solutions have yet to emerge, many common pitfalls can be avoided by informed experimental design. We have thus attempted to provide an introduction to the methods of micro-/nanoelectrochemistry and, in particular, to make the reader aware of non-idealities which are not necessarily obvious when extrapolating from the macro domain. We hope that linking some of the concepts addressed in this paper will be beneficial to the fluidic sensor community and will help to stimulate further exploration of the rich field of miniature sensing technologies.

Acknowledgments: The authors acknowledge financial support from the European Research Council (ERC) under project number 278801.

Author Contributions: Sahana Sarkar drafted the manuscript. Stanley C. S. Lai and Serge G. Lemay provided the framework of the review and revised the contents according to the scope of this work.

References

1. Oja, S.M.; Fan, Y.S.; Armstrong, C.M.; Defnet, P.; Zhang, B. Nanoscale electrochemistry revisited. *Anal. Chem.* **2016**, *88*, 414–430. [PubMed]
2. Oja, S.M.; Wood, M.; Zhang, B. Nanoscale electrochemistry. *Anal. Chem.* **2013**, *85*, 473–486. [CrossRef] [PubMed]
3. Watkins, J.J.; Zhang, B.; White, H.S. Electrochemistry at nanometer-scaled electrodes. *J. Chem. Educ.* **2005**, *82*, 712–719. [CrossRef]
4. Rackus, D.G.; Shamsi, M.H.; Wheeler, A.R. Electrochemistry, biosensors and microfluidics: A convergence of fields. *Chem. Soc. Rev.* **2015**, *44*, 5320–5340. [CrossRef] [PubMed]
5. Rassaei, L.; Singh, P.S.; Lemay, S.G. Lithography-based nanoelectrochemistry. *Anal. Chem.* **2011**, *83*, 3974–3980. [CrossRef] [PubMed]
6. Meier, J.; Schiotz, J.; Liu, P.; Norskov, J.K.; Stimming, U. Nano-scale effects in electrochemistry. *Chem. Phys. Lett.* **2004**, *390*, 440–444. [CrossRef]
7. Arrigan, D.W.M. Nanoelectrodes, nanoelectrode arrays and their applications. *Analyst* **2004**, *129*, 1157–1165. [CrossRef] [PubMed]
8. Kang, S.; Lemay, S.G. Nanoelectrochemical methods. In *Nanoelectrochemistry*; Mirkin, M.V., Amemiya, S., Eds.; CRC Press: Boca Ratan, FL, USA, 2015; pp. 573–600.
9. Micheal, M.V.; Amemiya, A. *Nanoelectrochemistry*; CRC Press: Boca Ratan, FL, USA, 2015.
10. Lee, T.M.H. Over-the-counter biosensors: Past, present, and future. *Sensors* **2008**, *8*, 5535–5559. [CrossRef]
11. Bakker, E.; Qin, Y. Electrochemical sensors. *Anal. Chem.* **2006**, *78*, 3965–3983. [CrossRef] [PubMed]
12. Grieshaber, D.; MacKenzie, R.; Voros, J.; Reimhult, E. Electrochemical biosensors—Sensor principles and architectures. *Sensors* **2008**, *8*, 1400–1458. [CrossRef]
13. Ino, K. Microchemistry- and MEMS-based integrated electrochemical devices for bioassay applications. *Electrochemistry* **2015**, *83*, 688–694. [CrossRef]
14. Wei, D.; Bailey, M.J.A.; Andrew, P.; Ryhanen, T. Electrochemical biosensors at the nanoscale. *Lab Chip* **2009**, *9*, 2123–2131. [CrossRef] [PubMed]
15. Gencoglu, A.; Minerick, A.R. Electrochemical detection techniques in micro- and nanofluidic devices. *Microfluid. Nanofluid.* **2014**, *17*, 781–807. [CrossRef]
16. Bard, A.J.; Faulkner, L.R. *Electrochemical Methods: Fundamentals and Applications*; Wiley: New York, NY, USA, 2000.
17. Brett, C.M.A.; Brett, A.M.O. *Electrochemistry: Principles, Methods, and Applications*; Oxford University Press: Oxford, UK, 1993.
18. Moore, A.M.; Weiss, P.S. Functional and spectroscopic measurements with scanning tunneling microscopy. *Annu. Rev. Anal. Chem.* **2008**, *1*, 857–882. [CrossRef] [PubMed]
19. Mirkin, M.V.; Nogala, W.; Velmurugan, J.; Wang, Y.X. Scanning electrochemical microscopy in the 21st century. Update 1: Five years after. *Phys. Chem. Chem. Phys.* **2011**, *13*, 21196–21212. [CrossRef] [PubMed]
20. Sun, P.; Laforge, F.O.; Mirkin, M.V. Scanning electrochemical microscopy in the 21st century. *Phys. Chem. Chem. Phys.* **2007**, *9*, 802–823. [CrossRef] [PubMed]
21. Macpherson, J.V.; Unwin, P.R. Noncontact electrochemical imaging with combined scanning electrochemical atomic force microscopy. *Anal. Chem.* **2001**, *73*, 550–557. [CrossRef] [PubMed]
22. Shinwari, M.W.; Zhitomirsky, D.; Deen, I.A.; Selvaganapathy, P.R.; Deen, M.J.; Landheer, D. Microfabricated reference electrodes and their biosensing applications. *Sensors* **2010**, *10*, 1679–1715. [CrossRef] [PubMed]
23. Hickling, A. Studies in electrode polarisation part IV—The automatic control of the potential of a working electrode. *Trans. Faraday Soc.* **1942**, *38*, 27–33. [CrossRef]
24. Brezinski, D.P. Kinetic, static and stirring errors of liquid junction reference electrodes. *Analyst* **1983**, *108*, 425–442. [CrossRef]
25. Mousavi, M.P.S.; Buhlmann, P. Reference electrodes with salt bridges contained in nanoporous glass: An underappreciated source of error. *Anal. Chem.* **2013**, *85*, 8895–8901. [CrossRef] [PubMed]
26. Kim, T.Y.; Hong, S.A.; Yang, S. A solid-state thin-film Ag/AgCl reference electrode coated with graphene oxide and its use in a pH sensor. *Sensors* **2015**, *15*, 6469–6482. [CrossRef] [PubMed]
27. Webster, T.A.; Goluch, E.D. Electrochemical detection of pyocyanin in nanochannels with integrated palladium hydride reference electrodes. *Lab Chip* **2012**, *12*, 5195–5201. [CrossRef] [PubMed]

28. Matsumoto, T.; Ohashi, A.; Ito, N. Development of a micro-planar Ag/AgCl quasi-reference electrode with long-term stability for an amperometric glucose sensor. *Anal. Chim. Acta* **2002**, *462*, 253–259. [CrossRef]
29. Uludag, Y.; Olcer, Z.; Sagiroglu, M.S. Design and characterisation of a thin-film electrode array with shared reference/counter electrodes for electrochemical detection. *Biosens. Bioelectron.* **2014**, *57*, 85–90. [CrossRef] [PubMed]
30. Rivas, I.; Puente, D.; Ayerdi, I.; Castano, E. Ag/AgI quasi-reference microelectrodes. In Proceedings of the 2005 Spanish Conference on Electron Devices, Tarragona, Spain, 2–4 Febuary 2005; pp. 465–468.
31. Zhou, J.H.; Ren, K.N.; Zheng, Y.Z.; Su, J.; Zhao, Y.H.; Ryan, D.; Wu, H.K. Fabrication of a microfluidic Ag/AgCl reference electrode and its application for portable and disposable electrochemical microchips. *Electrophoresis* **2010**, *31*, 3083–3089. [CrossRef] [PubMed]
32. Polk, B.J.; Stelzenmuller, A.; Mijares, G.; MacCrehan, W.; Gaitan, M. Ag/AgCl microelectrodes with improved stability for microfluidics. *Sens. Actuator B Chem.* **2006**, *114*, 239–247. [CrossRef]
33. Da Silva, E.T.S.G.; Miserere, S.; Kubota, L.T.; Merkoci, A. Simple on-plastic/paper inkjet-printed solid-state Ag/AgCl pseudoreference electrode. *Anal. Chem.* **2014**, *86*, 10531–10534. [CrossRef] [PubMed]
34. Desmond, D.; Lane, B.; Alderman, J.; Glennon, J.D.; Diamond, D.; Arrigan, D.W.M. Evaluation of miniaturised solid state reference electrodes on a silicon based component. *Sens. Actuator B Chem.* **1997**, *44*, 389–396. [CrossRef]
35. Suzuki, H.; Shiroishi, H.; Sasaki, S.; Karube, I. Microfabricated liquid junction Ag/AgCl reference electrode and its application to a one-chip potentiometric sensor. *Anal. Chem.* **1999**, *71*, 5069–5075. [CrossRef]
36. Suzuki, H. Advances in the microfabrication of electrochemical sensors and systems. *Electroanalysis* **2000**, *12*, 703–715. [CrossRef]
37. Du, J.L.; Chen, Z.F.; Chen, C.C.; Meyer, T.J. A half-reaction alternative to water oxidation: Chloride oxidation to chlorine catalyzed by silver ion. *J. Am. Chem. Soc.* **2015**, *137*, 3193–3196. [CrossRef] [PubMed]
38. Fritz, J.J. Thermodynamic properties of chloro-complexes of silver-chloride in aqueous-solution. *J. Solut. Chem.* **1985**, *14*, 865–879. [CrossRef]
39. Da Silva, R.A.B.; de Almeida, E.G.N.; Rabelo, A.C.; da Silva, A.T.C.; Ferreira, L.F.; Richter, E.M. Three electrode electrochemical microfluidic cell: Construction and characterization. *J. Braz. Chem. Soc.* **2009**, *20*, 1235–1241. [CrossRef]
40. Simonis, A.; Dawgul, M.; Luth, H.; Schoning, M.J. Miniaturised reference electrodes for field-effect sensors compatible to silicon chip technology. *Electrochim. Acta* **2005**, *51*, 930–937. [CrossRef]
41. Franklin, R.K.; Johnson, M.D.; Scott, K.A.; Shim, J.H.; Nam, H.; Kipke, D.R.; Brown, R.B. Iridium oxide reference electrodes for neurochemical sensing with MEMS microelectrode arrays. In Proceedings of IEEE Sensors 2005, Irvine, CA, USA, 31 October–3 November 2005; pp. 1400–1403.
42. Beati, A.A.G.F.; Reis, R.M.; Rocha, R.S.; Lanza, M.R.V. Development and evaluation of a pseudoreference Pt//Ag/AgCl electrode for electrochemical systems. *Ind. Eng. Chem. Res.* **2012**, *51*, 5367–5371. [CrossRef]
43. Kasem, K.K.; Jones, S. Platinum as a reference electrode in electrochemical measurements. *Platin. Met. Rev.* **2008**, *52*, 100–106. [CrossRef]
44. Dacuna, B.; Zaragoza, G.; Blanco, M.C.; Quintela, A.L.; Mira, J.; Rivas, J. Electrochemical synthesis of Fe/Ag and Co/Ag granular thin films. *Mater. Sci. Forum* **1998**, *269–272*, 307–312. [CrossRef]
45. Yang, H.S.; Kang, S.K.; Choi, C.A.; Kim, H.; Shin, D.H.; Kim, Y.S.; Kim, Y.T. An iridium oxide reference electrode for use in microfabricated biosensors and biochips. *Lab Chip* **2004**, *4*, 42–46. [CrossRef] [PubMed]
46. Xiong, J.W.; White, H.S. The I-V response of an electrochemical cell comprising two polarizable microelectrodes and the influence of impurities on the cell response. *J. Electroanal. Chem.* **2013**, *688*, 354–359. [CrossRef]
47. Sarkar, S.; Mathwig, K.; Kang, S.; Nieuwenhuis, A.F.; Lemay, S.G. Redox cycling without reference electrodes. *Analyst* **2014**, *139*, 6052–6057. [CrossRef] [PubMed]
48. Oja, S.M.; Zhang, B. Imaging transient formation of diffusion layers with fluorescence-enabled electrochemical microscopy. *Anal. Chem.* **2014**, *86*, 12299–12307. [CrossRef] [PubMed]
49. Bradley, J.C.; Chen, H.M.; Crawford, J.; Eckert, J.; Ernazarova, K.; Kurzeja, T.; Lin, M.D.; McGee, M.; Nadler, W.; Stephens, S.G. Creating electrical contacts between metal particles using directed electrochemical growth. *Nature* **1997**, *389*, 268–271. [CrossRef]
50. Loget, G.; Li, G.Z.; Fabre, B. Logic gates operated by bipolar photoelectrochemical water splitting. *Chem. Commun.* **2015**, *51*, 11115–11118. [CrossRef] [PubMed]

51. Loget, G.; Zigah, D.; Bouffier, L.; Sojic, N.; Kuhn, A. Bipolar electrochemistry: From materials science to motion and beyond. *Acc. Chem. Res.* **2013**, *46*, 2513–2523. [CrossRef] [PubMed]
52. Loget, G.; Roche, J.; Gianessi, E.; Bouffier, L.; Kuhn, A. Indirect bipolar electrodeposition. *J. Am. Chem. Soc.* **2012**, *134*, 20033–20036. [CrossRef] [PubMed]
53. Loget, G.; Kuhn, A. Bipolar electrochemistry for cargo-lifting in fluid channels. *Lab Chip* **2012**, *12*, 1967–1971. [CrossRef] [PubMed]
54. Fattah, Z.; Loget, G.; Lapeyre, V.; Garrigue, P.; Warakulwit, C.; Limtrakul, J.; Bouffier, L.; Kuhn, A. Straightforward single-step generation of microswimmers by bipolar electrochemistry. *Electrochim. Acta* **2011**, *56*, 10562–10566. [CrossRef]
55. Loget, G.; Kuhn, A. Propulsion of microobjects by dynamic bipolar self-regeneration. *J. Am. Chem. Soc.* **2010**, *132*, 15918–15919. [CrossRef] [PubMed]
56. Scida, K.; Sheridan, E.; Crooks, R.M. Electrochemically-gated delivery of analyte bands in microfluidic devices using bipolar electrodes. *Lab Chip* **2013**, *13*, 2292–2299. [CrossRef] [PubMed]
57. Chang, B.Y.; Chow, K.F.; Crooks, J.A.; Mavre, F.; Crooks, R.M. Two-channel microelectrochemical bipolar electrode sensor array. *Analyst* **2012**, *137*, 2827–2833. [CrossRef] [PubMed]
58. Sheridan, E.; Hlushkou, D.; Anand, R.K.; Laws, D.R.; Tallarek, U.; Crooks, R.M. Label-free electrochemical monitoring of concentration enrichment during bipolar electrode focusing. *Anal. Chem.* **2011**, *83*, 6746–6753. [CrossRef] [PubMed]
59. Dumitrescu, I.; Anand, R.K.; Fosdick, S.E.; Crooks, R.M. Pressure-driven bipolar electrochemistry. *J. Am. Chem. Soc.* **2011**, *133*, 4687–4689. [CrossRef] [PubMed]
60. Sequeira, C.A.C.; Cardoso, D.S.P.; Gameiro, M.L.F. Bipolar electrochemistry, a focal point of future research. *Chem. Eng. Commun.* **2016**, *203*, 1001–1008. [CrossRef]
61. Wang, Y.; Hernandez, R.M.; Bartlett, D.J.; Bingham, J.M.; Kline, T.R.; Sen, A.; Mallouk, T.E. Bipolar electrochemical mechanism for the propulsion of catalytic nanomotors in hydrogen peroxide solutions. *Langmuir* **2006**, *22*, 10451–10456. [CrossRef] [PubMed]
62. Kline, T.R.; Paxton, W.F.; Mallouk, T.E.; Sen, A. Catalytic nanomotors: Remote-controlled autonomous movement of striped metallic nanorods. *Angew. Chem. Int. Ed.* **2005**, *44*, 744–746. [CrossRef] [PubMed]
63. Paxton, W.F.; Kistler, K.C.; Olmeda, C.C.; Sen, A.; St Angelo, S.K.; Cao, Y.Y.; Mallouk, T.E.; Lammert, P.E.; Crespi, V.H. Catalytic nanomotors: Autonomous movement of striped nanorods. *J. Am. Chem. Soc.* **2004**, *126*, 13424–13431. [CrossRef] [PubMed]
64. Ma, C.X.; Zaino, L.P.; Bohn, P.W. Self-induced redox cycling coupled luminescence on nanopore recessed disk-multiscale bipolar electrodes. *Chem. Sci.* **2015**, *6*, 3173–3179. [CrossRef]
65. Oleinick, A.I.; Battistel, D.; Daniele, S.; Svir, I.; Amatore, C. Simple and clear evidence for positive feedback limitation by bipolar behavior during scanning electrochemical microscopy of unbiased conductors. *Anal. Chem.* **2011**, *83*, 4887–4893. [CrossRef] [PubMed]
66. Richter, M.M. Electrochemiluminescence (ECL). *Chem. Rev.* **2004**, *104*, 3003–3036. [CrossRef] [PubMed]
67. Zevenbergen, M.A.G.; Wolfrum, B.L.; Goluch, E.D.; Singh, P.S.; Lemay, S.G. Fast electron-transfer kinetics probed in nanofluidic channels. *J. Am. Chem. Soc.* **2009**, *131*, 11471–11477. [CrossRef] [PubMed]
68. Arora, A.; Eijkel, J.C.T.; Morf, W.E.; Manz, A. A wireless electrochemiluminescence detector applied to direct and indirect detection for electrophoresis on a microfabricated glass device. *Anal. Chem.* **2001**, *73*, 5633–5633. [CrossRef]
69. Leinweber, F.C.; Eijkel, J.C.T.; Bower, J.G.; van den Berg, A. Continuous flow microfluidic demixing of electrolytes by induced charge electrokinetics in structured electrode arrays. *Anal. Chem.* **2006**, *78*, 1425–1434. [CrossRef] [PubMed]
70. Bakker, E.; Pretsch, E. Modern potentiometry. *Angew. Chem. Int. Ed.* **2007**, *46*, 5660–5668. [CrossRef] [PubMed]
71. Malon, A.; Vigassy, T.; Bakker, E.; Pretsch, E. Potentiometry at trace levels in confined samples: Ion-selective electrodes with subfemtomole detection limits. *J. Am. Chem. Soc.* **2006**, *128*, 8154–8155. [CrossRef] [PubMed]
72. Bakker, E.; Pretsch, E. Nanoscale potentiometry. *Trends Anal. Chem* **2008**, *27*, 612–618. [CrossRef] [PubMed]
73. Pungor, E.; Toth, K. Ion-selective membrane electrodes—A review. *Analyst* **1970**, *95*, 625–648. [CrossRef]
74. Singh, P.S. From sensors to systems: CMOS-integrated electrocheimcal biosensors. *IEEE Access* **2015**, *3*, 249–259. [CrossRef]

75. Zhu, X.S.; Choi, J.W.; Ahn, C.H. A new dynamic electrochemical transduction mechanism for interdigitated array microelectrodes. *Lab Chip* **2004**, *4*, 581–587. [CrossRef] [PubMed]
76. Zhu, X.S.; Ahn, C.H. Electrochemical determination of reversible redox species at interdigitated array micro/nanoelectrodes using charge injection method. *IEEE Trans. Nanobiosci.* **2005**, *4*, 164–169. [CrossRef]
77. Zhu, X.S.; Ahn, C.H. On-chip electrochemical analysis system using nanoelectrodes and bioelectronic CMOS chip. *IEEE Sens. J.* **2006**, *6*, 1280–1286. [CrossRef]

6

High Throughput Studies of Cell Migration in 3D Microtissues Fabricated by a Droplet Microfluidic Chip

Xiangchen Che [1,†], Jacob Nuhn [2,†], Ian Schneider [2,3,*] and Long Que [1,*]

1 Department of Electrical and Computer Engineering, Iowa State University; Ames, IA 50011, USA; che@iastate.edu
2 Department of Chemical and Biological Engineering, Iowa State University; Ames, IA 50011, USA; januhn@iastate.edu
3 Department of Genetics, Development and Cell Biology, Iowa State University; Ames, IA 50011, USA
* Correspondence: ians@iastate.edu (I.S.); lque@iastate.edu (L.Q.)

† These authors contributed equally to this work.

Academic Editors: Manabu Tokeshi and Kiichi Sato

Abstract: Arrayed three-dimensional (3D) micro-sized tissues with encapsulated cells (microtissues) have been fabricated by a droplet microfluidic chip. The extracellular matrix (ECM) is a polymerized collagen network. One or multiple breast cancer cells were embedded within the microtissues, which were stored in arrayed microchambers on the same chip without ECM droplet shrinkage over 48 h. The migration trajectory of the cells was recorded by optical microscopy. The migration speed was calculated in the range of 3–6 μm/h. Interestingly, cells in devices filled with a continuous collagen network migrated faster than those where only droplets were arrayed in the chambers. This is likely due to differences in the length scales of the ECM network, as cells embedded in thin collagen slabs also migrate slower than those in thick collagen slabs. In addition to migration, this technical platform can be potentially used to study cancer cell-stromal cell interactions and ECM remodeling in 3D tumor-mimicking environments.

Keywords: cell motility; autocrine; paracrine; 3D micro-sized tissue; microfluidic droplet device

1. Introduction

It is very difficult to uncover how cells respond to the extracellular matrix (ECM) and how cells communicate using traditional cell culture systems. Traditional cell culture systems offer only two-dimensional (2D) substrates and lack the ability to isolate single cells or groups of cells [1,2].

An ideal platform for high throughput studies of cell-ECM interactions and cell-cell communication must have the following characteristics: (1) The platform is capable of realizing the encapsulation of cells in an ECM similar to that in the body, and the ECM should be three-dimensional (3D). (2) The platform can realize the isolation of single cells or groups of cells in order to control the cell-cell communication. This implies confining cells by providing barriers between the cell environment and the surroundings. (3) The platform must allow one to build microenvironments that are sufficiently small such as microtissues. Cells often communicate through the secretion of soluble molecules, so volumes between 10- to 1000-fold larger than the cell are appropriate to ensure that the secreted molecule concentration is sufficiently high. (4) Lastly, the platform is capable of rapidly generating a large number of cell-encapsulated microtissues in parallel in a cost-effective manner for high throughput studies.

2. Design, Operational Principle and Fabrication of the Droplet Microfluidic Device

Toward this goal, a droplet microfluidic chip for generating arrayed cell-encapsulated microtissues has been developed [3–5]. A schematic of a chip is given in Figure 1a. Filtered silicone oil is used as the continuous flow phase and the carrier fluid. Along the flowing direction of the fluids, as illustrated in Figure 1a, this device consists of a T-shape droplet generator, a liquid-droplet merger, a serpentine control channel (c-channel), and the droplet storage chambers (chambers). The droplet generator forms cell-laden collagen droplets. The c-channel is designed to prevent any air bubbles or non-uniform droplets from entering and occupying the chambers at the beginning of the operation of the device [6,7]. Once the uniform droplet generation is established, the c-channel is closed, and the outlet of the chambers is open. As a result, the droplets will flow toward the chambers, thereby entering and occupying them one by one. It should be noted that while this type of chip has been used for other applications [6], it is for the first time to be used to generate 3D microtissues and study the migration of the cells.

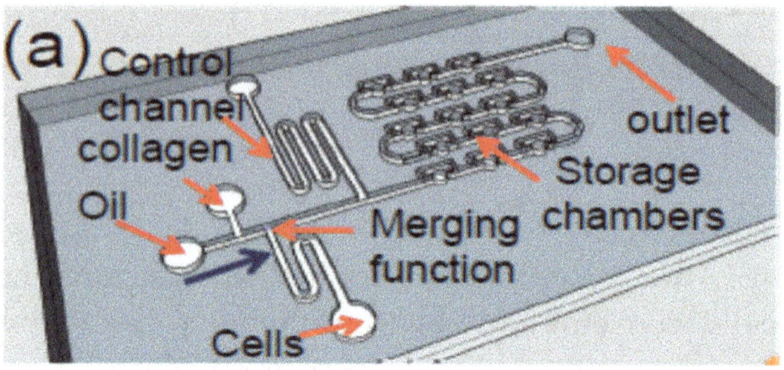

Figure 1. (**a**) Sketch of the droplet microfluidic chip for generating 3D microtissues (*not to scale*): Each storage chamber (a cylinder with a radius of 60 μm and height of 50 μm) has one 3D microtissue containing single or multiple cells; (**b**) Photo of a fabricated chip with 75 storage chambers.

A large scale of arrayed 3D microtissues, formed by polymerized collagen and cells, can be manufactured and stored in microchambers on the chip. To the best of our knowledge, this is the first demonstration of the fabrication of 3D microtissues using a droplet microfluidic chip to study cell migration in 3D microenvironments for an extended period of time.

The chip is fabricated using a soft lithography process [6,7]. Briefly, a 50-μm-thick SU-8 mold of the device is formed on a silicon substrate using conventional optical lithography. Polydimethylsiloxane (PDMS) is then casted on the mold, followed by 1.5 h of curing at the temperature of 65 °C. Finally, the PDMS microfluidic layer is peeled off from the mold, and then is bonded with a glass substrate after oxygen plasma treatment for 10 s. The input and output holes are made in the PDMS layer for the delivery of the samples to the chip, followed by assembling input and output tubing (Upchurch Scientific, Inc., Oak Harbor, DC, USA), and being connected with syringes controlled by several syringe pumps (KD Scientific, Inc., Holliston, MA, USA). A photo of a fabricated chip is shown in Figure 1b.

3. Materials and Methods

Breast cancer MDA-MB-231 cells were subcultured in Dulbecco's modified Eagle's medium with 10% fetal bovine serum, 2% Glutamax, and 1% penicillin/streptomycin. Imaging media was the same except it lacked phenol red and was supplemented with 12 mM HEPES. On the day the chambers were loaded, cells were trypsinized and suspended in 2 mg/mL collagen solution (rat tail-CORNING-354249) neutralized with imaging media at a cell density of 2×10^6 cells/mL. Chips were either loaded with a continuous collagen phase or with droplets in the storage chambers. Cells were allowed to spread over 24 h. Phase contrast images were then taken every 0.5–2 min over 4–8 h. Slabs of collagen were generated between two microscope slides with 60 μm (thin) or 360 μm (thick) spacers. Cells were prepared and imaged in the same way as for the devices with the exception that 3×10^5 cells/mL were used.

In order to mitigate or even eliminate the droplet shrinking issue due to evaporation, the fabricated chip was firstly soaked in PBS buffer solution (pH-7.4) in incubator (FISHER SCIENTIFIC-ISOTEMP 3530) overnight before use to ensure that PDMS was saturated with PBS. Silicone oil (SIGMA-ALDRICH) was used as the fluid carrier. Harvard syringe pump (70–4500) was connected with syringes for flowing the oil and the collagen/cells. In the experiments, the cell loading in the collagen droplets was based on Poisson distribution without any attempt to control the loading process. In addition, no surfactant was used to facilitate the droplet stability. During the collagen droplet generation and storage process, the collagen flowing input tube and syringe were submerged into a cold water tank (0~2 °C) to avoid fast polymerization since the polymerization rate is highly depended on temperature. After the droplets were stored in the chambers, the device was flipped over every minute within 10 min until the collagen was fully polymerized in the storage chamber, and to make sure the cells were in the middle of the storage chamber (along the z-axis), thereby ensuring the cells to stay in the 3D-matrix. For the experiments, the droplet microtissues remain surrounded by silicone oil. Experiments on the cell behaviors after the oil is replaced by cell culture media are in progress.

Confocal reflectance microscope (LEICA LAS-AF, Weltzlar, Germany) was used to image the 3D-matrix system. Standard incubator (FISHER SCINTIFIC-ISOTEMP 3530, FISHER SCINTIFIC, Waltham, MA, USA) was used to incubate the chip overnight in order to make cells accommodate to 3D-matrix system for cells' optimum behavior. OLYMPUS IX73 (OLYMPUS, Tokyo, Japan) with camera DP73 (OLYMPUS, Tokyo, Japan) was used to track the cell migration.

During the cell tracking process, the chip was submerged into a glass petri dish filled with PBS buffer at 37 °C to prevent drying problem. A heating stage (HARVARD APPARATUS-c-11842, HARVARD APPARATUS, Holliston, MA, USA) was applied to supply continuous heat. Finally, image J (National Institutes of Health, Bethesda, MD, USA) with a cell tracker model was used to track and plot the cell migration diagram. Experiments found that oxygen depletion was not a problem, even in our relatively small microtissues with PDMS and media above. Cell death did not occur over the period of about two days in the chamber, particularly if it was kept under proper pH buffering and temperature conditions. The oxygen consumption rate (OCR) for cancer cells is no higher than 30 $pmol \cdot s^{-1} \cdot 10^{-6} \cdot$ cells [8]. The volume of each microtissue is ~6.0×10^{-10} L and no more than 10 cells occupy a microtissue. Consequently, the OCR for one microtissue is 500 nM/s. If no oxygen transfer occurs, it would take over a day for the cells in each microtissue to decrease the oxygen concentration from 260 μM, the saturated level of media in equilibrium with air in the incubator, to 200 μM, a value still well above hypoxic conditions. However, there is oxygen transport across the liquid and PDMS, and the transport is governed by the following equation at steady state: $OCR = (D/h)A(C^* - C)$, where OCR is the oxygen consumption rate (0.3 fmol/s), D is the diffusion coefficient of oxygen in PDMS or water (3×10^{-5} cm^2/s) [9], A is the cross-sectional area of each microtissue (1.2×10^{-4} cm^2), C^* is the equilibrium concentration of oxygen in fluid (260 μM), C is the local oxygen concentration around the cells and h is the height of the PDMS and fluid above the microtissue. At a height of 0.8 cm, the steady-state oxygen concentration is about 200 μM. While there is little information on whether cell function is altered at this concentration, it is well above that which is considered hypoxic (<6 μM). Furthermore, because media is initially at an equilibrium concentration of 260 μM oxygen, it takes time for the oxygen concentration to reach this steady state. At the time that experiments are conducted, the oxygen level is 200–210 μM. Consequently, the 0.8 cm of PDMS and media is thin enough to support the relatively low rate of oxygen consumption within the microtissues.

4. Results and Discussion

The optical image of the fabricated arrayed microtissues inside the storage chambers is given in Figure 2a. Following the procedure described in Section 2, it has been demonstrated that the uniform microtissues can be formed and stored in the storage chambers on the chip routinely. However, it should be emphasized that care should be taken to avoid the polymerization of the collagen in the flowing channels on the chip; otherwise, the storage chambers cannot be occupied by microtissues properly. In Figure 2b, a close-up optical image of a droplet shows a cell inside a polymerized collagen fiber. In order to show the collagen fiber more clearly, a confocal image in Figure 2c has been taken on the droplet, showing one cell embedded in the polymerized collagen fiber.

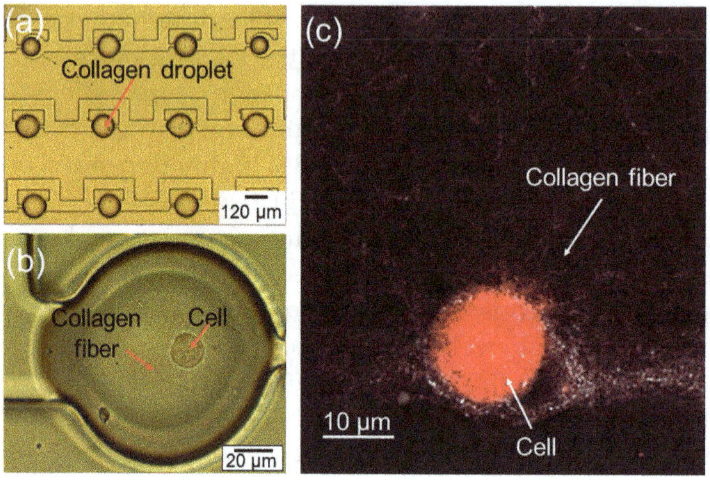

Figure 2. (**a**) Photo of arrayed microtissues stored in storage chambers; (**b**) close-up of one microtissue containing one cell; (**c**) confocal image of one cell inside polymerized collagen fiber, forming a microtissue.

In order to confirm that the cell is indeed surrounded by a 3D extracellular matrix (ECM), which is made up of polymerized collagens, some confocal images of the microtissues have been taken. A topside view, cross-section view and the stacked images from the bottom to the top of a microtissue are obtained in Figure 3. Given that the nominal height of the fabricated storage chambers is ~50 μm, the cell is roughly ~20 μm above the bottom of the microtissue and ~20 μm below the top of the microtissue. Basically, the cell is embedded inside the collagen fibers. Note that the gap of the cell from the top and bottom of the microtissue can be readily increased by increasing the height of the storage chambers.

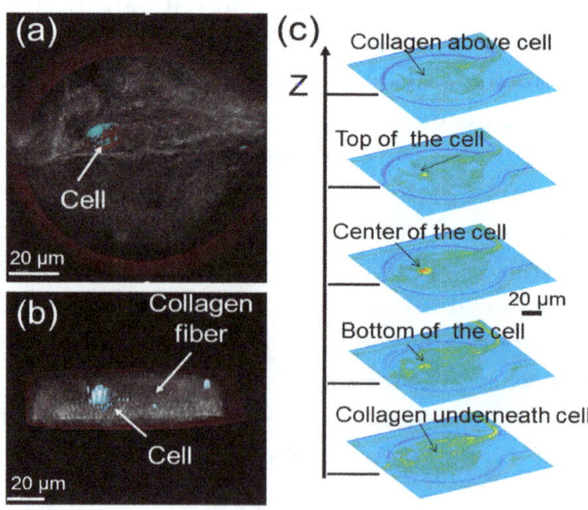

Figure 3. Confocal images showing one cell inside a 3D microtissue in a storage chamber: (**a**) topside view; (**b**) cross-section view; (**c**) stacked confocal images of a microtissue showing one cell inside a 3D microtissue.

It has been found that as long as the silicone oil does not directly contact the cells, it will not affect cell viability. In the experiments, only the cells embedded within the polymerized collagen have been studied. These cells are not directly exposed to oil. The total time for the cells inside the polymerized collagen for the experiments was up to 32 h, and no clear effect on cell viability was observed during this time period, suggesting that the oil does not diffuse into the microtissue droplets.

It has also been observed that the polymer gel structure has some differences at the interior *versus* the edges of the microtissue droplets. Interactions with surfaces could potentially nucleate collagen fiber assembly or simply act as an adherent surface for collagen fibers. The typical time for the polymerization of the collagen is ~15 min at room temperature, similar to that for collagen polymerization on a glass cover slip.

The real-time migration videos (in the supplementary) of the cells inside microtissue have been recorded using an optical microscope. The representative images in Figure 4a,b shows the migration of three congregated cells inside microtissue in a 7 h period of time, while the representative images in Figure 4c,d shows the migration of one cell inside a microtissue during the same period of time. These experiments demonstrate that the chip can provide a platform to study the migration of one single cell or multiple cells in a microtissue environment. In addition, since the cells are confined in a small volume (~600 pL), the communication among them may be easily studied.

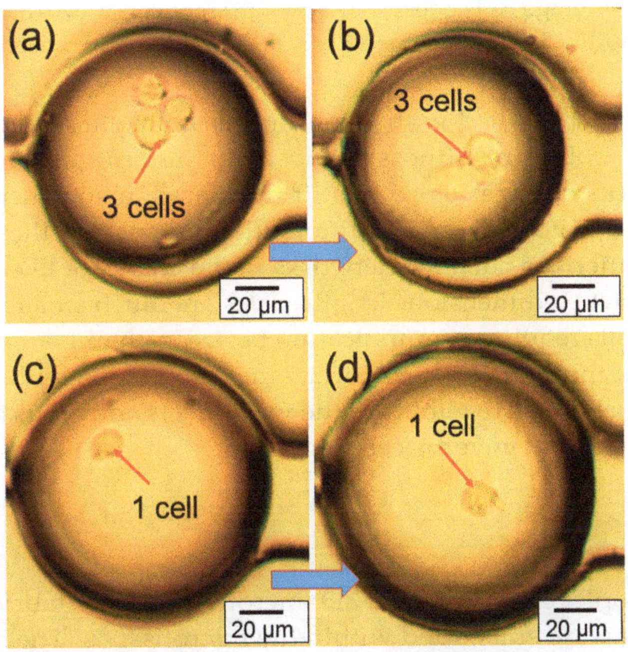

Figure 4. Representative optical images showing (**a**,**b**) the migration of three cells inside 3D microtissue during a 7 h period at 37 °C; (**c**,**d**) the migration of one cell inside 3D microtissue during a 7 h period at 37 °C.

Based on the recorded videos (in the supplementary), the cell migration speed has been calculated under two conditions. The first condition includes chips that are filled with cells embedded in collagen, generating a continuous collagen network. This increases the volume of the environment, decreasing the opportunity for the depletion of nutrients or accumulation of waste. Also, cells in different chambers may communicate. The second condition includes devices that only contain cells embedded in collagen in droplets within the chambers (Figure 4). These droplets have relatively small volumes and cells in a particular droplet cannot communicate with cells in other droplets. These conditions were compared to cell migration in thin (60 μm) and thick (360 μm) slabs of collagen. Representative migration trajectories are shown in Figure 5a,b. Cells in the continuous collagen gels migrate similarly to those in the thin collagen slabs and slower than those in the thick collagen slabs (Figure 5c). Cells in droplets migrated much slower than any other condition (Figure 5c).

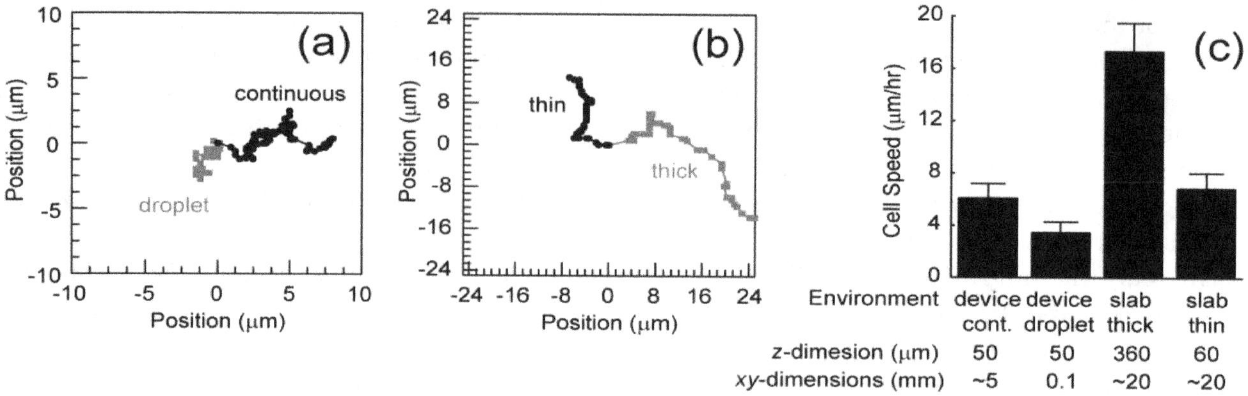

Figure 5. Representative trajectories of cells embedded in collagen (2 mg/mL) in the chip (**a**) and embedded in a collagen (2 mg/mL) slab between two coverslips (thick: grey, thin: black); (**b**) The chip is either filled with a continuous polymerized collagen network (grey) or droplets of collagen within the chambers (black); (**c**) Average cell speed under the different conditions as well as the length scales associated with each condition. Error bars are 95% confidence intervals.

It is interesting that the chip filled with a continuous collagen network and a thin slab results in similar migration rates. Collagen stiffness is known to alter migration speeds and the observed stiffness of flexible networks changes close to stiff interfaces, a so-called wall effect. The similar *z*-dimension length scales between these conditions likely generate the similar migration speeds. Consequently, thicker polymerized collagen networks in the chips are likely needed to observe faster migration. Finally, the droplet *xy*-dimensions length scales are much smaller than the other conditions suggesting that either (1) cells require communication between chambers or (2) small volumes in this first generation chip inhibit migration. The second generation chips with storage chambers that are both thicker and larger will allow us to eliminate the wall effects and focus on cell-cell communication within and between chambers that governs cell migration.

5. Conclusions

Using microfluidic droplet chips, arrayed 3D microtissues were fabricated successfully. One or multiple breast cancer cells were embedded within the microtissues. The migration trajectory of the cells was recorded and analyzed. The migration speed inside 3D microtissues was in the range of 3–6 μm/h. It was found that cells in chips filled with a continuous collagen network migrated faster than those where only isolated droplets were arrayed in the chambers. Besides being used for studying the cell migration, this technical platform can be also potentially useful for studying cancer cell-stromal cell interactions and ECM remodeling in 3D tumor-mimicking environments.

Acknowledgments: The authors acknowledge Margaret Carter at the Confocal and Multiphoton Facility under the Office of Biotechnology at Iowa State University (ISU). Ian Schneider acknowledges support from National Institute of Health/National Cancer Institute (R03CA184575) and the Engineering Research Institute in the College of Engineering at ISU for general project funding. Long Que acknowledges financial support from National Science Foundation (NSF) and a seed grant from Engineering Research Institute at Iowa State University. No funds were received to cover the costs to publish in open access journals.

Author Contributions: Ian Schneider and Long Que conceived and designed the experiments; Xiangchen Che and Jacob Nuhn performed the experiments; Long Que, Ian Schneider, Jacob Nuhn and Xiangchen Che analyzed the data; Long Que and Ian Schneider wrote the paper.

References

1. Pampaloni, F.; Emmanuel, G.; Ernst, H. The third dimension bridges the gap between cell culture and live tissue. *Nat. Rev. Mol. Cell Biol.* **2007**, *8*, 839–845. [CrossRef] [PubMed]
2. Fischbach, C.; Chen, R.; Matsumoto, T.; Schmelze, T.; Brugge, J.S.; Polverini, P.J.; Mooney, D.J. Engineering tumors with 3D scaffolds. *Nat. Methods* **2007**, *4*, 855–860. [CrossRef] [PubMed]
3. Teh, S.-Y.; Lin, R.; Hung, L.-H.; Lee, A.P. Droplet microfluidics. *Lab Chip* **2008**, *8*, 198–220. [CrossRef] [PubMed]
4. Li, C.Y.; Wood, D.K.; Huang, J.H.; Bhatia, S.N. Flow-based pipeline for systematic modulation and analysis of 3D tumor microenvironments. *Lab Chip* **2013**, *13*, 1969–1978. [CrossRef] [PubMed]
5. Hong, S.M.; Hsu, H.J.; Kaunas, R.; Kameoka, J. Collagen microsphere production on a chip. *Lab Chip* **2012**, *12*, 3277–3280. [CrossRef] [PubMed]
6. Cheng, W.; He, Y.; Chang, A.; Que, L. A microfluidic chip for controlled release of drugs from microcapsules. *Biomicrofluidics* **2013**, *7*, 064102. [CrossRef] [PubMed]
7. Cheng, W.; He, Y.; Que, L. Controlled drug release in a microfluidic device with droplet merging and storage functions. In Proceeding of 2013 IEEE Sensors, Baltimore, MD, USA, 3–6 November 2013; pp. 1815–1818.
8. Herst, P.M.; Berridge, M.V. Cell surface oxygen consumption: a major contributor to cellular oxygen consumption in glycolytic cancer cell lines. *Biochim. Biophys. Acta. Bioenerg.* **2007**, *1767*, 170–177. [CrossRef] [PubMed]
9. Cox, M.E.; Dunn, B. Oxygen diffusion in poly(dimethyl siloxane) using fluorescence quenching. I. Measurement technique and analysis. *J. Polym. Sci. Part A Polym. Chem.* **1986**, *24*, 621–636. [CrossRef]

7

Three-Dimensional Electro-Sonic Flow Focusing Ionization Microfluidic Chip for Mass Spectrometry

Cilong Yu [1,‡], Xiang Qian [1,*,‡], Yan Chen [2], Quan Yu [1], Kai Ni [1] and Xiaohao Wang [1,3,*]

1 Division of Advanced Manufacturing, Graduate School at Shenzhen, Tsinghua University, Shenzhen 518055, China; yu-cl12@mails.tsinghua.edu.cn (C.Y.); yu.quan@sz.tsinghua.edu.cn (Q.Y.); ni.kai@sz.tsinghua.edu.cn (K.N.)

2 Shenzhen Institutes of Advanced Technology, Chinese Academy of Sciences, Shenzhen 518055, China; yan.chen@siat.ac.cn

3 The State Key Laboratory of Precision Measurement Technology and Instruments, Tsinghua University, Beijing 100084, China

* Correspondence: qian.xiang@sz.tsinghua.edu.cn (X.Q.); wang.xiaohao@sz.tsinghua.edu.cn (X.W.)

† This paper is an extended version of our paper presented in the 17th Annual Conference of the Chinese Society of Micro-Nano Technology, Shanghai, China, 11–14 October 2015.

‡ These authors contributed equally to this work.

Academic Editors: Manabu Tokeshi and Kiichi Sato

Abstract: Increasing research efforts have been recently devoted to the coupling of microfluidic chip-integrated ionization sources to mass spectrometry (MS). Considering the limitations of microfluidic chips coupled with MS such as liquid spreading, dead volume, and manufacturing troubles, this paper proposed a new three-dimensional (3D) flow focusing (FF)-based microfluidic ionizing source. This source was fabricated by using the two-layer soft lithography method with the nozzle placed inside the chip. The proposed FF microfluidic chip can realize two-phase FF with liquid in air regardless of the viscosity ratio of the continuous and dispersed phases. MS results indicated that the proposed FF microfluidic chip can work as a typical electrical ionization source when supplied with high voltage and can serve as a sonic ionization source without high voltage. The electro-sonic FF ionization microfluidic chip is expected to have various applications, particularly in the integrated and portable applications of ionization sources coupling with portable MS in the future.

Keywords: microfluidic chip; ionization; electro-sonic flow focusing; two-phase flow; soft lithography; mass spectrometry

---

1. Introduction

Electrospray ionization (ESI) mass spectrometry (MS) is vital to biological analysis because of its excellent ability to detect a great number of analytes with high sensitivity while also identifying the structural information of detected species [1–3]. Given the rapid development of miniaturization technology, microfluidic chips have an important function in metabolomics, proteomics, and other biochemical analyses owing to their efficient and fast separations [4,5] in integrating complex sample pretreatment functions and their automatic manipulation of small sample volumes [6–10]. Therefore, the coupling of microfluidic chips with MS has received considerable research interest [11], particularly the design and improvement of chip-based microfluidic ionization sources coupled with MS, which has been comprehensively reviewed by a large number of research groups [12–15].

Chip-based ESI sources are generally divided into three types. The first type is the monolithic source, which involves direct spraying from the edge of a microfluidic chip, and was earlier reported by Karger *et al.* [16] and Ramsey *et al.* [17]. Although this approach is relatively simple, the ionization source encountered liquid spreading problems along the edge of the chip, thus resulting in the formation of a large Taylor cone. Tapered fused-silica capillaries, which served as electrospray tips, were inserted into the end of the channels in the microfluidic chips to overcome this problem [18]. However, the fabrication process of this approach was complicated. Moreover, large dead volumes were generated at the interfaces between the capillaries and micro-channels, thus causing a possible degradation of electrospray performance. In recent years, an increasing number of research efforts adopted the design of integrating a nozzle in the microfluidic chip during the fabrication process [6,14]. This approach undoubtedly possesses distinct advantages compared with previous methods. Moreover, this integrated nozzle was also developed from one to multi-nozzles for high-throughput analysis [19,20]. However, all of these microfluidic chip ionization sources have only one channel of liquid for spraying and a lack air for atomization which is usually used in macro-ionization [21].

Various materials have been employed to fabricate microfluidic chips, such as silicon [22,23], polymers (SU-8) [24–26], polymethyl methacrylate [27], glass [28–31] and poly (dimethylsiloxane) (PDMS) [1,6,32,33]. Koster *et al.* [14] conducted a thorough summary of the materials used for microfluidic chips. The fabrication process of integrated nozzles on microfluidic chips was generally easier with polymers than with glass [34,35]. Among these polymers, PDMS was a widely used material for microfluidic chips because of its chemical inertness, low cost, and rapid fabrication process by soft lithography [36]. Furthermore, compared with the hydrophilic surface of glass, the hydrophobic surface of PDMS can effectively prevent the solution from wetting the nozzle; otherwise, the wetted surface may preclude the operation at nano-ESI flow rates and lead to unstable electrospray [1,37]. However, forming an excellent electrospray nozzle integrated on the microfluidic chip was difficult because the PDMS was soft and the tip was too small to cut at the two sides of the micro-channel front end. A raised layer cutting method by two-layer soft lithography was proposed in our previous work [38]. Furthermore, the electrospray nozzle was exposed outside and was prone to damage [1,6,32,33,39,40].

Inspired by flow focusing (FF) technology [41–49], we designed an inner electrospray nozzle that has a Taylor cone at the front end of the liquid (dispersed phase) channel focused by the air (continuous phase) inside the microfluidic chip. This water in the air FF regime was first implemented by using a coaxial capillary tube closed to a small hole in a thin plate [50–54]. However, the fabrication processes were complicated and coupling such devices with other microfluidic modules was difficult. For microfluidic FF, two-dimensional (2D) microfluidic FF chips always require dispersed and continuous phases with low viscosity ratios (approximately 0.1) [55], such as water in oil [41] and air in water [47]. The high viscosity ratio is also barely implemented because the 2D micro-channel will form a shear flow and the dispersed phase will not break [56–59]. Alternatively, the three-dimensional (3D) microfluidic FF chip, in which the dispersed phase was suspended in the continuous phase without contacting the micro-channel wall, can guarantee the formation of extensional flow. This 3D microfluidic FF chip is suitable for applications with low viscosity ratios, such as generating submicron emulsion droplets and cell counting [49,60,61] and high viscosity ratios. Although the 3D microfluidic FF chip was recently implemented by Trebbin *et al.* [62] at a high viscosity ratio to produce liquid jets and droplets, we conceived and designed a similar 3D microfluidic FF chip independently and differently. Compared with the three-layer structure of Trebbin *et al.*, we fabricated the microfluidic chip by two soft lithography layers with the nozzle inside the microfluidic chip to simplify the fabrication and alignment processes and protect the nozzle from damage. Furthermore, the 3D microfluidic FF chip implemented in the current paper was adopted as an ionization source. MS data were collected to verify the ionization performances with different applied voltages. In the rest of the paper, we mainly demonstrated the fabrication process of the proposed microfluidic chip by two-layer soft lithography. A corresponding 2D microfluidic FF chip was also fabricated, and the spray effects of the 2D and the 3D microfluidic chips were compared. The jet diameter of the dispersed phase from the 3D microfluidic chip was measured with the changes of the dispersed phase flow rate and the continuous phase pressure. Furthermore, we also show the simple application of the proposed microfluidic chip coupled with MS. The results indicated that such a microfluidic chip can obtain stable MS signals with and without high voltage supply. We called this microfluidic chip with an ion source the electro-sonic FF ionization (ESFFI) microfluidic chip, which has a potential for portable and on-site applications in the future.

2. Materials and Methods

2.1. Materials and Equipment

HPLC-grade methanol and acetic acid were purchased from Merck KGaA (Darmstadt, Germany). PDMS elastomer base and curing agent (Sylgard 184) were purchased from Dow Corning (Midland, MI, USA). SU-8 photoresist was obtained from Microchem Co. (Naton, MA, USA). All dispersed and continuous phases were supplied to the microfluidic chip through short stainless steel tubes embedded in the reservoirs using a pneumatic pressure controller (MFCS, Fluigent, Paris, France). This pneumatic pressure controller allowed almost non-fluctuating flow, which was essential to form a steady Taylor cone. The high voltage generated by a power supply module (Dongwen High Voltage Power Supply Co., Ltd, Tianjin, China) was applied on the stainless steel tube of the dispersed phase. A high-speed camera (ORCA-flash, Hamamatsu, Shizuoka, Japan) mounted on an inverted optical microscope (Eclipse TE 2000-U, Nikon, Tokyo, Japan) was used to observe the experiments. An ion trap mass spectrometer (Thermo Fisher Scientific Inc., Waltham, MA, USA) was coupled to the microfluidic chip, and MS data were collected by the computer.

2.2. ESFFI Microfluidic Chip Design

In most cases, the top and bottom halves of the nozzle tend to separate at the tip, thus seriously affecting the spray effect. A Taylor cone was generated inside the microfluidic chip in this paper to avoid this drawback. Moreover, the proposed structure and corresponding fabrication process avoided the trouble of cutting along the edge of the nozzle outlet. The trumpet-shaped outlet was instead cut far from the nozzle (Figure 1a), thus greatly improving the craftwork and allowing the rapid mass production of microfluidic chips. The ESFFI microfluidic chip was designed by using AutoCAD (Autodesk, Inc., San Rafael, CA, USA). The photo mask and nozzle size of the ESFFI microfluidic chip are presented in Figure 1. In this paper, two photo masks (patterns illustrated in Figure 1a) were manufactured by Qingyi Precision Mask Making Co., Ltd (Shenzhen, China). The heights of the liquid and air channels were approximately 25 and 320 μm, respectively. An air channel higher than the liquid channel leads to a better focus effect because a higher air channel is beneficial for the dispersed phase to suspend in the continuous phase without coming into contact with micro-channel walls.

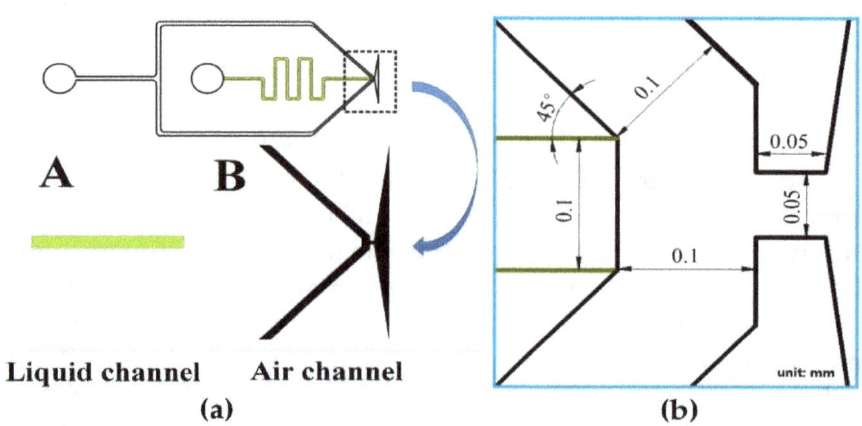

Figure 1. Photo mask and nozzle size of the ESFFI microfluidic chip. (**a**) Photo mask of the ESFFI microfluidic chip. (**b**) Nozzle size of the ESFFI microfluidic chip.

2.3. Fabrication of the ESFFI Microfluidic Chip

The ESFFI microfluidic chip was fabricated by using standard multilayer soft lithography techniques [63,64]. The fabrication process is shown in Supplementary Materials Figure S1. First, a 3 inches silicon wafer template was treated in oxygen plasma (PDC-M, Chengdu Mingheng Science & Technology Co., Ltd, Chengdu, China) to prevent SU-8 photoresist from spalling. The negative photoresist (SU-8 2025) was then poured on the silicon wafer. After spinning and soft-baking, the photoresist was exposed via photo mask A, which served as the liquid channel layer, providing an orifice and a channel for the dispersed phase. After post-baking and cooling, a second layer of negative photoresist (SU-8 2100) was applied at the top of the liquid channel layer without developing uncross-linked photoresist. After spinning and soft-baking, photo mask B, which was aligned with the liquid channel layer by a UV aligner, was placed on the second layer photoresist for exposure. This second layer served as the air channel layer with the channels and orifice for the continuous phase. After post-baking and cooling, the SU-8 photoresist layers were developed in propylene glycol methyl ether acetate and were then placed on the thermostatic platform for hard-baking. This SU-8 master mold served as the top layer of our micro-channel structure. Another SU-8 master mold with patterns of photo mask B only was prepared on another 3 inches silicon wafer template for the bottom PDMS micro-channel half-devices. This SU-8 master mold only had an air channel; thus, the fabrication process needed exposure only once. The final structures of the SU-8 master mold are illustrated in Figure 2.

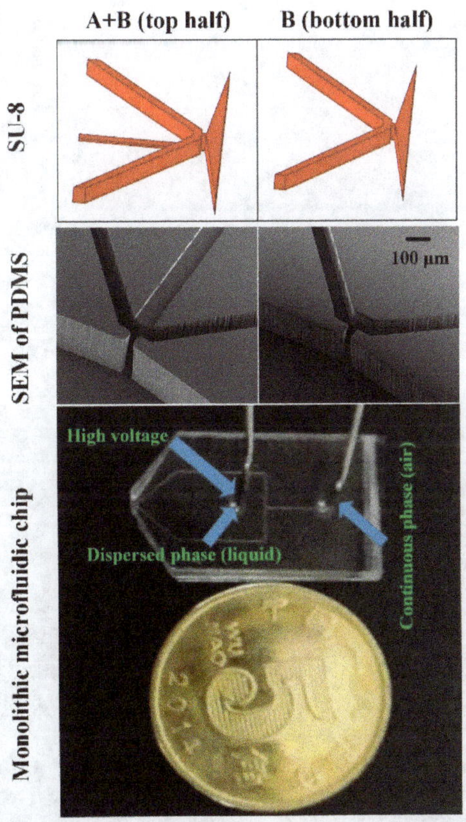

Figure 2. (**Top**)Top and bottom SU-8 master molds; (**Middle**) Top and bottom structures of PDMS under SEM; (**Bottom**) monolithic microfluidic chip.

The SU-8 master molds were modified with vapor-phase TMCS (Chlorotrimethylsilane) to assist the release of PDMS membranes. PDMS base monomer and curing agent were mixed at 10:1 and 5:1 weight ratios and then poured on the top and bottom SU-8 master molds, respectively. After degassing under vacuum, these two half-pieces were cured in an oven at 80 °C for 2 h. Subsequently, two PDMS slabs were peeled off from the two master molds and the inlet holes were drilled at the top by using a punch (tip diameter of 0.75 mm). Figure 2 also shows the two PDMS slabs under a scanning electron microscope (SEM). In general, great attention should be paid in removing the excess PDMS along the nozzle tips with a razor blade. However, in this paper, a razor blade was only required to cut off the excess PDMS along the expansion trumpet-shaped outlet far from the nozzle. Both PDMS slabs, which were treated in oxygen plasma (PDC-M, Chengdu Mingheng Science & Technology Co., Ltd, Chengdu, China), were then bonded together by using an xyz-manipulator (Beijing Optical Century Instrument Co., Ltd., Beijing, China). Following assembly, the PDMS microfluidic chip was cured at 80 °C for 72 h to enhance the strength of the bonding and to eliminate the MS background from PDMS. The final monolithic microfluidic chip is shown in Figure 2.

3. Results and Discussions

For the 2D micro-channel, most research efforts have focused on the low viscosity ratio of the dispersed phase and continuous phase, such as water in oil [41] and air in water [47]. For comparison, a corresponding 2D micro-channel was fabricated in this paper and the high viscosity ratio of the dispersed phase and continuous phase was performed. The spray effect of the 2D microfluidic chip in Figure 3 shows that the dispersed phase was separated into two layers (arrow pointed). The shear force caused by the viscosity of the walls prevented the dispersed phase from separation.

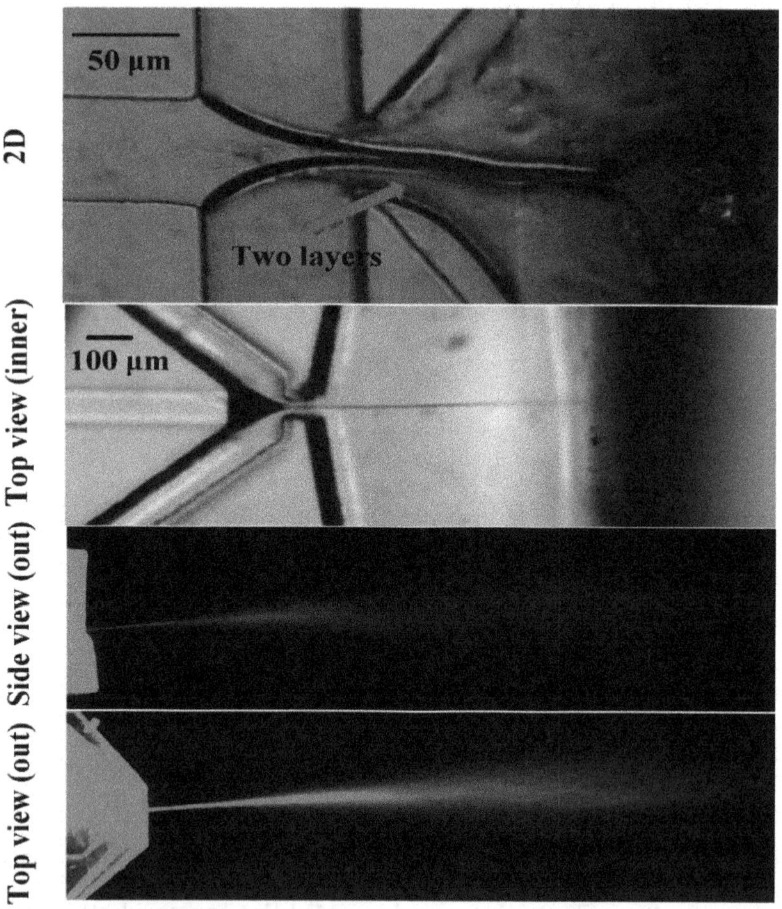

Figure 3. Spray effect of the 2D and 3D microfluidic chips. Pressures on the continuous and dispersed phases were 300 and 210 mbar, respectively. High voltage (2 kV) was added on the dispersed phase. The arrow pointed to the two layers of the liquid separated by the shear force of the walls.

However, for the 3D microfluidic FF chip proposed in this paper, the dispersed phase was suspended in the continuous phase without contacting the micro-channel walls, thus allowing this phase to separate at the high viscosity ratio. Figure 3 shows the perfect spray effect of the 3D microfluidic chip, in which a steady thin liquid jet was smoothly emitted from the Taylor cone inside the microfluidic chip and extended over several millimeters out of the microfluidic chip; this effect was similar to the liquid jet from the macrostructure device [54] and microfluidic chip [62] for FF. Under this circumstance, dead volume and liquid spreading problems, which often occur in microfluidic chips for MS, were successfully settled. Figure 4 shows the details of the jet diameters with the changes of the dispersed phase flow rates. With increasing dispersed phase flow rates, the liquid jet diameters increased in power (Figure 5). The dispersed phase flow rates varied from 120 to 8400 μL/h, and the pressures on the continuous phase were 150 and 300 mbar. However, the jet diameters in our study were larger than the data calculated from the theoretical formula [54] as shown in Equation (1):

$$d_j = \left(\frac{8\rho_l}{\pi^2 P_g}\right)^{1/4} Q^{1/2} \tag{1}$$

where ρ_l is the density of the dispersed phase, P_g is the pressure drop of the continuous phase, and Q is the dispersed phase flow rate; these phases presented similar power function relationships, thus revealing that such a microfluidic liquid jet system may share the same underlying physics with the well-studied orifice plate configuration [62]. The differences between the rectangular channels and circular pipelines might contribute to the error. Moreover, the selected position of the jet diameter and measurement error might also introduce data dissimilarity. In general, the liquid jet diameter was proportional to the ratio of the pressure applied on the dispersed phase and continuous phase.

This microfluidic chip can be widely used in various fields in which micron or sub-micron liquid jets are generated, such as inkjet printing, pharmaceutical formulations [62] and MS. As an example application of this ESFFI microfluidic chip, we presented the experiment of coupling this microfluidic chip with MS. Figure 6 displays the configuration of the ESFFI microfluidic chip with the mass spectrometer. The microfluidic chip was held by a laboratory-built platform and coupled to the ion trap mass spectrometer. The distance between the microfluidic chip emitter and the MS inlet orifice is about 2–10 mm, which was adjusted by a xyz-manipulator.

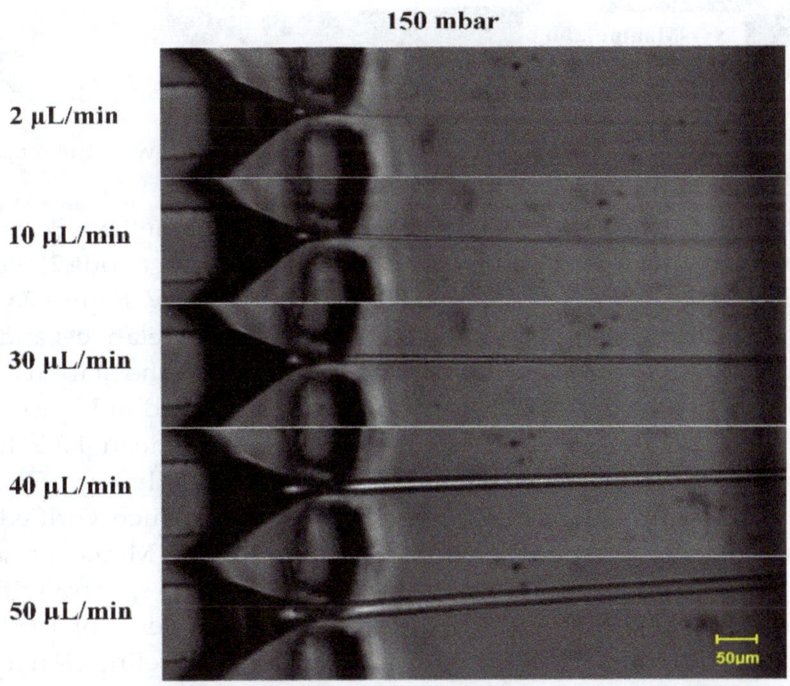

Figure 4. Jet diameter changes with dispersed phase flow rates under the gas pressure of 150 mbar. The red box presents the position of the liquid jet diameter.

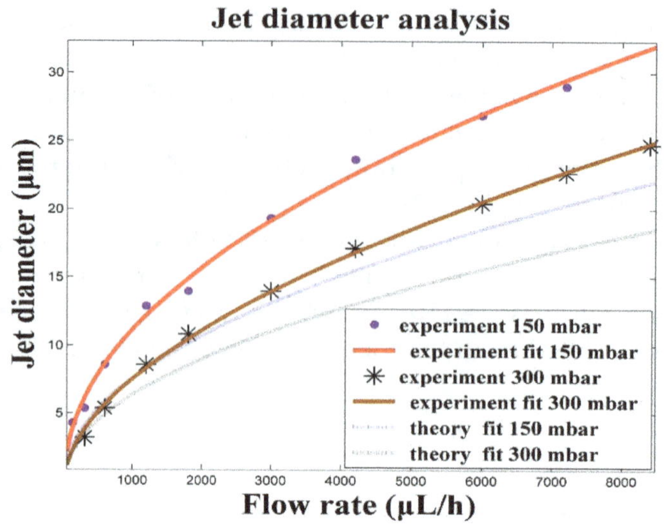

Figure 5. Comparison of experimental data with theoretical predictions for jet diameter and breakup transition analyses.

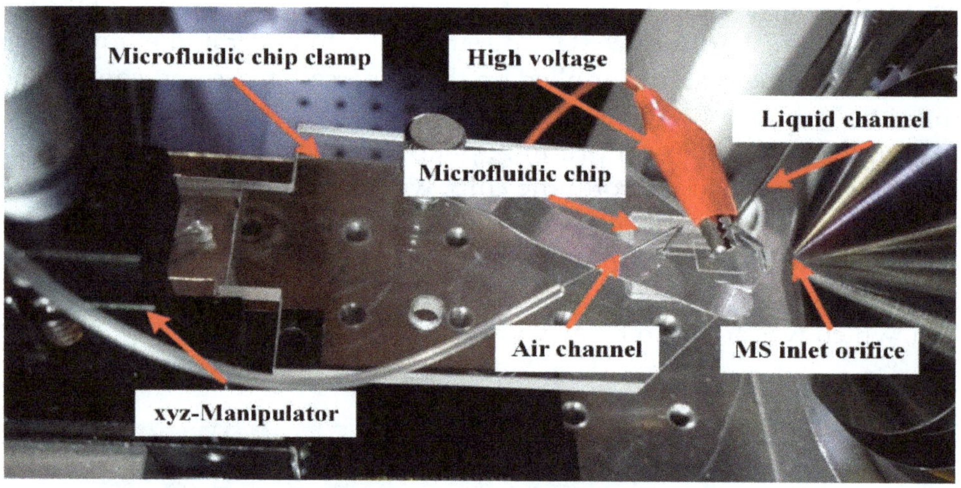

Figure 6. The configuration of the xyz-manipulator microfluidic chip with the mass spectrometer.

Figure 7 shows the MS signal of 41.5 μM Reserpine in 3/1 (v/v) methanol/water with 0.2% formic acid, the gas pressures for the liquid channel and air channel were about 250 mbar and 300 mbar, respectively, the high voltage added on the liquid channel was 5 kV. Figure 7a is the 10 min mean signal of reserpine and the signal stability is shown in Figure 7b; the relative standard deviation (RSD) for the total ion current (TIC) was 5.02%. To further demonstrate the stability of this microfluidic chip, the batch-to-batch reproducibility of three chips were displayed in Figure 8. The experimental conditions were the same as the signal shown in Figure 7. The mean TICs for chip 1 to 3 were 1.17×10^6, 8.87×10^5, and 8.4×10^5, respectively; chip 1 had a slightly high TIC. However, the base peak mean intensity of these three chips was about 5.32×10^4, which verified the batch-to-batch reproducibility and stability of the proposed ESFFI microfluidic chip. Moreover, several samples with lower-concentration Rhodamine B solution were tested, and the MS spectra with a 0.01 μM analyte is shown in Figure 9. The Rhodamine B signal intensity is at the level of 10^2 ion counts, and can be estimated to be more than three times larger than the nearby baseline signal (although there are existing several miscellaneous peaks with the same level). Thus, we can speculate that, though not rigorously, the limit of detection of such an electro-sonic flow focusing ionization microfluidic chip for Rhodamine B can be lower than 0.01 μM.

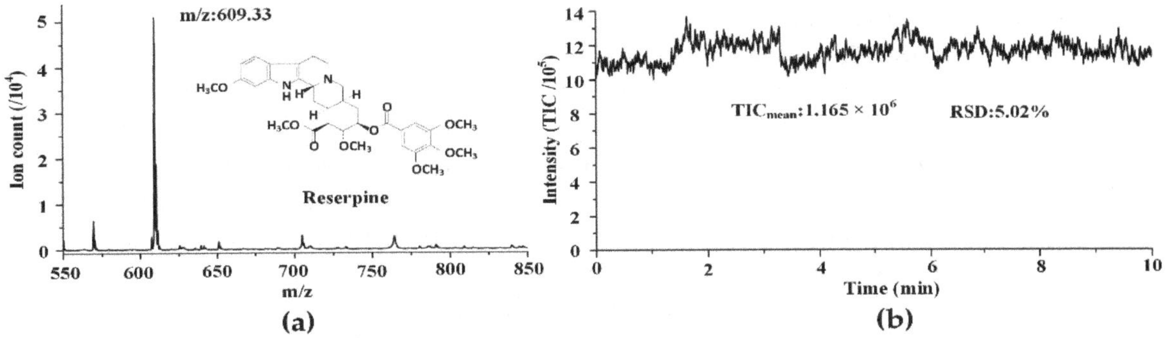

Figure 7. MS signal of 41.5 μM Reserpine in 3/1 (*v*/*v*) methanol/water with 0.2% formic acid, the gas pressures for the liquid channel and air channel were about 250 mbar and 300 mbar, respectively, and the high voltage added on liquid channel was 5 kV. (**a**) Reserpine ion counts of the 10 min mean signal. (**b**) Reserpine signal stability of 10 min; TIC mass range from 550 to 850.

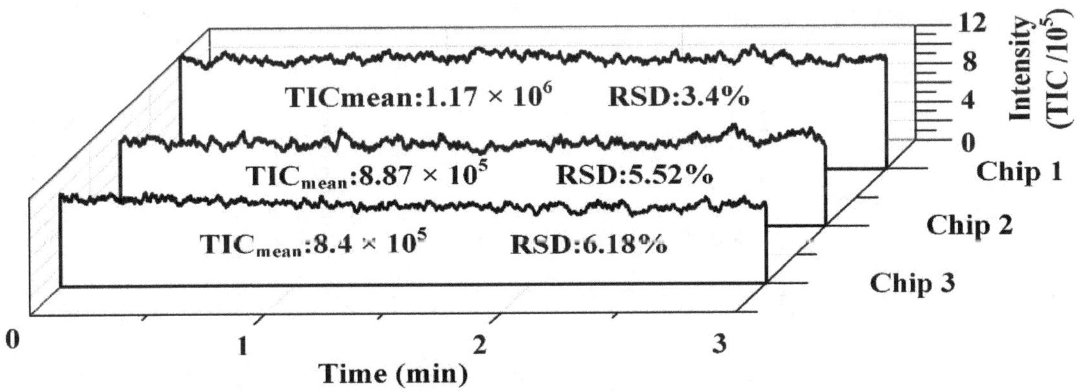

Figure 8. The batch-to-batch reproducibility of three chips. The conditions for three chips were the same: 41.5 μM Reserpine in 3/1 (*v*/*v*) methanol/water with 0.2% formic acid, the gas pressures for the liquid channel and air channel were about 250 and 300 mbar, respectively, and the high voltage added on the liquid channel was 5 kV. TIC mass range from 550 to 850.

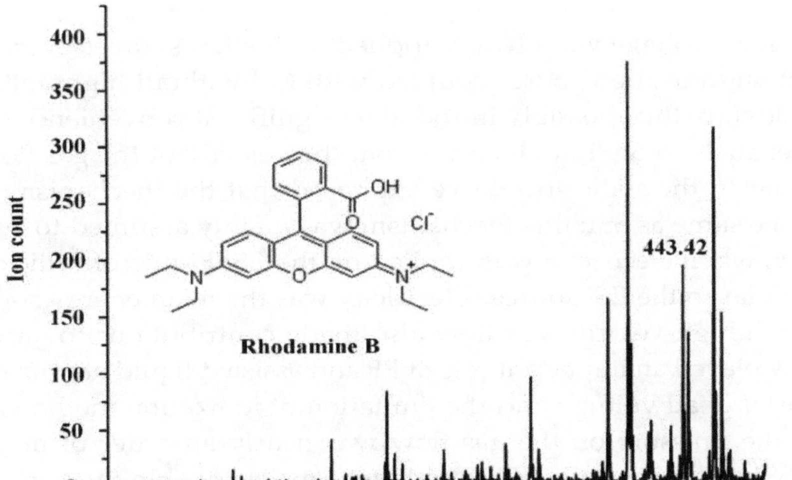

Figure 9. MS signal of 0.01 μM Rhodamine B in methanol with 0.2% formic acid, the gas pressures for the liquid channel and air channel were about 250 and 300 mbar, respectively, and the high voltage added on the liquid channel was 5 kV.

Figure 10 shows the MS signal of 6 μM Rhodamine B in methanol with 0.2% formic acid, which was the mean signal within 30 s. Figure 10a shows the signal of the ESFFI microfluidic chip without high voltage (*i.e.*, the sonic FF ionization (SFFI) mode). Figure 10b demonstrates the signal of the ESFFI microfluidic chip with high voltage (4 kV) (*i.e.*, the ESFFI mode). Figure 10c shows the signal of the commercial ESI source with high voltage (4 kV). The signal-to-noise ratios (SNR) of the SFFI mode, the ESFFI mode, and the commercial ESI source were 54, 31, and 14, respectively. The SNR of the SFFI mode was nearly two times larger than that of the ESFFI mode and three times larger than that of the commercial ESI source. Although the signal intensity can be enhanced greatly with high voltage (Figure 10b,c), the relative intensity without high voltage (Figure 10a) was enough to analyze the samples because of the high SNR. The characteristics of the SFFI mode with high SNR and low signal intensity were consistent with sonic spray ionization (SSI) [65,66].

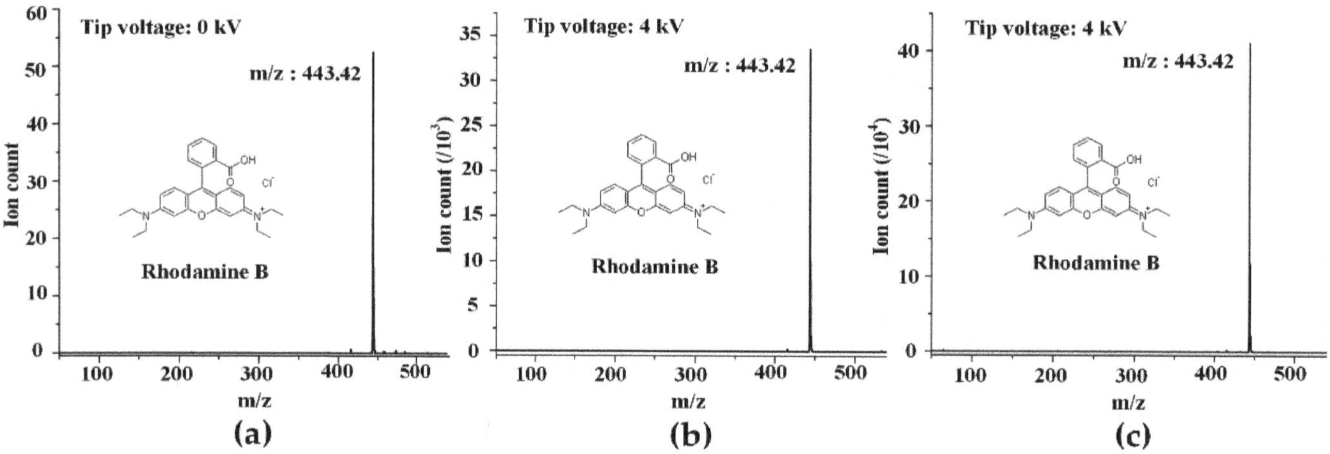

Figure 10. MS spectra of the 6 μM solution of Rhodamine B molecules in methanol with 0.2% formic acid. The flow rate of the solution was 20 μL/min, and the gas pressure for the ESFFI microfluidic chip was 300 mbar. (**a**) ESFFI microfluidic chip without high voltage; (**b**) ESFFI microfluidic chip with 4 kV; (**c**) Commercial ESI sources with 4 kV.

In most cases, a high voltage was always applied on the ESI sources except for the SSI [65,66]. In the present paper, an ionization source coupled with MS without high voltage was realized in the ESFFI microfluidic chip, thus possibly introducing significant convenience when a high voltage is unavailable. Under such an SFFI mode, given that the velocity of the gas flow in the outlet was estimated to be similar to the sonic speed, we supposed that the mechanism of this gaseous ion formation might be the same as SSI; this mechanism was mainly assumed to be the charge residue model [67]. However, when electricity was applied on the ESFFI microfluidic chip, such an ESFFI mode was mainly similar to the ESI source. Electricity was the main energy source for gaseous ion formation even though high-velocity gas flow also made contributions to gaseous ion formation. In both cases, gas flow played an important role of FF and assisted liquid atomizing, thus successfully settling the problems of dead volume and the limitation of ionization methods in the microfluidic chip. Furthermore, the pressure on the gas flow was nearly an order of magnitude lower than the macrostructure SSI. Therefore, the SFFI mode was more suitable for portable and integration applications on MS in the future, whereas the ESFFI mode could be applied to analyze various samples. For detailed applications, interested readers could refer to the macrostructure of electro-SSI [68]. Although the intensity of the SFFI mode was low in this report, recent works [69,70] have alleviated the ion suppression effect. On the basis of the current work, efforts will be performed in our future work to optimize the structure to realize self-aspirated samples by using the negative pressure caused by the high-velocity gas flow in the microfluidic chip. The driven forces for the liquid channel might be ignored, which will further simplify the accessory equipment for portable applications.

4. Conclusions

A new 3D ESFFI microfluidic chip structure was proposed to successfully realize steady water in air FF with the nozzle inside the microfluidic chip by simple fabrication craft. This ESFFI microfluidic chip combined the liquid and air in one channel; this process was beneficial for liquid atomizing. The measurement results demonstrated that this approach fully avoided the disadvantages of the microfluidic chips coupled with MS including liquid spreading, dead volume, and manufacturing troubles. This approach also realized sample ionization in the microfluidic chip without the assistance of high voltage. These properties might make significant contributions to the integration and portable applications of ionization sources coupled with MS. In addition to the MS field, this microfluidic chip might also be widely used in other fields such as inkjet printing and microfiber spinning in which micron liquid jets are needed.

Acknowledgments: Acknowledgments: This work is supported by the National Natural Science Foundation of China (Grant No. 81201165) and the Interdiscipline Research and Innovation Fund of Graduate School at Shenzhen of Tsinghua University (Grant No. JC20140005).

Author Contributions: Author Contributions: Xiang Qian designed the experiments and revised the paper; Cilong Yu performed the experiments and wrote the paper; Yan Chen optimized the fabrication process and revised the paper; Quan Yu analyzed the mass spectrometer data; Kai Ni designed the homemade electrical and mechanical setup; Xiaohao Wang organized and revised the paper. All authors were involved in the preparation of this manuscript.

References

1. Sun, X.; Kelly, R.T.; Tang, K.; Smith, R.D. Membrane-Based Emitter for Coupling Microfluidics with Ultrasensitive Nanoelectrospray Ionization-Mass Spectrometry. *Anal. Chem.* **2011**, *83*, 5797–5803. [CrossRef] [PubMed]
2. Liu, T.; Belov, M.E.; Jaitly, N.; Qian, W.J.; Smith, R.D. Accurate mass measurements in proteomics. *Chem. Rev.* **2007**, *107*, 3621–3653. [CrossRef] [PubMed]
3. Aebersold, R.; Mann, M. Mass spectrometry-based proteomics. *Nature* **2003**, *422*, 198–207. [CrossRef] [PubMed]
4. Culbertson, C.T.; Jacobson, S.C.; Ramsey, J.M. Microchip devices for high-efficiency separations. *Anal. Chem.* **2000**, *72*, 5814–5819. [CrossRef] [PubMed]
5. Jacobson, S.C.; Culbertson, C.T.; Daler, J.E.; Ramsey, J.M. Microchip structures for submillisecond electrophoresis. *Anal. Chem.* **1998**, *70*, 3476–3480. [CrossRef]
6. Sun, X.; Kelly, R.T.; Tang, K.; Smith, R.D. Ultrasensitive nanoelectrospray ionization-mass spectrometry using poly(dimethylsiloxane) microchips with monolithically integrated emitters. *Analyst* **2010**, *135*, 2296–2302. [CrossRef] [PubMed]
7. Huang, B.; Wu, H.K.; Bhaya, D.; Grossman, A.; Granier, S.; Kobilka, B.K.; Zare, R.N. Counting low-copy number proteins in a single cell. *Science* **2007**, *315*, 81–84. [CrossRef] [PubMed]
8. Lazar, I.M.; Trisiripisal, P.; Sarvaiya, H.A. Microfluidic liquid chromatography system for proteomic applications and biomarker screening. *Anal. Chem.* **2006**, *78*, 5513–5524. [CrossRef] [PubMed]
9. Qu, H.Y.; Wang, H.T.; Huang, Y.; Zhong, W.; Lu, H.J.; Kong, J.L.; Yang, P.Y.; Liu, B.H. Stable microstructured network for protein patterning on a plastic microfluidic channel: Strategy and characterization of on-chip enzyme microreactors. *Anal. Chem.* **2004**, *76*, 6426–6433. [CrossRef] [PubMed]
10. Sakai-Kato, K.; Kato, M.; Toyo'oka, T. Creation of an on-chip enzyme reactor by encapsulating trypsin in sol-gel on a plastic microchip. *Anal. Chem.* **2003**, *75*, 388–393. [CrossRef] [PubMed]
11. Oleschuk, R.D.; Harrison, D.J. Analytical microdevices for mass spectrometry. *TrAC Trend. Anal. Chem.* **2000**, *19*, 379–388. [CrossRef]

12. Gao, D.; Liu, H.X.; Jiang, Y.Y.; Lin, J.M. Recent advances in microfluidics combined with mass spectrometry: Technologies and applications. *Lab Chip* **2013**, *13*, 3309–3322. [CrossRef] [PubMed]
13. Lin, S.; Bai, H.; Lin, T.; Fuh, M. Microfluidic chip-based liquid chromatography coupled to mass spectrometry for determination of small molecules in bioanalytical applications. *Electrophoresis* **2012**, *33*, 635–643. [CrossRef] [PubMed]
14. Koster, S.; Verpoorte, E. A decade of microfluidic analysis coupled with electrospray mass spectrometry: An overview. *Lab Chip* **2007**, *7*, 1394–1412. [CrossRef] [PubMed]
15. Sung, W.C.; Makamba, H.; Chen, S.H. Chip-based microfluidic devices coupled with electrospray ionization-mass spectrometry. *Electrophoresis* **2005**, *26*, 1783–1791. [CrossRef] [PubMed]
16. Xue, Q.F.; Foret, F.; Dunayevskiy, Y.M.; Zavracky, P.M.; McGruer, N.E.; Karger, B.L. Multichannel microchip electrospray mass spectrometry. *Anal. Chem.* **1997**, *69*, 426–430. [CrossRef] [PubMed]
17. Ramsey, R.S.; Ramsey, J.M. Generating electrospray from microchip devices using electroosmotic pumping. *Anal. Chem.* **1997**, *69*, 1174–1178. [CrossRef]
18. Li, J.; Thibault, P.; Bings, N.H.; Skinner, C.D.; Wang, C.; Colyer, C.; Harrison, J. Integration of Microfabricated Devices to Capillary Electrophoresis-Electrospray Mass Spectrometry Using a Low Dead Volume Connection: Application to Rapid Analyses of Proteolytic Digests. *Anal. Chem.* **1999**, *71*, 3036–3045. [CrossRef] [PubMed]
19. Mao, P.; Gomez-Sjoberg, R.; Wang, D. Multinozzle Emitter Array Chips for Small-Volume Proteomics. *Anal. Chem.* **2013**, *85*, 816–819. [CrossRef] [PubMed]
20. Mao, P.; Wang, H.; Yang, P.; Wang, D. Multinozzle Emitter Arrays for Nanoelectrospray Mass Spectrometry. *Anal. Chem.* **2011**, *83*, 6082–6089. [CrossRef] [PubMed]
21. Covey, T.R.; Thomson, B.A.; Schneider, B.B. Atmospheric pressure ion sources. *Mass Spectrom. Rev.* **2009**, *28*, 870–897. [CrossRef] [PubMed]
22. Su, S.; Gibson, G.T.T.; Mugo, S.M.; Marecak, D.M.; Oleschuk, R.D. Microstructured Photonic Fibers as Multichannel Electrospray Emitters. *Anal. Chem.* **2009**, *81*, 7281–7287. [CrossRef] [PubMed]
23. Mery, E.; Ricoul, F.; Sarrut, N.; Constantin, O.; Delapierre, G.; Garin, J.; Vinet, F. A silicon microfluidic chip integrating an ordered micropillar array separation column and a nano-electrospray emitter for LC/MS analysis of peptides. *Sens. Actuator B Chem.* **2008**, *134*, 438–446. [CrossRef]
24. Arscott, S. SU-8 as a material for lab-on-a-chip-based mass spectrometry. *Lab Chip* **2014**, *14*, 3668–3689. [CrossRef] [PubMed]
25. Sikanen, T.; Tuomikoski, S.; Ketola, R.A.; Kostiainen, R.; Franssila, S.; Kotiaho, T. Analytical characterization of microfabricated SU-8 emitters for electrospray ionization mass spectrometry. *J. Mass Spectrom.* **2008**, *43*, 726–735. [CrossRef] [PubMed]
26. Sikanen, T.; Heikkilä, L.; Tuomikoski, S.; Ketola, R.A.; Kostiainen, R.; Franssila, S.; Kotiaho, T. Performance of SU-8 Microchips as Separation Devices and Comparison with Glass Microchips. *Anal. Chem.* **2007**, *79*, 6255–6263. [CrossRef] [PubMed]
27. Lee, J.; Soper, S.A.; Murray, K.K. Development of an efficient on-chip digestion system for protein analysis using MALDI-TOF MS. *Analyst* **2009**, *134*, 2426–2433. [CrossRef] [PubMed]
28. Batz, N.G.; Mellors, J.S.; Alarie, J.P.; Ramsey, J.M. Chemical vapor deposition of aminopropyl silanes in microfluidic channels for highly efficient microchip capillary electrophoresis-electrospray ionization-mass spectrometry. *Anal. Chem.* **2014**, *86*, 3493–3500. [CrossRef] [PubMed]
29. Mellors, J.S.; Black, W.A.; Chambers, A.G.; Starkey, J.A.; Lacher, N.A.; Ramsey, J.M. Hybrid Capillary/Microfluidic System for Comprehensive Online Liquid Chromatography-Capillary Electrophoresis-Electrospray Ionization-Mass Spectrometry. *Anal. Chem.* **2013**, *85*, 4100–4106. [CrossRef] [PubMed]
30. Chambers, A.G.; Ramsey, J.M. Microfluidic Dual Emitter Electrospray Ionization Source for Accurate Mass Measurements. *Anal. Chem.* **2012**, *84*, 1446–1451. [CrossRef] [PubMed]
31. Mellors, J.S.; Jorabchi, K.; Smith, L.M.; Ramsey, J.M. Integrated Microfluidic Device for Automated Single Cell Analysis Using Electrophoretic Separation and Electrospray Ionization Mass Spectrometry. *Anal. Chem.* **2010**, *82*, 967–973. [CrossRef] [PubMed]
32. Sun, X.F.; Tang, K.Q.; Smith, R.D.; Kelly, R.T. Controlled dispensing and mixing of pico- to nanoliter volumes using on-demand droplet-based microfluidics. *Microfluid. Nanofluid.* **2013**, *15*, 117–126. [CrossRef] [PubMed]
33. Kelly, R.T.; Tang, K.; Irimia, D.; Toner, M.; Smith, R.D. Elastomeric microchip electrospray emitter for stable cone-jet mode operation in the nanoflow regime. *Anal. Chem.* **2008**, *80*, 3824–3831. [CrossRef] [PubMed]

34. Hoffmann, P.; Haeusig, U.; Schulze, P.; Belder, D. Microfluidic glass chips with an integrated nanospray emitter for coupling to a mass spectrometer. *Angew. Chem. Int. Ed.* **2007**, *46*, 4913–4916. [CrossRef] [PubMed]
35. Yue, G.E.; Roper, M.G.; Jeffery, E.D.; Easley, C.J.; Balchunas, C.; Landers, J.P.; Ferrance, J.P. Glass microfluidic devices with thin membrane voltage junctions for electrospray mass spectrometry. *Lab Chip* **2005**, *5*, 619–627. [CrossRef] [PubMed]
36. Barbier, V.; Tatoulian, M.; Li, H.; Arefi-Khonsari, F.; Ajdari, A.; Tabeling, P. Stable modification of PDMS surface properties by plasma polymerization: Application to the formation of double emulsions in microfluidic systems. *Langmuir* **2006**, *22*, 5230–5232. [CrossRef] [PubMed]
37. Rohner, T.C.; Rossier, J.S.; Girault, H.H. Polymer microspray with an integrated thick-film microelectrode. *Anal. Chem.* **2001**, *73*, 5353–5357. [CrossRef] [PubMed]
38. Qian, X.; Xu, J.; Yu, C.; Chen, Y.; Yu, Q.; Ni, K.; Wang, X. A Reliable and Simple Method for Fabricating a Poly(Dimethylsiloxane) Electrospray Ionization Chip with a Corner-Integrated Emitter. *Sensors* **2015**, *15*, 8931–8944. [CrossRef] [PubMed]
39. Lindberg, P.; Dahlin, A.P.; Bergstrom, S.K.; Thorslund, S.; Andren, P.E.; Nikolajeff, F.; Bergquist, J. Sample pretreatment on a microchip with an integrated electrospray emitter. *Electrophoresis* **2006**, *27*, 2075–2082. [CrossRef] [PubMed]
40. Dahlin, A.P.; Wetterhall, M.; Liljegren, G.; Bergstrom, S.K.; Andren, P.; Nyholm, L.; Markides, K.E.; Bergquist, J. Capillary electrophoresis coupled to mass spectrometry from a polymer modified poly(dimethylsiloxane) microchip with an integrated graphite electrospray tip. *Analyst* **2005**, *130*, 193–199. [CrossRef] [PubMed]
41. Kim, S.H.; Kim, B. Controlled formation of double-emulsion drops in sudden expansion channels. *J. Colloid Interface Sci.* **2014**, *415*, 26–31. [CrossRef] [PubMed]
42. Zhou, T.; Liu, Z.; Wu, Y.; Deng, Y.; Liu, Y.; Liu, G. Hydrodynamic particle focusing design using fluid-particle interaction. *Biomicrofluidics* **2013**, *7*, 54104. [CrossRef] [PubMed]
43. Zhao, C.X. Multiphase flow microfluidics for the production of single or multiple emulsions for drug delivery. *Adv. Drug Deliver. Rev.* **2013**, *65*, 1420–1446. [CrossRef] [PubMed]
44. Wörner, M. Numerical modeling of multiphase flows in microfluidics and micro process engineering: A review of methods and applications. *Microfluid. Nanofluid.* **2012**, *12*, 841–886. [CrossRef]
45. Vladisavljevic, G.T.; Kobayashi, I.; Nakajima, M. Production of uniform droplets using membrane, microchannel and microfluidic emulsification devices. *Microfluid. Nanofluid.* **2012**, *13*, 151–178. [CrossRef]
46. Zhao, C.; Middelberg, A.P.J. Two-phase microfluidic flows. *Chem. Eng. Sci.* **2011**, *66*, 1394–1411. [CrossRef]
47. Park, J.I.; Nie, Z.H.; Kumachev, A.; Kumacheva, E. A microfluidic route to small CO_2 microbubbles with narrow size distribution. *Soft Matter* **2010**, *6*, 630–634. [CrossRef]
48. Taylor, J.K.; Ren, C.L.; Stubley, G.D. Numerical and experimental evaluation of microfluidic sorting devices. *Biotechnol. Prog.* **2008**, *24*, 981–991. [CrossRef] [PubMed]
49. Chang, C.C.; Huang, Z.X.; Yang, R.J. Three-dimensional hydrodynamic focusing in two-layer polydimethylsiloxane (PDMS) microchannels. *J. Micromech. Microeng.* **2007**, *17*, 1479–1486. [CrossRef]
50. Gañán-Calvo, A.; Montanero, J. Revision of capillary cone-jet physics: Electrospray and flow focusing. *Phys. Rev. E* **2009**, *79*, 066305.
51. Gañán-Calvo, A. Unconditional jetting. *Phys. Rev. E* **2008**, *78*, 026304. [CrossRef] [PubMed]
52. Ganan-Calvo, A.M. Electro-flow focusing: The high-conductivity low-viscosity limit. *Phys. Rev. Lett.* **2007**, *98*, 239904. [CrossRef]
53. Gañán-Calvo, A.M.; López-Herrera, J.M.; Riesco-Chueca, P. The combination of electrospray and flow focusing. *J. Fluid Mech.* **2006**, *566*, 421–445. [CrossRef]
54. Gañán-Calvo, A.M. Generation of Steady Liquid Microthreads and Micron-Sized Monodisperse Sprays in Gas Streams. *Phys. Rev. Lett.* **1998**, *80*, 285–288. [CrossRef]
55. Lee, W.; Walker, L.M.; Anna, S.L. Competition Between Viscoelasticity and Surfactant Dynamics in Flow Focusing Microfluidics. *Macromol. Mater. Eng.* **2011**, *296*, 203–213. [CrossRef]
56. Stone, H.A. Dynamics of Drop Deformation and Breakup in Viscous Fluids. *Annu. Rev. Fluid Mech.* **1994**, *26*, 65–102. [CrossRef]
57. Debruijn, R.A. Tipstreaming of drops in simple shear flows. *Chem. Eng. Sci.* **1993**, *48*, 277–284. [CrossRef]
58. Bentley, B.J.; Leal, L.G. An experimental investigation of drop deformation and breakup in steady, two-dimensional linear flows. *J. Fluid Mech.* **1986**, *167*, 241–283. [CrossRef]

59. Taylor, G.I. The formation of emulsions in definable fields of flow. *Proc. R. Soc. Lond. Ser. A* **1934**, *146*, 501–523. [CrossRef]
60. Jeong, W.C.; Lim, J.M.; Choi, J.H.; Kim, J.H.; Lee, Y.J.; Kim, S.H.; Lee, G.; Kim, J.D.; Yi, G.R.; Yang, S.M. Controlled generation of submicron emulsion droplets via highly stable tip-streaming mode in microfluidic devices. *Lab Chip* **2012**, *12*, 1446–1453. [CrossRef] [PubMed]
61. Golden, J.P.; Kim, J.S.; Erickson, J.S.; Hilliard, L.R.; Howell, P.B.; Anderson, G.P.; Nasir, M.; Ligler, F.S. Multi-wavelength microflow cytometer using groove-generated sheath flow. *Lab Chip* **2009**, *9*, 1942–1950. [CrossRef] [PubMed]
62. Trebbin, M.; Krüger, K.; DePonte, D.; Roth, S.V.; Chapman, H.N.; Förster, S. Microfluidic liquid jet system with compatibility for atmospheric and high-vacuum conditions. *Lab Chip* **2014**, *14*, 1733–1745. [CrossRef] [PubMed]
63. Unger, M.A.; Chou, H.P.; Thorsen, T.; Scherer, A.; Quake, S.R. Monolithic microfabricated valves and pumps by multilayer soft lithography. *Science* **2000**, *288*, 113–116. [CrossRef] [PubMed]
64. Xia, Y.; Whitesides, W.G. Soft Lithography. *Annu. Rev. Mater. Sci.* **1998**, *28*, 153–184. [CrossRef]
65. Santos, V.G.; Regiani, T.; Dias, F.F.G.; Romao, W.; Jara, J.L.P.; Klitzke, C.F.; Coelho, F.; Eberlins, M.N. Venturi Easy Ambient Sonic-Spray Ionization. *Anal. Chem.* **2011**, *83*, 1375–1380. [CrossRef] [PubMed]
66. Hirabayashi, A.; Sakairi, M.; Koizumi, H. Sonic spray ionization method for atmospheric-pressure ionization mass-spectrometry. *Anal. Chem.* **1994**, *66*, 4557–4559. [CrossRef]
67. Takats, Z.; Nanita, S.C.; Cooks, R.G.; Schlosser, G.; Vekey, K. Amino acid clusters formed by sonic spray ionization. *Anal. Chem.* **2003**, *75*, 1514–1523. [CrossRef] [PubMed]
68. Takáts, Z.; Wiseman, J.M.; Gologan, B.; Cooks, R.G. Electrosonic Spray Ionization. A Gentle Technique for Generating Folded Proteins and Protein Complexes in the Gas Phase and for Studying Ion-Molecule Reactions at Atmospheric Pressure. *Anal. Chem.* **2004**, *76*, 4050–4058. [CrossRef] [PubMed]
69. Kanaki, K.; Pergantis, S.A. Use of 3-nitrobenzonitrile as an additive for improved sensitivity in sonic-spray ionization mass spectrometry. *Rapid Commun. Mass Spectrom.* **2014**, *28*, 2661–2669. [CrossRef] [PubMed]
70. Li, G.; Huang, G. Alleviation of ion suppression effect in sonic spray ionization with induced alternating current voltage. *J. Mass Spectrom.* **2014**, *49*, 639–645. [CrossRef] [PubMed]

8

Quasi-Optical Terahertz Microfluidic Devices for Chemical Sensing and Imaging

Lei Liu [1,*], Zhenguo Jiang [1], Syed Rahman [1], Md. Itrat Bin Shams [1], Benxin Jing [2], Akash Kannegulla [3] and Li-Jing Cheng [3]

1 Department of Electrical Engineering, University of Notre Dame, 275 Fitzpatrick, Notre Dame, IN 46556, USA; zjiang@nd.edu (Z.J.); srahman@nd.edu (S.R.); mshams@nd.edu (M.I.B.S.)
2 Department of Chemical Engineering and Material Science, Wayne State University, 5050 Anthony Wayne Dr., Detroit, MI 48202, USA; Jingbenxin@gmail.com
3 School of Electrical Engineering and Computer Science, Oregon State University, Corvallis, OR 97330, USA; kannegua@oregonstate.edu (A.K.); chengli@eecs.oregonstate.edu (L.-J.C.)
* Correspondence: lliu3@nd.edu

Academic Editor: Manabu Tokeshi

Abstract: We first review the development of a frequency domain quasi-optical terahertz (THz) chemical sensing and imaging platform consisting of a quartz-based microfluidic subsystem in our previous work. We then report the application of this platform to sensing and characterizing of several selected liquid chemical samples from 570–630 GHz. THz sensing of chemical mixtures including isopropylalcohol-water (IPA-H_2O) mixtures and acetonitrile-water (ACN-H_2O) mixtures have been successfully demonstrated and the results have shown completely different hydrogen bond dynamics detected in different mixture systems. In addition, the developed platform has been applied to study molecule diffusion at the interface between adjacent liquids in the multi-stream laminar flow inside the microfluidic subsystem. The reported THz microfluidic platform promises real-time and label-free chemical/biological sensing and imaging with extremely broad bandwidth, high spectral resolution, and high spatial resolution.

Keywords: terahertz; microfluidic; quasi-optical; frequency domain; chemical sensing and imaging; laminar flow; label free; molecule diffusion

1. Introduction

Chemical and biochemical sensing has been increasingly more important in security, environmental, medical and clinical applications [1–4]. However, most current sensing and probing techniques are time-consuming, and require specific expertise and expensive equipment [5]. In addition, many of these techniques require chemical alteration of samples or labeling with fluorescent chromophores prior to detection and analysis [6]. For example, current DNA hybridization detection is mainly based on fluorescent labeling, which introduces unwanted preliminary processing steps and eventually modifies the DNA sample under test, resulting in system inefficiency and low accuracy [7]. Alternative label-free methods, such as mass sensitive [8], electrochemical [9], and acoustic wave [10], have been intensively studied, but no approach has been mature enough to provide performance as competitive as standard fluorescent-based systems.

With the emerging advances in device and circuit technologies in the terahertz (THz) regime (0.1 to 10 THz), electromagnetic waves in this frequency range have found many promising applications in noninvasive, label-free and remote sensing for many substances of interest (e.g., chemicals, explosives, drugs) owing to the strong interaction between THz waves and low-energy events (e.g., molecular rotation, torsion, vibration, as well as inter- and intra-molecular hydrogen-bonding) in chemical samples [11–16].

However, the THz sensing capability for liquid samples has been greatly limited due to strong undesired THz absorption introduced by hydrogen-bonding in the aqueous media or matrices (e.g., water) [17–20]. In view of this, researchers have started to combine THz technology with microfluidic devices to reduce the wave traveling path in liquid samples for leveraging THz absorption [21–24]. Although THz time-domain spectroscopy (THz-TDS) systems are widely employed for the above purposes, their spectral and spatial resolution for chemical sensing and imaging are generally inferior [13,25].

In this paper, we first review the development of a frequency-domain quasi-optical THz sensing and imaging platform consisting of a quartz-based microfluidic subsystem supporting four-stream laminar flow. We then applied the platform to sensing and imaging of a variety of chemicals, mixtures as well as molecular diffusion at liquid-liquid interface. The employed THz frequency-domain spectroscopy (THz-FDS) offers much improved spectral (better than ~10 kHz) and spatial resolution while the microfluidic device exhibits significant advantages including low analyte consumption, well-confined multi-stream flow, fast dynamics and autonomous operation [26], making the platform an extremely versatile system for high performance chemical/biological sensing and imaging.

2. Experimental Section

2.1. Experimental Setup

Figure 1a schematically illustrates the THz sensing and imaging platform we developed for this work. The quasi-optical THz-FDS system was reported in our previous work [26–28]. In this system, an amplifier multiplier chain (AMC, Virginia Diodes, Charlottesville, VA, USA) is employed as the THz emitter to provide THz radiation from 570 to 630 GHz with an average power level of ~1 mW. Much broader frequency coverage (e.g., 0.1–1.0 THz) can be achieved by using multiple THz emitters for different THz bands. The THz wave emitted from the frequency multiplier chain is first collimated and then focused onto the microfluidic device by the off-axis parabolic mirrors M1 and M2. The transmitted THz signal is then collimated and focused onto the quasi-optical THz detector for broadband and room-temperature operation [29]. In order to enable two-dimensional (2-D) THz mapping and imaging, the microfluidic device is mounted on a computer-controlled *X-Y-Z* positioning stage (not shown). As pointed out in our previous work, the spatial resolution of the THz microfluidic system was designed to be approximately 0.5 mm at the sampling position [26].

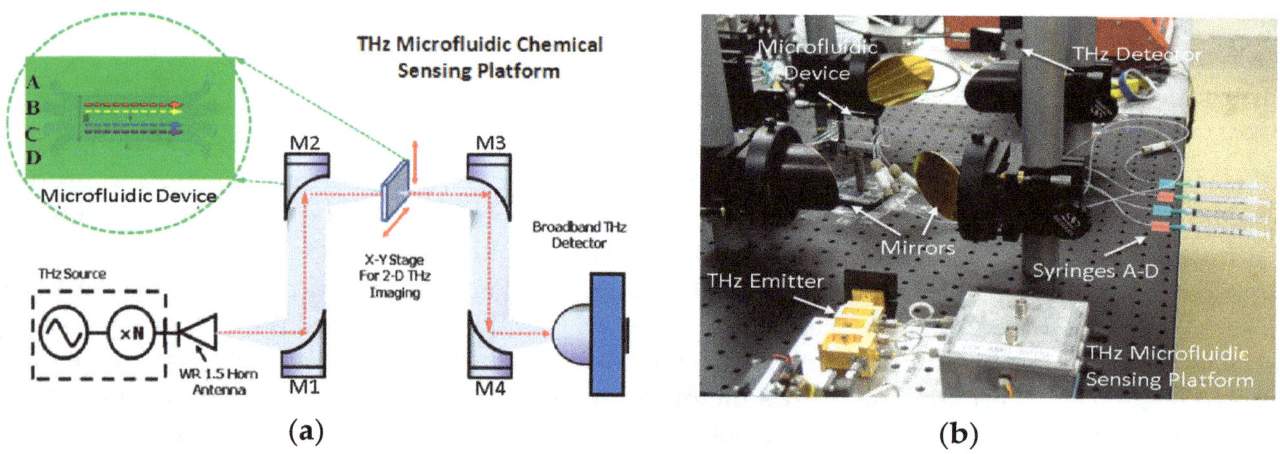

Figure 1. The THz microfluidic chemical sensing and imaging platform: (**a**) a schematic of the system comprising a quasi-optical THz-FDS spectroscopy and a four-channel microfluidic subsystem; (**b**) a photo showing the actual experimental setup. Liquid samples are delivered to the microfluidic chip through the four syringes A–D [26].

2.2. *Quartz-Based Microfluidic Device*

As seen in Figure 1b, the key device in this sensing platform is the four-channel (with multiple inlets/outlets) quartz-based microfluidic subsystem. The microfluidic subsystem was fabricated by bonding two fused quartz substrates (1 mm-thick, 1 in. × 3 in. in dimension) as shown in Figure 2a. Quartz substrates were chosen for building this microfluidic chip because of their low insertion loss (THz), low dielectric constant and optical transparency. The main channel of the device was etched to have a depth of 50 μm and a width of 1 cm, leading to an inner volume of ~20 μL. In order to support four-stream laminar flow, four through holes were drilled on the top quartz at both ends of the main channel. Two polydimethylsiloxane (PDMS) slabs were then bonded to the top slide with corresponding through holes aligned with those on the quartz substrate. Polytetrafluoroethylene (PTFE) tubings were then inserted to the through holes for delivery and drainage of the fluid, respectively. In order to reduce the "standing-wave" (or "cavity") effect due to the relatively thick quartz substrates (*i.e.*, 2 mm thickness or 80 GHz in period), thinner and wedged THz window configuration can be micromachined on both the top and bottom quartz slides as shown in Figure 2a. During the chemical sensing and imaging experiments, four syringes (A–D) as seen in Figure 1b were used to inject liquid chemical samples into the main channel with one sample at a time or multiple samples simultaneously to form a multi-stream laminar flow (*i.e.*, 2–4 streams). As shown in Figure 2b, an initial test by injecting red and blue dyes simultaneously into the inlets A and D respectively at a rate of 100 μL/min clearly showed the formation of a stable two-stream laminar flow, demonstrating the success of the design and fabrication of the microfluidic subsystem. Finally, the operation of the entire system including frequency scan for THz spectroscopy and two-dimensional (2D) spatial scan for imaging were fully controlled and automated by a computer using a home-written LabView interface.

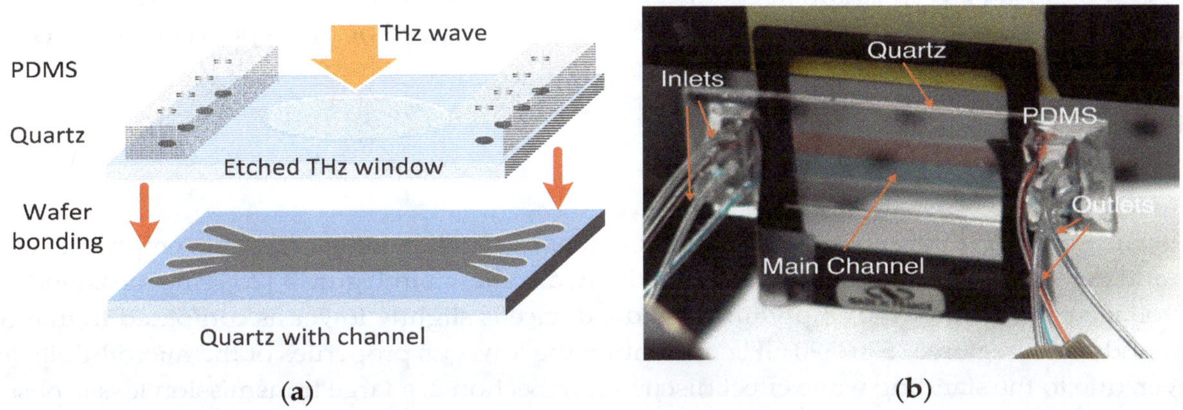

Figure 2. (**a**) The fabrication of the quartz-based microfluidic device using wafer bonding; (**b**) A two-stream laminar flow (with red and blue dyes at an injection rate of 100 μL/min) formed inside the device main channel demonstrating that the design and fabrication of the device were successful.

2.3. *Sensing Platform Characterization and Data Acquisition*

During the operation of the sensing platform, the THz emitter is modulated with an optical chopper at 1 kHz. The modulated DC voltage output from the quasi-optical THz detector is fed into a preamplifier (×100) and a lock-in amplifier for processing. Before applying this THz microfluidic platform to chemical sensing and imaging, the system's performance was examined by measuring dynamic range and Mylar thin films. Figure 3a shows the measured maximum signal level and noise floor from 570 to 630 GHz. An average system dynamic range of ~50 dB has been demonstrated which is considered enough for many applications. Figure 3b shows the measurement results for a 1-mil thick Mylar thin film (solid line) and a 3-mil Mylar thin film (dashed line) [27]. Both curves were obtained by normalizing the measured data to that of background signal. The expected lower transmittance (~73%) for the 3-mil Mylar as compared to that of the 1-mil Mylar (~92%), as well as the flat linear frequency response observed from both samples clearly show the system's capability for sensing different samples with a measurement accuracy better than 2% [27].

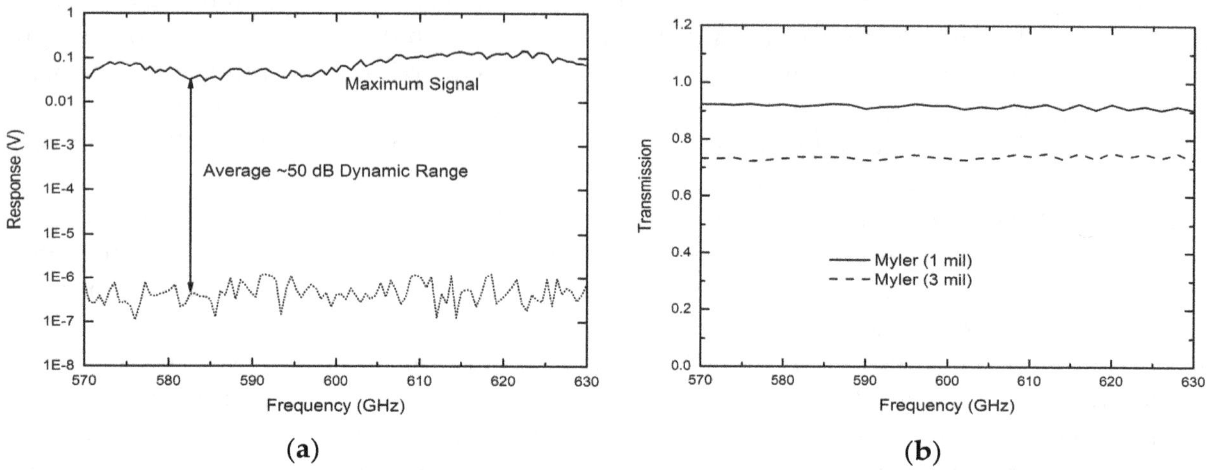

Figure 3. (**a**) Measured system dynamic range showing an average of 50 dB over the frequency range of 570–630 GHz; (**b**) measured transmission spectrum of Mylar thin films [27].

For a prototype demonstration, the developed frequency domain quasi-optical THz microfluidic platform has been applied to chemical sensing and imaging by performing frequency scan in the band of 570–630 GHz. The transmission spectra of background, the empty microfluidic device, as well as a variety of microfluidic-chip confined chemicals and their mixtures were acquired automatically by performing frequency scan at a resolution of 0.6 GHz with a speed of 10 ms per data point using the LabView interface described in Section 2.2.

3. Results and Discussion

3.1. THz Sensing of Chemicals

The measured THz transmission spectral without normalization for background, empty microfluidic device and deionized water filled device, respectively, are shown in Figure 4 [26]. The measured signal level (voltage response) for the empty microfluidic device is slightly lower as compared to that of the background in the region of 570–590 GHz, indicating the low-loss properties of the microfluidic device. However, due to the standing wave effect discussed in Section 2, a large transmission loss is observed from 590 to 630 GHz (*i.e.*, half period of ~40 GHz). This undesired standing wave effect can be effectively reduced by incorporating thin and wedged THz window on the quartz microfluidic device (see Figure 2a). The THz transmission signal level for the water-filled device was further reduced. However, the detected signal level falls well within the system's dynamic range, demonstrating that the quartz microfluidic device with a 50-μm THz transmission path is a suitable design for sensing chemicals in a well-controlled aqueous environment.

In our previous work, we focused on sensing and imaging of only isopropylalcohol (IPA), water and their mixtures [26]. In this paper, we extend our research by applying the sensing platform to characterize a variety of chemicals. In Figure 5a, we first compare the measured THz transmission responses (raw data without normalization) of IPA, methanol and water. The THz absorption increases in the following order: IPA < methanol < water, which is consistent with the findings from previous research [25]. This phenomenon can be explained using the different hydrogen bond densities in these three liquids. As discussed in Reference [26], the hydrogen bond densities of water, methanol, and isopropanol are calculated to be 1.2×10^{23}, 2.7×10^{22} and 1.4×10^{22} cm^{-3}, respectively [30–33]. Different from infrared spectroscopy, THz waves interact significantly with hydrogen bonds—liquids with higher hydrogen bond density tend to yield high THz absorption. This explains the different

THz transmission levels observed in the three liquid samples in Figure 5a. To better compare the sensing results for different chemicals, we normalized the transmission response raw data for each chemical to that of water. Figure 5b shows the normalized THz transmission features for Benzyl (BEZ) alcohol, IPA, methanol and acetonitrile (ACN) in the same frequency band, *i.e.*, 570–630 GHz. A response peak is observed at ~625 GHz for all chemicals and this is believed to be introduced by the well-know standing-wave effect [26]. Different from BEZ alcohol, IPA and methanol that have a lower THz absorption due to lower hydrogen bond densities, pure ACN shows larger THz absorption than water due to the relatively strong resonance between THz waves and the vibration mode of individual ACN cluster.

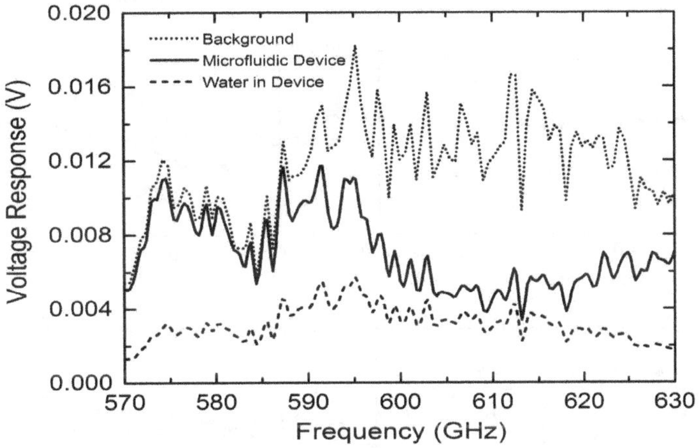

Figure 4. Measured THz responses (raw data without normalization) for background (ambient), empty microfluidic device and water-filled microfluidic device, respectively [26].

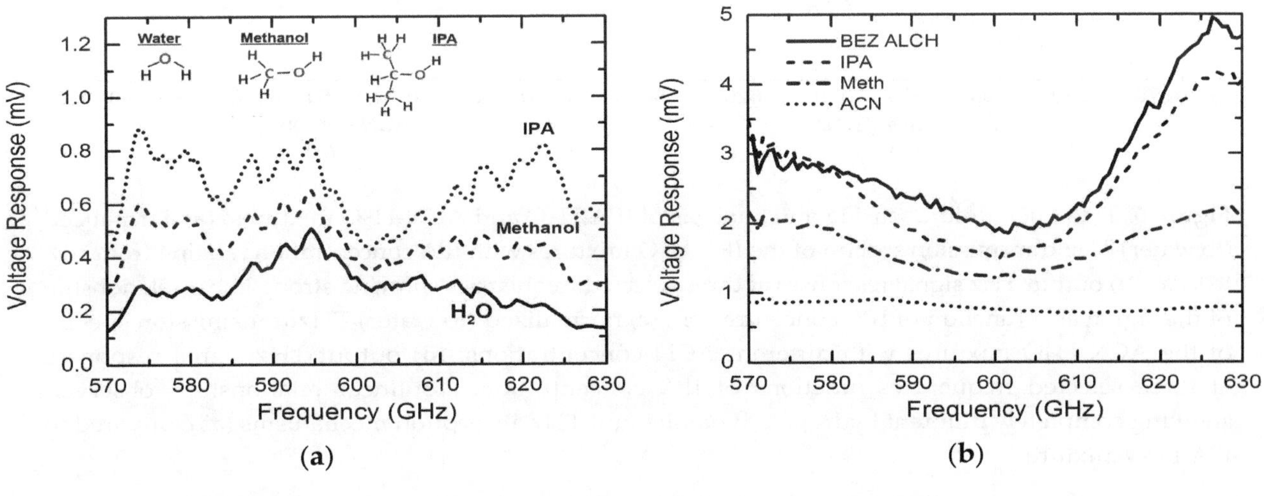

Figure 5. (**a**) Raw data of the THz spectra responses for isopropylalcohol (IPA), Methanol and water; (**b**) Comparison of normalized (to water) THz spectra responses for a variety of chemicals (Benzyl alcohol (BEZ ALCH), isopropylalcohol (IPA), methanol (Meth) and acetonitrile (ACN)), demonstrating the system's capability for discriminating different chemicals.

3.2. *THz Sensing of Chemical Mixtures*

In order to reveal the details of the above observation, we further studied and compared the THz transmission spectra for IPA-H_2O and ACN-H_2O mixtures with various concentrations. In this study, all spectra data were processed by normalizing to water, and the known cavity effect was removed numerically. As seen in Figure 6a, the THz signal intensity increases when the IPA concentration is increased because the decreasing in hydrogen bond density leading to lower THz absorption based on what was observed in Figure 5. The transmission signals for IPA-H_2O mixtures at three different frequencies, *i.e.*, 580, 590 and 600 GHz, were plotted as functions of the IPA concentration as shown in Figure 6b. Although slight nonlinearity is found, all these three plots show monotonous relationship between THz response and IPA concentration.

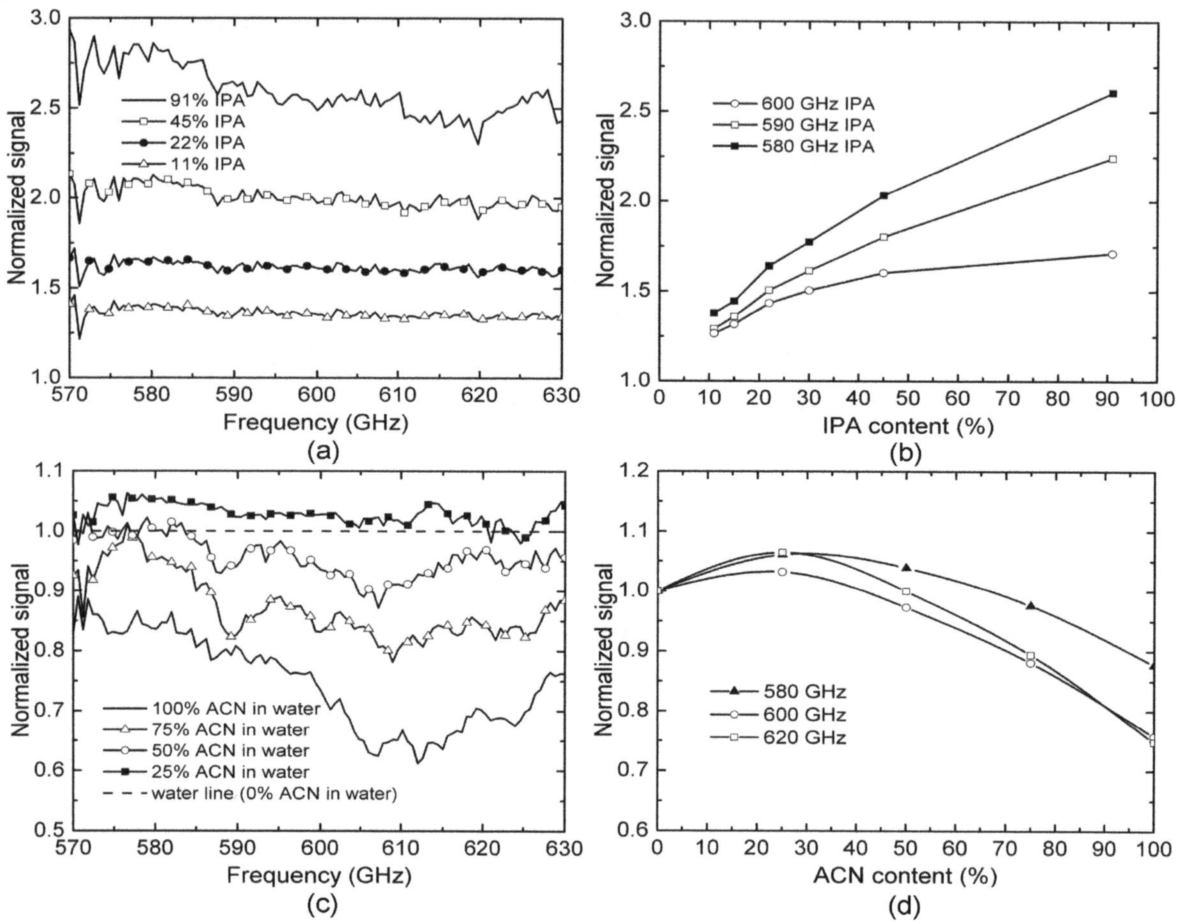

Figure 6. THz microfluidic sensing and analysis of IPA-H_2O and ACN-H_2O mixtures: (**a**) normalized (to water) THz transmission spectra of the IPA-H_2O mixtures with IPA concentration ranging from 10% to 91%; (**b**) output THz signal responses at three selected frequencies showing strong linear relationship of the signal as a function of IPA concentration; (**c**) normalized (to water) THz transmission spectra of the ACN-H_2O mixtures with different ACN concentrations; (**d**) output THz signal responses at three selected frequencies functions of IPA concentration. Nonlinear relationships observed showing completely different hydrogen dynamics and THz absorption mechanisms (as compared to IPA-H_2O mixtures).

Similar experiments have been also performed to characterize ACN-H_2O mixtures, as shown in Figure 6c, with ACN concentrations ranging from 0% to 100%. Although pure AN has higher THz absorption than water (*i.e.*, lower normalized signal intensity is expected), surprisingly, the normalized signal intensity increases first for smaller ACN concentration (<25%) and then decreases for larger concentrations. Figure 6d shows the normalized THz transmission response as a function of ACN concentration at three selected frequencies, *i.e.*, 580, 600 and 620 GHz. Different from the monotonous relation that has been observed for IPA-H_2O mixtures (see Figure 6b), THz responses for ACN-H_2O mixtures show a strong nonlinear and non-monotonous relation to ACN concentration, with the lowest THz absorption (maximum signal intensity) occurring for an ACN concentration around

25%. This observation reveals completely different hydrogen bond dynamics and THz absorption mechanisms in IPA-H_2O and ACN-H_2O mixtures.

Similar nonlinearity for ACN-H_2O mixtures has been reported in studies using infrared spectroscopy [34,35]. The absorption of THz energy by liquid samples can be attributed to the resonance between THz waves and vibration modes of molecular clusters formed by either a hydrogen bond or dispersion force.

From Figure 5, we can see that higher hydrogen bond density leads to more molecular clusters with the resonating vibration mode. However, when ACN is mixed with water, the situation is different. In the low ACN concentration regime, each ACN molecule only forms one hydrogen bond with water and thus behaves as an end-cap agent that suppresses hydrogen bond number/density between water molecules.

The increase of ACN in a mixture could lead to the decreasing concentration of water cluster resulting in lower THz absorption [34]. However, once the fraction of ACN in water is high enough (>25%), ACN clusters will be formed through dipole-dipole interaction. The vibration of the individual ACN cluster leads to higher THz absorption. As a result, in the high ACN concentration regime, THz absorption increases with ACN concentration, as seen in Figure 6d [36].

3.3. *THz Imaging of Molecular Diffusion*

In order to apply this THz microfluidic platform to study molecular diffusion and potential chemical reactions between two liquid chemical samples, we attempted THz mapping for a two-stream laminar flow situation, initially tested by using red and blue dyes (see Figure 7a, lower inset). For the above purpose, water and IPA-H_2O mixtures having variant IPA concentrations were injected into the microfluidic device at an injection rate of 100 μL/min to form a two-stream laminar flow.

THz image with 20 × 200 ($X \times Y$) pixels was taken at 580 GHz by performing 2-D scanning of the device [26]. As shown in Figure 7a (upper inset), the interface between the two streams was clearly seen. Figure 7b shows the measurement results of one-dimensional (1-D) THz scanning across the microfluidic device (Y-direction) at X = 30 mm. As expected, the distinguishable THz transmission levels were detected across the entire device region in response to different chemical streams (Stream-I (H_2O) region and Stream-II (mixture) region).

It is clearly observed that the signal level in the Stream-II region changed as expected when the IPA concentration for Stream-II region was increased from 0% to 100%, demonstrating that the approach of using multi-stream microfluidic device indeed enables THz sensing of molecular diffusion and potential chemical reactions at the interfaces between adjacent liquid samples [26].

For a prototype demonstration of this imaging capability to study molecular diffusion, we performed 1-D THz scan at the same position (*i.e.*, across the channel at X = 30 mm) for laminar flows formed by water and IPA, for different injection rates from 40 to 1 μL/min as shown in Figure 7c. With decreasing of the flow speed, the transition width at the liquid-liquid interface was observed to increase from ~3 to ~6 mm. The transition area almost doubled at a lower flow speed, indicating higher diffusion level between the two liquids was detected. We then kept the injection rate at 0.5 μL/min and performed THz 1-D scan across the microfluidic device at different positions from X = 10 mm to X = 35 mm. As seen in Figure 7d, the transition width at the interface changes from 5.8 mm (at X = 10 mm) to 6.2 mm (at X = 35 mm), also showing stronger diffusion at the outlets side of the device than that at the inlets side of the device, as expected. This same approach can be adopted to study and visualize chemical reactions between two or more chemicals at their laminar flow interfaces.

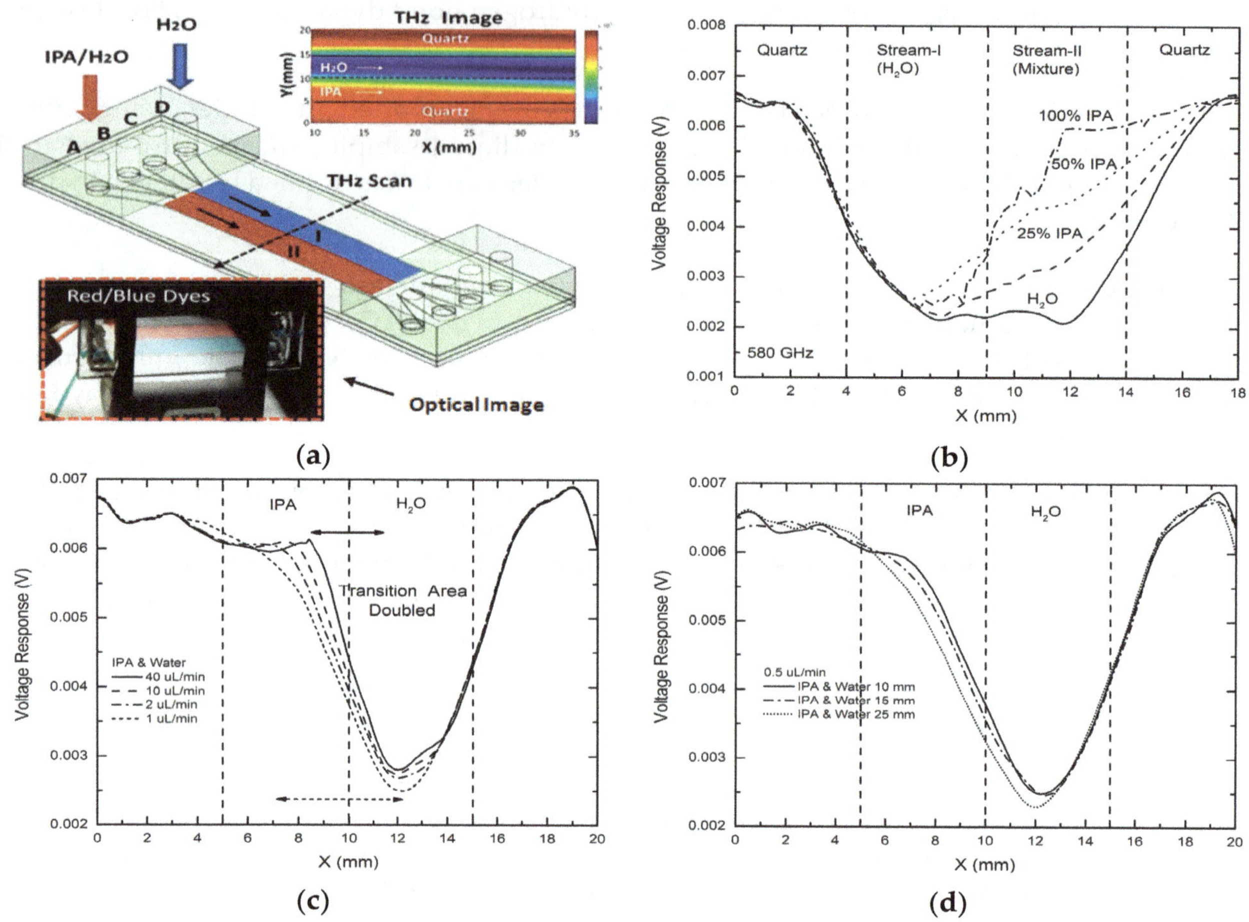

Figure 7. THz chemical sensing and imaging (580 GHz) for studying molecular diffusion at liquid-liquid interfaces of two-stream laminar flows: (**a**) schematic showing the THz imaging of two-stream laminar flow inside the device (lower left inset shows an optical image of a two-stream laminar flow with red and blue dyes at an injection rate of 100 μL/min; upper right inset shows a THz 2-D scanning image of a laminar flow formed by water and IPA; (**b**) 1-D THz scanning results for two-stream laminar flows formed by water and IPA-H_2O mixtures [26]; (**c**) 1-D THz scanning across the device at $X = 30$ mm for laminar flows formed by water and IPA at different injection rate from 40 to 0.5 μL/min. The transition region at the interface is nearly doubled; (**d**) 1-D THz scanning at different positions of the device ($X = 10$–25 mm) for a laminar flow by water and IPA at a rate of 0.5 μL/min. The transition at the liquid-liquid interface changes from 5.8 to 6.2 mm when X changes from 10 to 25 mm, showing stronger diffusion at the outlets side of the microfluidic device.

4. Conclusions

A frequency domain quasi-optical THz chemical sensing and imaging platform consisting of a quartz-based microfluidic device supporting multi-stream laminar flow has been developed. The performance of this sensing platform has been fully characterized. This system has been successfully applied to sensing several selected liquid chemical samples from 570 to 630 GHz. THz spectroscopic sensing of chemical mixtures including IPA-H_2O and AN-H_2O mixtures with different concentrations have been successfully demonstrated, revealing different hydrogen dynamics and absorption mechanisms in different mixture systems. 2-D mapping and imaging of two-stream laminar flows as well as molecule diffusion at the liquid-liquid interface has been performed and discussed. The reported THz microfluidic platform promises real-time and label-free chemical/biological sensing and imaging with extremely broad bandwidth, high spectral resolution, and high spatial resolution.

Acknowledgments: The authors acknowledge partial supports from National Science Foundation (NSF) grants ECCS-1002088, ECCS-1102214 and ECCS-1508057, the Advanced Diagnostics and Therapeutics (AD&T) and the Center for Nano Science and Technology (NDnano) at the University of Notre Dame. The authors also thank Hsueh-Chia Chang, Patrick Fay and Huili (Grace) Xing for helpful discussions.

Author Contributions: L.L. and L.C. conceived and designed the experiments; Z.J., S.R. and M.I.B. performed the experiments; L.L., L.C., B.J., Z.J., S.R., M.I.B. and A. K. analyzed the data; L.L. and L.C. wrote the paper.

References

1. Janata, J. *Principles of Chemical Sensors*, 2nd ed.; Springer: Berlin/Heidelberg, Germany, 2009.
2. Pejcic, B.; Eadington, P.; Ross, A. Environmental monitoring of hydrocarbons: A chemical sensor perspective. *Environ. Sci. Technol.* **2007**, *41*, 6333–6342. [CrossRef] [PubMed]
3. Li, C.M.; Dong, H.; Cao, X.; Luong, J.H.T.; Zhang, X. Implantable electrochemical sensors for biomedical and clinical applications: progress, problems, and future possibilities. *Curr. Med. Chem.* **2007**, *14*, 937–951. [PubMed]
4. Wang, J. Survey and summary: From DNA biosensors to gene chips. *Nucleic Acids Res.* **2000**, *16*, 3011–3016. [CrossRef]
5. Willis, R.C. Soring out the mess. *Modern Drug Discov.* **2004**, *7*, 30–32.
6. Schena, M.; Shalon, D.; Heller, R.; Chai, A.; Brown, P.O.; Davis, R.W. Parallel human genome analysis: Microarray-based expression monitoring of 1000 genes. *Proc. Natl. Acad. Sci. USA* **1996**, *93*, 10614–10619. [CrossRef] [PubMed]
7. Ozaki, H.; Mclaughlin, L.W. The estimation of distances between specific backbone-labeled sites in DNA using fluorescence resonance energy transfer. *Nucleic Acids Res.* **1992**, *20*, 5205–5214. [CrossRef] [PubMed]
8. Okahata, Y.; Kawase, M.; Niikura, K.; Ohtake, F.; Furusawa, H.; Ebara, Y. Kinetic measurements of DNA hybridization on an oligonucleotide-immobilized 27 MHz quartz-crystal microbalance. *Anal. Chem.* **1998**, *70*, 1288–1296. [CrossRef] [PubMed]
9. Larramendy, M.L.; El-Rifai, W.; Knuutila, S. Comparison of fluorescein isothiocyanate- and Texas red-conjugated nucleotides for direct labeling in comparative genomic hybridization. *Cytometry* **1998**, *31*, 174–179. [CrossRef]
10. Zhang, H.; Tan, H.; Wang, R.; Wei, W.; Yao, S. Immobilization of DNA on silver surface of bulk acoustic wave sensor and its application to the study of UV-C damage. *Anal. Chim. Acta* **1998**, *374*, 31–38. [CrossRef]
11. Ajito, K.; Ueno, Y. THz chemical imaging for biological applications. *IEEE Trans. Terahertz Sci. Technol.* **2011**, *1*, 293–300. [CrossRef]
12. Tanaka, M.; Hirori, H.; Nagai, M. THz nonlinear spectroscopy of solids. *IEEE Trans. Terahertz Sci. Technol.* **2011**, *1*, 301–312. [CrossRef]
13. Walther, M.; Fischer, B.M.; Ortner, A.; Bitzer, A.; Thoman, A.; Helm, H. Chemical sensing and imaging with pulsed terahertz radiation. *Anal. Bioanal. Chem.* **2010**, *397*, 1009–1017. [CrossRef] [PubMed]
14. Woolard, D.; Brown, E.; Pepper, M.; Kemp, M. Terahertz frequency sensing and imaging: A time of reckoning future applications? *Proc. IEEE* **2005**, *93*, 1722–1743. [CrossRef]
15. Plusquellic, D.F.; Siegrist, K.; Heilweil, E.J.; Esenturk, O. Applications of terahertz spectroscopy in biosystems. *ChemPhysChem* **2007**, *8*, 2412–2431. [CrossRef] [PubMed]
16. Globus, T.; Norton, M.L.; Lvovska, M.I.; Gregg, D.A.; Khromova, T.B.; Gelmont, B.L. Reliability analysis of THz characterization of modified and unmodified vector sequences. *IEEE Sensors J.* **2010**, *10*, 410–418. [CrossRef]
17. Globus, T.; Bykhovski, A.; Khromova, T.; Gelmont, B.; Tamm, L.K.; Salay, L.C. Low terahertz spectroscopy of liquid water. *Proc. SPIE* **2007**, *6772*, 67720S.
18. Xu, J.; Plaxco, K.; Allen, S. Absorption spectra of liquid water and aqueous buffers between 0.3 and 3.72 THz. *J. Chem. Phys.* **2006**, *124*, 036101. [CrossRef] [PubMed]
19. Dietlein, C.; Popovic, Z.; Grossman, E.N. Aqueous blackbody calibration source for millimeter-wave/terahertz metrology. *Appl. Opt.* **2008**, *47*, 5604–5615. [CrossRef] [PubMed]
20. Ogawa, Y.; Cheng, L.; Hayashi, S.; Fukunaga, K. Attenuated total reflection spectra of aqueous glycine in the terahertz region. *IEICE Electron. Expr.* **2009**, *6*, 117–121. [CrossRef]

21. Baragwanath, A.J.; Swift, G.P.; Dai, D.; Gallant, A.J.; Chamberlain, J.M. Silicon based microfluidic cell for terahertz frequencies. *J. Appl. Phys.* **2010**, *108*, 013102. [CrossRef]
22. Mendis, R.; Astley, V.; Liu, J.; Mittleman, D.M. Terahertz microfluidic sensor based on a parallel-plate waveguide resonant cavity. *Appl. Phys. Lett.* **2009**, *95*, 171113. [CrossRef]
23. George, P.A.; Hui, W.; Rana, F.; Hawkins, B.G.; Smith, A.E.; Kirby, B.J. Microfluidic devices for terahertz spectroscopy of biomolecules. *Opt. Express* **2008**, *16*, 1577–1582. [CrossRef] [PubMed]
24. Kiwa, T.; Oka, S.; Kondo, J.; Kawayama, I.; Yamada, H.; Tonouchi, M.; Tsukada, K. A terahertz chemical microscope to visualize chemical concentrations in microfluidic chips. *Jpn. J. Appl. Phys.* **2007**, *46*, L1052–L1054. [CrossRef]
25. Merbold, H.P. Terahertz Time-Domain Spectroscopy of Aqueous Systems in Reflection Geometry and Construction of Polarization-Sensitive Photoconductive Terahertz Antennas. Master's Thesis, University of Freiburg, Freiburg, Germany, 2006.
26. Liu, L.; Pathak, R.; Cheng, L.J.; Wang, T. Real-time frequency-domain terahertz sensing and imaging of isopropyl alcohol-water mixtures on a microfluidic chip. *Sensors Actuators B* **2013**, *184*, 228–234. [CrossRef]
27. Liu, L.; Hesler, J.; Weikle, R.; Wang, T.; Fay, P.; Xing, H. A 570–630 GHz frequency domain spectroscopy system based on a broadband quasi-optical Schottky diode detector. *Int. J. High Speed Electron. Syst.* **2011**, *20*, 629–638. [CrossRef]
28. Das, A.; Magaridis, C.M.; Liu, L.; Wang, T.; Biswas, A. Design and synthesis of superhydrophobic carbon nanofiber composite coatings for terahertz frequency shielding and attenuation. *Appl. Phys. Lett.* **2011**, *98*, 174101. [CrossRef]
29. Liu, L.; Hesler, J.; Xu, H.; Lichtenberger, A.; Weikle, R. A broadband quasi-optical terahertz detector utilizing a zero bias Schottky diode. *IEEE Microw. Wirel. Compon. Lett.* **2010**, *20*, 504–506. [CrossRef]
30. Soper, A. The radial distribution functions of water and ice from 220 to 673 K and at pressures up to 400 MPa. *Chem. Phys.* **2000**, *258*, 121–137. [CrossRef]
31. Blumberg, R.; Stanley, H.; Geiger, A.; Mausbach, P. Connectivity of hydrogen-bonds in liquid water. *J. Chem. Phys.* **1984**, *80*, 3387–3391. [CrossRef]
32. Narten, A.; Habenschuss, A. Hydrogen-bonding in liquid methanol and ethanol determined by X-ray diffraction. *J. Chem. Phys.* **1984**, *80*, 3387–3391. [CrossRef]
33. Takamuku, T.; Saisho, K.; Aoki, S.; Yamaguchi, T. Large-angle X-ray scattering investigation of the structure of 2-propanol-water mixtures. *Naturforsch Z.* **2002**, *A57*, 982–994. [CrossRef]
34. Sakurai, M. Partial molar volumes for acetonitrile + water. *J. Chem. Eng. Data* **1992**, *37*, 358–362. [CrossRef]
35. Moreau, C.; Douheret, G. Thermodynamic and physical behavior of water + acetonitrile mixtures. *J. Chem. Thermodyn.* **1976**, *8*, 403–410. [CrossRef]
36. Takamuku, T.; Tabata, M.; Yamaguchi, A.; Nishimoto, J.; Kumamoto, M.; Wakita, H.; Yamaguchi, T. Liquid structure of acetonitrile-water mixtures by X-ray diffraction and infrared spectroscopy. *J. Phys. Chem.* **1998**, *102*, 8880–8888. [CrossRef]

9

Advances in Microfluidic Paper-Based Analytical Devices for Food and Water Analysis

Lori Shayne Alamo Busa [1,2], Saeed Mohammadi [1], Masatoshi Maeki [3], Akihiko Ishida [3], Hirofumi Tani [3] and Manabu Tokeshi [3,4,5,6,*]

1 Graduate School of Chemical Sciences and Engineering, Hokkaido University, Kita 13 Nishi 8, Kita-ku, Sapporo 060-8628, Japan; lorishayne_busa@eis.hokudai.ac.jp (L.S.A.B.); drsaeedmoh@ec.hokudai.ac.jp (S.M.)
2 Physical Sciences Department, Nueva Vizcaya State University, Bayombong, Nueva Vizcaya 3700, Philippines
3 Division of Applied Chemistry, Faculty of Engineering, Hokkaido University, Kita 13 Nishi 8, Kita-ku, Sapporo 060-8628, Japan; m.maeki@eng.hokudai.ac.jp (M.M.); ishida-a@eng.hokudai.ac.jp (A.I.); tani@eng.hokudai.ac.jp (H.T.)
4 ImPACT Research Center for Advanced Nanobiodevices, Nagoya University, Furo-cho, Chikusa-ku, Nagoya 464-8603, Japan
5 Innovative Research Center for Preventive Medical Engineering, Nagoya University, Furo-cho, Chikusa-ku, Nagoya 464-8601, Japan
6 Institute of Innovation for Future Society, Nagoya University, Furo-cho, Chikusa-ku, Nagoya 464-8601, Japan
* Correspondence: tokeshi@eng.hokudai.ac.jp

Academic Editor: Joost Lötters

Abstract: Food and water contamination cause safety and health concerns to both animals and humans. Conventional methods for monitoring food and water contamination are often laborious and require highly skilled technicians to perform the measurements, making the quest for developing simpler and cost-effective techniques for rapid monitoring incessant. Since the pioneering works of Whitesides' group from 2007, interest has been strong in the development and application of microfluidic paper-based analytical devices (μPADs) for food and water analysis, which allow easy, rapid and cost-effective point-of-need screening of the targets. This paper reviews recently reported μPADs that incorporate different detection methods such as colorimetric, electrochemical, fluorescence, chemiluminescence, and electrochemiluminescence techniques for food and water analysis.

Keywords: μPADs; food analysis; water analysis; point-of-need

1. Introduction

Ensuring the safety and quality of food is an incessant concern. Hamburg's editorial in *Science* entitled "Advancing regulatory science" [1] states the relevance of this matter, and indeed, one of the key points of food analysis is to ensure food safety [2]. In order to meet this goal, there is a constant search for new and more practical methods for food monitoring. Food is after all the source of nutrition and energy of every human. Similarly, water safety and quality is of great importance. With water being the major constituent of the human body, it is natural that enough water must be consumed to regulate bodily functions [3]. However, failure to warrant the safety and quality of food and water brings risks that often lead to illnesses and sometimes fatalities.

The safety of food and water is often affected by several factors, including the presence of pathogens, pesticides and herbicides, metals and other toxic materials generally borne to the food and water through agricultural and industrial processes. Another influencing factor is the amount of food additives used to provide food preservation, coloring and sweetening [4]. Such food additives have to be controlled due to the potential risks that these substances pose to human health. Some have even become prohibited due to their toxicity such as furylfuramide (AF-2), which was used as food preservative in Japan from 1965 or earlier; it was later banned due to its carcinogenicity in experimental animals [5].

This review discusses the recent progress in microfluidic paper-based analytical device (μPAD) technology for food and water safety monitoring, specifically μPAD applications to the detection of different target compounds and pathogens that are either borne naturally to food and water, or caused by unmonitored industrial and agricultural processing and waste contamination to both. Lateral-flow immunoassays (also known as immunochromatographic assays) are excluded as they have been reviewed elsewhere [6,7]. This review also covers the types of paper substrates that have been utilized in the μPAD fabrication and the detection methods that were incorporated into the μPAD for specific target detection for food and water analysis.

2. Paper in Microfluidics

Microfluidics as defined by Whitesides [8] in his article published in *Nature* in 2006 is the science and technology of systems that process and manipulate small amounts of fluid up to 10^{-9} to 10^{-18} L using fluidic channels with dimensions ranging from tens to hundreds of micrometers. Microfluidics has undergone rapid growth with notable impacts to the analytical chemistry community due to a number of capabilities including its ability to utilize small amounts of samples and reagents and to perform separation and detection with high resolution and sensitivity, at low cost and rapidly [9]. Some of the early reports on microfluidic fabrication involved the use of glass [10,11], silicon [12,13], and polymers such as poly(dimethylsiloxane) (PDMS) [14,15] as substrates. Though these microfluidic devices miniaturize the conventional methods for specific target separation and detection, they have some drawbacks such as the expense of the substrate materials, and the need for power supply and fluid transport instruments.

Paper on the other hand is a very promising substrate material for microfluidic device fabrication for a number of reasons. The properties of paper and the many advantages that it provides as a low-cost platform for diagnostics have been well-discussed [16–18]: It is easily printed, coated and impregnated; its cellulose composition is particularly compatible with proteins and biomolecules; it is environment-compatible as it is easily disposed of by incineration; and it is accessible almost everywhere. With paper as its main substrate, the cellulose membrane network of the microfluidic paper-based analytical devices (μPADs) provide instrument-free liquid transport by capillary action, a high surface area to volume ratio that enhances detection limits for colorimetric assays, and the ability to store chemical components in their active form within the paper fiber network [19]. Although μPADs lack the high resolution and sensitivity that the silicon, glass or plastic-based devices offer, the application of μPADs is highly suitable to point-of-need monitoring that requires inexpensive analysis for constant testing especially in less industrialized countries where complex instrumentation and analytical laboratories and experts are limited. Hence, μPADs have emerged as an attractive alternative to highly sophisticated instrumentation in analytical research applications particularly in food and water monitoring and safety.

To date, much analytical research has focused on the development and application of μPADs for food and water safety and quality monitoring; including fabrication procedures of the μPADs and suitable methods of detection for qualitative or quantitative interpretation of measurements. Fabrication usually entails the selection of a type of paper substrate before subjecting it to fabrication techniques such as cutting [20–25], inkjet printing [26,27], wax patterning [28,29], wax pencil drawing [30], wax printing [31–40], screen printing [29,41,42], contact stamping [43–45], and photolithography [46–48]. Examples of μPADs fabricated using various methods and paper substrates are shown in Figure 1. Among the various cellulose-based paper substrates that have been used, Whatman chromatography paper grade 1 was the first type to be utilized in 2007 [17] and it has been subsequently used in many reported μPAD fabrication and detection methods [28,29,33,37,38,47,49,50].

Whatman filter paper grade 1, on the other hand, has been the most commonly used paper substrate for μPAD fabrication in food and water analysis [25,30,32,34–36,41,45,51–54]. Paper substrates that have been similarly utilized include Whatman chromatography paper 3 MM Chr [20,21], Whatman filter paper grade 4 [42,55], Whatman RC60 regenerated cellulose membrane filter [56], Millipore MCE membrane filter [57], Canson paper [58], Fisherbrand P5 filter paper [59], JProLab JP 40 filter paper [44], Advantec 51B chromatography paper [48], and Ahlstrom 319 paper [39]. Although comparing the capabilities of each paper substrate is inappropriate when different fabrication methods and detection methods are employed among the studies, some comparisons of substrates have been made. Liu *et al.* [20], for instance, investigated paper substrates including nitrocellulose membrane, filter paper, quantitative filter paper, qualified filter paper and Whatman 3 mm chromatography paper for the μPAD chemiluminescence (CL) detection of dichlorvos (DDV) in vegetables. With the filter paper, quantitative filter paper and qualified filter paper, a high CL signal of the blank sample and poor repeatability for sample detection were observed due to the non-uniform thickness of the substrates (from 10 to 250 μm) affecting the optical path length, scattering, assay sensitivity, and volume of fluid required for an assay. However, Whatman 3 mm chromatography paper, which has high quality, purity and consistency, provided good repeatability.

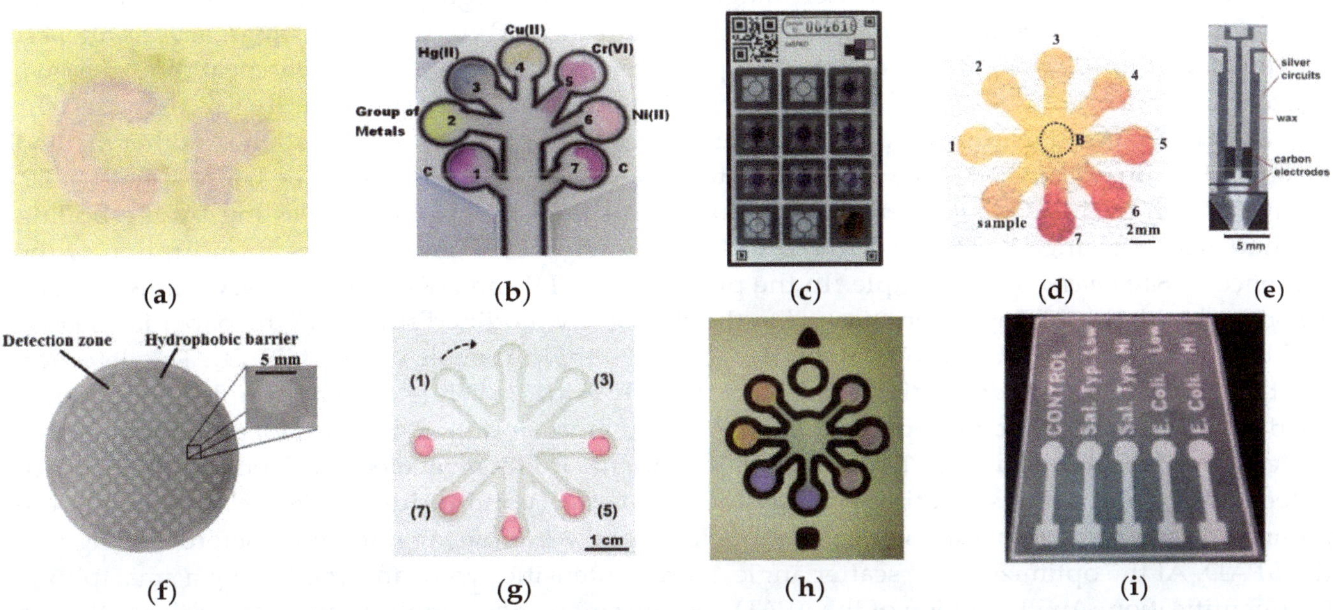

Figure 1. Examples of μPADs fabricated using different methods and paper substrates: (**a**) Wax patterning, WCP1. Reprinted with permission from reference [28]. Copyright 2015 American Chemical Society. (**b**) Wax printing, WP1. Reprinted with permission from reference [31]. Copyright 2011 American Chemical Society. (**c**) Wax printing, AP319. Reprinted with permission from reference [39]. Copyright 2015 American Chemical Society. (**d**) Alkylsilane self-assembling and UV/O_3-patterning, WFP1. Reprinted with permission from reference [52]. Copyright 2013 American Chemical Society. (**e**) Wax printing with screen-printed electrodes, WCP1. Reprinted with permission from reference [38]. Copyright 2010 The Royal Society of Chemistry. (**f**) Polymer screen printing, WFP4. Reprinted with permission from reference [42]. Copyright 2016 The Royal Society of Chemistry. (**g**) Contact stamping, JPFP40. Reprinted with permission from reference [44]. Copyright 2015 The Royal Society of Chemistry. (**h**) Contact stamping, WFP1. Reprinted with permission from reference [45]. Copyright 2014 American Chemical Society. (**i**) Photolithography, CP. Reprinted with permission from reference [46]. Copyright 2013 The Royal Society of Chemistry. WFP1, Whatman No. 1 filter paper; WCP1, Whatman No. chromatography paper; WP1, Whatman No. 1 paper; AP310, Ahlstrom 319 paper; WFP4, Whatman No. 4 filter paper; JPFP40, JProLab JP 40 filter paper; CP, chromatography paper.

3. Applications to Food and Water Contamination

3.1. Detection of Foodborne and Waterborne Pathogens

Paper-based approaches for food safety monitoring are attractive because simple, low-cost, and on-site detection of foodborne contaminants is achievable and they are also applicable as preventive measures. μPADs developed for pathogen detection in food have relied primarily on enzymatic assay-based optical methods where results are either confirmed visually by the naked eye or digitally converted and measured using image analysis software. Two of the most commonly used programs are ImageJ and Adobe Photoshop where RGB (red-green-blue) image intensities are measured relative to the image pixels or are first converted into CMYK (cyan-magenta-yellow-key) scale before intensity measurement. In a study reported by Jokerst *et al.* [32], a μPAD was developed for the microspot assay of *Escherichia coli* (*E. coli*) O157:H7, *Listeria monocytogenes* (*L. monocytogenes*) and *Salmonella* Typhimurium in ready-to-eat meat samples. The pathogens were collected from foods by a swab sampling technique and then cultured in media before adding to a chromogen-impregnated paper-based well device. A color change is observed indicating the presence of an enzyme associated with the pathogen of interest and detection is achieved. Although the detection limits determined for each of the live bacterial assays after ImageJ analysis were high (10^6 colony-forming unit (CFU) mL^{-1} for *E. coli*, 10^4 CFU mL^{-1} for *Salmonella* Typhimurium, and 10^8 CFU mL^{-1} for *L. monocytogenes*), the developed μPAD was capable of detecting pathogenic bacteria in ready-to-eat meat (bologna) at a concentration of as low as 10^1 CFU mL^{-1} within 12 h or less, which is significantly less time than the gold standard method (requires several days) for bacterial detection and enumeration. Another method presented by Jin *et al.* [33] was based on CL detection of *Salmonella* via adenosine triphosphate (ATP) quantification on μPAD. *Salmonella* was cultured and then lysed after harvesting by the boiling method. Color change is observed in the μPAD only when ATP is present as an indication of the presence of *Salmonella* in the sample. In the presence of ATP, the HRP-tagged DNA that is initially associated with the ATP aptamer attached to the chemically modified surface of the paper is released and later it allows the catalytic oxidation of 3-amino-9-ethylcarbazole by HRP/H_2O_2. The detection limit for *Salmonella* was determined to be 2×10^7 CFU mL^{-1}. While no real samples were tested, the developed μPAD could be applied for food and water monitoring. Park *et al.* [46] presented another optical-based technique using a highly angle-dependent and less wavelength-dependent method of detection through a Mie scattering strategy for *Salmonella* Typhimurium. *Salmonella* samples were pre-mixed with anti-*Salmonella* conjugated particles to allow immunoagglutination before loading into the μPAD. At the optimized Mie scatter angle, scatter intensities were analyzed using a smartphone for quantification. An illustration of the μPAD and the smartphone application used for the pathogen quantification are shown in Figure 2a,b, respectively. The detection limit of the smartphone-based μPAD assay was 10^2 CFU mL^{-1}. A one-step multiplexed fluorescence (FL) strategy for detecting pathogens was also developed by Zuo *et al.* [60] using a μPAD that was a hybrid of PDMS and glass. The paper substrate enabled the integration of the fluorescent aptamer-functionalized graphene oxide biosensor on the microfluidic device (Figure 2c). While the aptamer is adsorbed on the surface of the graphene oxide, the FL of the aptamer is quenched. In the presence of the target pathogen, the pathogen induced the liberation of the aptamer from the graphene oxide layer and thereby restored the FL of the aptamer for detection. The detection limits for the simultaneous detection of *S. aureus* and *S. enterica* were 800.0 CFU mL^{-1} and 61.0 CFU mL^{-1}, respectively. Other works on *E. coli* detection in water were reported by Burnham *et al.* [57] and Ma *et al.* [30]. Burnham *et al.* specifically demonstrated the use of bacteriophages as capture and sensing elements for the paper-based detection of the pathogen. The method was based on the detection of β-galactosidase released from the pathogenic cells following bacteriophage-mediated lysis. Colorimetric and bioluminescence methods were performed for *E. coli* detection using red-β-D-galactopyranoside chromogenic substrate and Beta-Glo® reagent (Promega Corporation, Madison, WI, USA) to produce the color and bioluminescence, respectively, for measurement with a detection limit of 4 CFU mL^{-1} for both methods. Ma *et al.*, on the other hand, presented a μPAD for the colorimetric determination of *E. coli* using AuNP-labeled detection antibodies via sandwich immunoassay with a silver enhancing step for signal amplification. The detection limit was 57 CFU mL^{-1}.

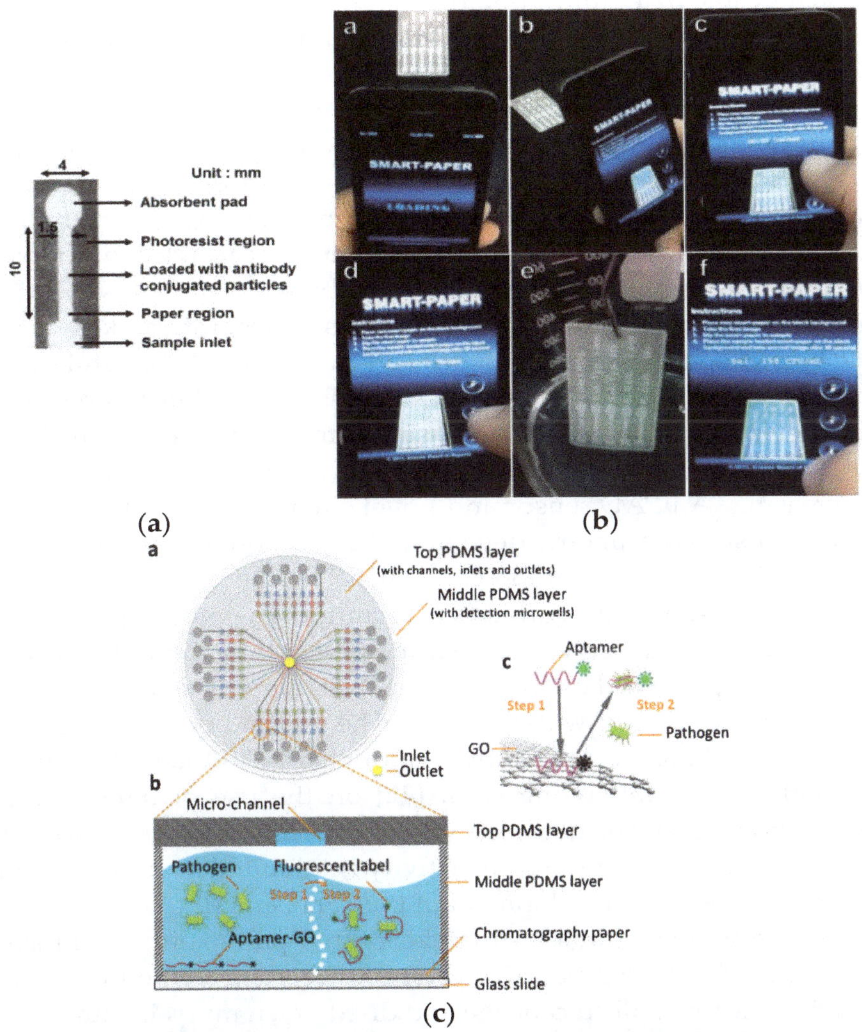

Figure 2. Detection methods for pathogens. (**a**) An image of a single-channel μPAD and (**b**) the smartphone application for *Salmonella* detection on a multi-channel μPAD. Reprinted with permission from reference [46]. Copyright 2013 The Royal Society of Chemistry. (**c**) Schematic layout of the PDMS/paper hybrid μPAD system and illustration of the one-step multiplexed FL detection principle on the μPAD during aptamer adsorption (Step 1) and liberation (Step 2) from the GO surface and the restoration of the FL for detection in the presence of the target pathogen. Reprinted with permission from reference [60].

3.2. *Detection of Pesticides and Herbicides*

Pesticides have been used for many years in agriculture and have significantly contributed to maintaining food quality and production. Simultaneously, however, these materials bring harmful effects on human health [61,62]. Wang *et al.* [49] developed a paper-based molecular imprinted polymer-grafted multi-disk micro-disk plate for CL detection of 2,4-dichlorophenoxyacetic acid (2,4-D). The MIP approach was proposed as an alternative to immunoassays, which rely on antibodies and have fundamental drawbacks such as the possible denaturation and instability of the antibodies during manufacture and transport. An indirect competitive assay was made with tobacco peroxidase (TOP)-labeled 2,4-D that was molecularly imprinted on the polymer-grafted device. An enzyme catalyzed CL emission was achieved from the luminol-TOP-H_2O_2 CL system with a detection limit of 1.0 pM. A simple paper-based luminol-H_2O_2 CL detection of DDV was reported by Liu *et al.* [20]. Paper chromatography was combined in the μPAD CL assay of DDV in fruits and vegetables and the separation was achievable in 12 min utilizing 100 μL of developing reagent. The method was successfully applied to the trace DDV detection on cucumber, tomato and cabbage by a spiking method with a detection limit of 3.6 ng· mL^{-1}. Liu *et al.* [21] also presented another MIP-based approach using a paper-based device with a molecularly imprinted polymer for the CL detection of DDV. The detection limit was 0.8 ng· mL^{-1} and the method was successfully applied to cucumber and tomato. A paper-based colorimetric approach has also been demonstrated for the detection of organophosphate and carbamate pesticides. Badawy *et al.* [58] developed a method that was based on the inhibition of acetylcholinesterase (AChE) on the degradation of acetylcholine molecules into choline and acetic acid by organophosphate (methomyl) and carbamate (profenos) pesticides. The degree of inhibition of the AChE indicates the toxicity of the pesticides; this makes the AChE a standard bioevaluator for the presence of organophosphates and carbamates [63]. While the method was not tested on real samples, the method could detect AChE inhibitors within 5 min response time.

With the goal to devise portable and easy measuring techniques and considering the increasing use of smartphones, the number of μPAD strategies that incorporate mobile or smartphones for target measurements is increasing. A μPAD sensor and novel smartphone application was developed by Sicard *et al.* [34] for the on-site colorimetric detection of organophosphate pesticides (paraoxon and malathion) based on the inhibition of immobilized AChE by the pesticides. AChE hydrolyzes the colorless indoxyl acetate substrate and converts it to an indigo-colored product in the absence of pesticides. The color intensity is reduced with increasing pesticide concentration owing to inhibition of AChE. The color produced is processed by the image analysis algorithm using a smartphone, allowing real time monitoring and mapping of water quality. The method is capable of detecting pesticide concentration of around 10 nM as evidenced by a color change in the μPAD. Another colorimetric approach was reported by Nouanthavong *et al.* [42] on the use of nanoceria-coated μPAD for colorimetric organophosphate pesticide detection via enzyme-inhibition assay with AChE and choline oxidase. In the presence of the pesticides, AChE activity is inhibited leading to no or less production of H_2O_2 and hence less yellow color development of the nanoceria (the color production mechanism is shown in Figure 3). The assay was able to analyze methyl-paraoxon and chlorpyrifos-oxon with detection limits of 18 ng· mL^{-1} and 5.3 ng· mL^{-1}, respectively. The method was successfully applied for methyl-paraoxon detection on spiked cabbage and dried green mussel, with ~95% recovery values for both samples.

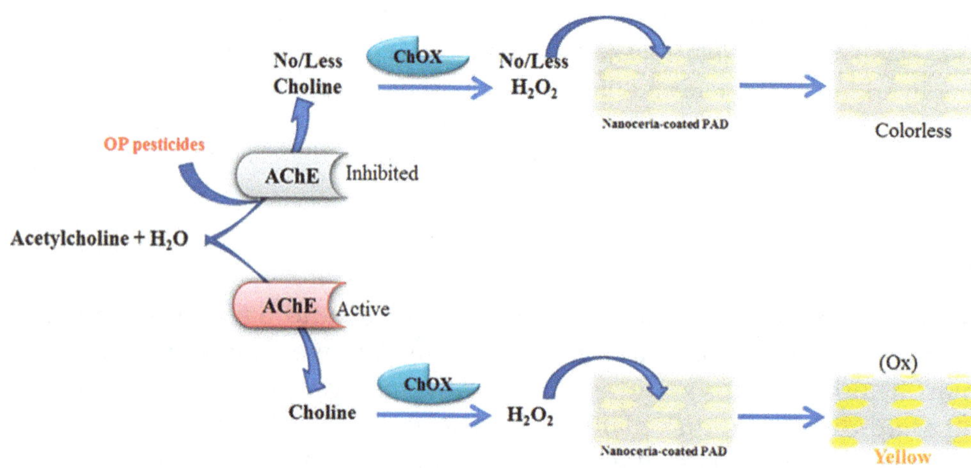

Figure 3. Colorimetric detection of pesticides based on the enzyme inhibition properties of the pesticide on nanoceria substrate. Reprinted with permission from reference [42].

Another pesticide causing a health concern is pentachlorophenol (PCP) [64–66]. PCP is a xenobiotic that accumulates in the body with carcinogenic and acute toxic effects. Sun *et al.* [50] developed a photoelectrochemical (PEC) sensor that utilized the MIP technique on a μPAD to detect PCP. The paper working electrode of the μPAD was covered with a layer of gold nanoparticles (AuNPs) and a layer of polypyrrole (Ppy)-functionalized ZnO nanoparticles. The photoelectrochemical mechanism involves the excitation of electrons from Ppy from its highest occupied molecular orbital to the lowest unoccupied molecular orbital of ZnO after being irradiated with visible light. Since the lowest unoccupied molecular orbital of ZnO and Ppy matched well, the transfer of the excited electrons to ZnO was allowed and the electrons subsequently reached the gold-paper working electrode (Au-PWE) surface, where photocurrent generation efficiency was improved leading to a sharp increase of the photocurrent. However, in the presence of the PCP, the steric hindrance toward the diffusion of the quencher molecules and/or photogenerated holes on the interface of the electrode increased, thereby leading to a decrease in generated photocurrent. The device was capable of measuring PCP down to a limit of 4 $pg \cdot mL^{-1}$.

The only paper-based approach applied to herbicide detection that has utilized FL as a method of detection for methyl viologen is presented by Su *et al.* [67]. The method was based on the integration of CdTe Qdots on the paper device and the CdTe quenching effect in the presence of the target methyl viologen. Presence of a higher methyl viologen concentration in the system gave a darker area on the μPAD as a result of the quenching of the methyl viologen on the CdTe Qdots. The detection limit of the CdTe-paper-based visual sensor was 0.16 $\mu mol \cdot L^{-1}$.

3.3. Detection of Food Additives

In food and beverage industries, wide use is made of food additives such as glucose, fructose and sucrose, which are specifically used as sweeteners, and other food additives, which are used to improve or enhance the flavor or color of the food or beverage. Though most of these food additives are essentially nontoxic, large intakes of them may promote unhealthy nutrition, and some become toxic above a certain amount. Hence, there is a strong demand for fast, highly sensitive and economical methods of analysis that can be provided by the easily accessible and portable point-of-need testing of μPAD technology. Kuek Lawrence *et al.* [51] reported on an amperometric detection of glucose on a screen-printed electrode μPAD. The assay involved the use of ferrocene monocarboxylic acid as a mediator for the catalytic oxidation of glucose on the μPAD by the immobilized glucose oxidase on the paper. The method was successfully applied to glucose detection in commercially marketed carbonated beverages with a limit of 0.18 mM. Adkins *et al.* [35] presented a μPAD that utilized microwire electrodes as an alternative to screen-printed electrodes for the non-enzymatic electrochemical detection of glucose, fructose and sucrose in beverage samples. A copper working electrode was used and the copper electrocatalytically reacted with glucose in the alkaline media, allowing the non-enzymatic electrochemical detection of the carbohydrates. A variety of commercial beverages were tested including Coca-Cola™, Orange Powerade™, Strawberry Lemonade Powerade™, Red Bull™ and Vitamin Water™. The detection limits were 270 nM, 340 nM and 430 nM for glucose, fructose and sucrose, respectively.

Colletes *et al.* [43] presented a study that utilized a paraffin-stamped paper substrate for the detection of glucose in hydrolysis of liquors (detection limit 2.77 $mmol \cdot L^{-1}$) by paper spray mass spectrometry (PS-MS). PS-MS is a fast, precise, accurate and cost-effective ionization method introduced by Crooks and co-workers in 2010 that provides complex analyses in a simple and economical way by mass spectrometry [68]. Although the paraffin-stamped paper substrate is not a μPAD *per se*, Colletes *et al.* explained the potential of the paper substrate for the combination of a microfluidic paper-based analytical device with mass spectrometry that used paper spray as the ionization method.

Nitrites are food additives used to prevent the growth of microorganisms as well as to inhibit lipid oxidation that causes rancidity [69]. Nitrite monitoring in food and water is essential due to the ability of nitrite to readily react with secondary and tertiary amines and produce carcinogenic nitrosamine compounds [70]. Several works on nitrite detection have involved the use of the Griess-color reaction mechanism to visually detect the presence of nitrite in food. For instance, He *et al.* [52] described a μPADusing the Griess-color nitrite assay, where, upon reaction of nitrite with the Griess reagent in the μPAD, a color developed with intensities depending on the amount of nitrite in the sample.

Image processing was done for quantification showing a dynamic range of 0.156–2.50 mM, and a successful application to nitrite detection in red cubilose (a traditional nutritious food and medicine in China) was achieved. Other works presented by Lopez-Ruiz *et al.* [45], Cardoso *et al.* [44] and Jayawardane *et al.* [53] similarly focused on the colorimetric detection of nitrite in water and food using the Griess method in μPADs. Lopez-Ruiz *et al.* presented a strategy using a mobile phone with a customized algorithm for image analysis and detection. As depicted in Figure 4a, the method allowed a multidetection of the μPAD sensing areas specific for pH detection simultaneously with nitrite detection in water samples.

The strategy involved capturing the μPAD image upon sample detection with the smartphone camera, and processing of the image in order to extract the colorimetric information for measurement, wherein, hue (H) and saturation (S) of the HSV color space were used for the determination of pH and nitrite concentration, respectively. The colorimetric assay for pH determination was based on the use of two pH indicators, phenol red and chlorophenol red.

A color transition of chlorophenol red from yellow to purple indicated a pH from 4 to 6, while a color transition of phenol red from yellow to pink indicated a pH from 6 to 9. The nitrite assay, on the other hand, involved a Griess-color reaction in which the color formation was quantitatively interpreted showing a detection limit of $0.52\ mg\cdot L^{-1}$.

Cardoso *et al.* similarly reported a μPAD strategy for nitrite detection in ham, sausage and the preservative water from a bottle of Vienna sausage using the Griess-color assay with a detection limit of 5.6 μM. The colorimetric analysis was performed by first taking the image of the detection device using a scanner, and later processing the magenta scale of the image after conversion to the CMYK using Corel Photo-Paint™ software.

Finally, Jayawardane *et al.* presented their work for nitrite and nitrate determination in different water samples using two μPADs, each specific for nitrate and nitrite, respectively. The image of the 2D and 3D μPADs used for detection are shown in Figure 4b.
The nitrite detection simply employed the Griess method for colorimetric measurements after image scanning and processing using ImageJ software. In the nitrate detection however, a conversion of the colorimetrically undetected species was first performed to the colorimetrically detected nitrite using a Zn reduction channel incorporated in the μPAD for nitrate detection.
After conversion, the Griess method was employed and image quantification was performed. The method was successfully applied to actual analysis of different water samples (tap water, mineral water, and pond water) with detection limits of 1.0 μM and 19 μM for nitrite and nitrate, respectively.

The addition of colorants to food has become a normal practice to enhance or change food color and make it more attractive to consumers. However, most of these colorants are potentially harmful to human health especially after excessive consumption. One μPAD design that has been developed for detecting colorants was presented in the work of Zhu *et al.* [22] where a poly(sodium 4-styrenesulfonate)-functionalized paper substrate was used for the rapid separation, preconcentration and detection of colorants in drinks with complex components via a surface-enhanced Raman spectroscopy (SERS) method.

Sunset yellow and lemon yellow were both detected in grape juice and orange juice with detection limits of 10^{-5} M and 10^{-4} M, respectively.

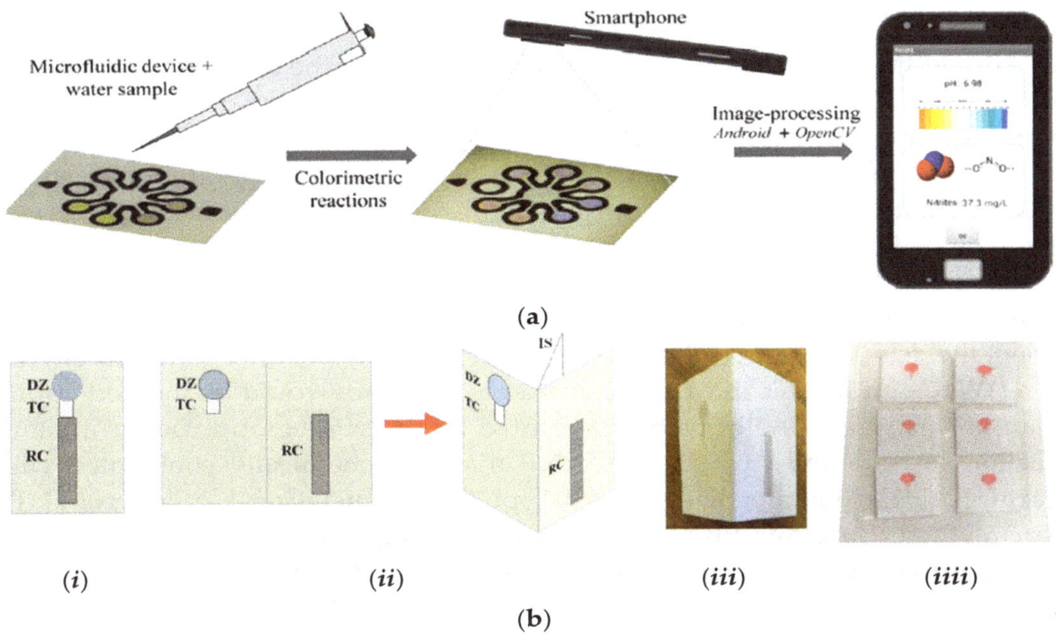

Figure 4. (**a**) Griess-color reaction assay-based detection methods for nitrite using a smartphone for image processing. Reprinted with permission from reference [45]. Copyright 2014 American Chemical Society. (**b**) Griess-color reaction assay-based detection methods for nitrite and nitrate using 2D (*i*) and 3D (*ii–iv*) µPADs. Reprinted with permission from reference [53].

3.4. Detection of Heavy Metals

Several µPADs have been developed for the detection of heavy metals in both food and water. The most common methods of detection integrated with the µPADs were colorimetric-based using silver or gold nanoparticles and nanoplates, but electrochemical and FL based methods were used as well. Nie *et al.* [47] developed a µPAD for the versatile and quantitative electrochemical detection of biological and inorganic analytes in aqueous solutions. Specifically, for water analysis, lead was investigated via square wave anodic stripping voltammetry using a µPAD with screen-printed electrodes as shown in Figure 5a. The measurements relied on the simultaneous plating of bismuth and lead onto the screen-printed carbon electrodes of the µPAD, which formed alloys, followed by anodic stripping of the metals from the electrode. The method showed a detection limit of 1.0 ppb in water medium. Similarly, Shi *et al.* [54] developed an electrochemical µPAD for Pb(II) and Cd(II) detection based on square wave anodic stripping voltammetry (SWASV) relying on *in situ* plating of bismuth film. The method was capable of detecting lead and cadmium ions simultaneously in carbonated electrolyte drink (salty soda water as described by the authors) samples with detection limits of 2.0 ppb and 2.3 ppb for Pb(II) and Cd(II), respectively.

Using silver nanoparticles (AgNP) self-assembled with aminothiol compounds on µPADs, Ratnarathorn *et al.* [25] reported on the colorimetric detection of copper in drinking water samples. In the presence of Cu^{2+}, the modified AgNP solution changed from yellow to orange and then green-brown due to nanoparticle aggregation. The method was tested on tap water and pond water samples with a detection limit of 7.8 nM or 0.5 $\mu g \cdot L^{-1}$. Two other applications of µPAD with colorimetric detection for Cu(II) were reported by Jayawardane *et al.* [55] and Chaiyo *et al.* [36]. In the former work, a polymer inclusion membrane (PIM) containing the chromophore (1-(2′-pyridylazo)-2-naphthol (PAN)) reactive to Cu(II) was incorporated in the µPAD and was used as the sensing element selective to the metal ion. The original yellow color of the membrane changed to red/purple as the Cu(II) formed a complex with PAN. The device was applied to Cu(II) determination in hot tap water samples with a detection limit of 0.6 $mg \cdot L^{-1}$. The latter work by Chaiyo *et al.*

on the other hand used silver nanoplates (AgNPls) modified with hexadecyltrimethyl-ammonium bromide (CTAB) for the colorimetric detection of Cu(II) based on the catalytic etching of the AgNPls with thiosulfate ($S_2O_3^{2-}$). The violet-red $S_2O_3^{2-}$/CTAB/AgNPl on the detection zone lost its color with increasing Cu^{2+} concentration. The method was applied for determination of Cu^{2+} in drinking water, ground water, tomato and rice with a detection limit of 1.0 $ng \cdot mL^{-1}$ by visual detection. Nath *et al.* [23] presented a sensing system that could detect As^{3+} ions using gold nanoparticles chemically conjugated with thioctic acid (TA) and thioguanine (TG) molecules on paper. During detection, a visible bluish-black color appeared on the paper due to nanoparticle aggregation through transverse diffusive mixing of the Au–TA–TG with As^{3+} ions. While no real water sample testing was performed, the detection limit (1.0 ppb) was lower than the reference standard of World Health Organization (WHO) for arsenic in drinking water, hence there would be method applicability to real water sample analysis. Another work presented by the same group used a similar approach for the detection of Pb^{2+} and Cu^{2+} using AuNP that was chemically conjugated with TA and dansylhydrazine [24]. The detection limit was $\leqslant$0.0 ppb for both metal ions. Apilux *et al.* [41] developed a colorimetric method using AgNPls for the detection of Hg(II) ion levels. A change in color from pinkish violet to pinkish yellow occurred with the Hg(II) ion detection, a phenomenon that can be attributed to a change in the surface plasmon resonance of the AgNPls, which is related to the AgNPl apparent color. At Hg(II) concentration levels above 25 ppm, the color of the AgNPls fades as observed by the naked eye. With digital imaging and software processing though, the quantitative capability of the system was improved and showed a detection limit of 0.12 ppm with successful applications to real sample analysis of drinking water and tap water. Another method via FL detection for the determination of Hg(II), Ag(I) and neomycin (NEO) for food analysis was presented by Zhang *et al.* [37]. The method used a Cy5-labeled single-stranded DNA (ssDNA)-functionalized graphene oxide (GO) sensor that generated FL in the presence of the target analytes, otherwise, the Cy5 was quenched while adsorbed on the GO surface. The detection limits were 121 nM, 47 nM and 153 nM for Hg(II), Ag(I) and NEO, respectively.

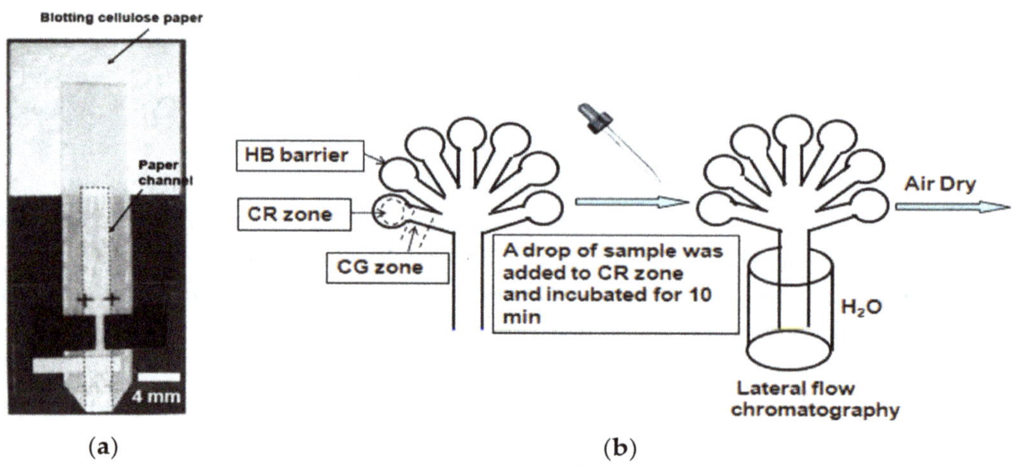

Figure 5. Detection methods for metals. (**a**) Electrochemical device for SWASV analysis of lead in water with screen-printed carbon working and counter electrodes and Ag/AgCl pseudo-reference electrode. Reprinted with permission from reference [47]. Copyright 2009 The Royal Society of Chemistry. (**b**) Multiplexed colorimetric detection of metals based on B-GAL and CPRG interaction in the presence of Hg^{2+}, Cu^{2+}, Cr^{6+} and Ni^{2+} mixture. Reprinted with permission from reference [31].

Hossain *et al.* [31] presented a multiplexed µPAD that is capable of detecting heavy metals simultaneously in a single µPAD. As shown in Figure 5b, the µPAD is composed of seven reaction zones, two of which are for control experiments, one for testing the mixture of metal ions via β-galactosidase

(B-GAL) assay, and four using colorimetric reagents specific for Hg(II), Cu(II), Cr(VI) and Ni(II), respectively. In the B-GAL assay, the chromogenic substrate, chlorophenol red β-galactopyranoside (CPRG), which is printed on a region upstream to the B-GAL zone, is transported into the detection zone by the sample solution through capillary action and it is hydrolyzed by the B-GAL enzyme to form the red-magenta product. In the presence of the metal ions, the red-magenta color produced upon CPRG hydrolysis is lost to a degree dependent on the concentration of the metal ions in the sample. For the assays specific for each metal ion, color appearance is observed in the presence of each metal ion on their respective detection zones, while the absence of any of the metal ions results in no color change on the respective zones. The detection limit of the device is ~0.5–1.0 ppm. Li *et al.* [28] demonstrated the use of a μPAD that enables easy detection of trace metals via text-reporting of results. Using the color-generating periodic table symbols of the specific trace metals fabricated on the μPAD as markers, even nonprofessional users can carry out handy detection and monitoring. The Cu(II) assay was based on the formation of an orange to brown complex by bathocuproine as the indicator with Cu(II). For the Cr(VI) assay, a magenta to purple complex formed in the presence of the metal ion with the indicator 1,5-diphenylcarbazide in acidic medium, while for the Ni(II) assay, a stable pink-magenta colored complex formed between dimethylglyoxime and Ni(II). The device was capable of colorimetric detection of Cu(II), Cr(VI) and Ni(II) in tap water with concentrations of $\geqslant 0.8~mg\cdot L^{-1}$, $>0.5~mg\cdot L^{-1}$ and $\geqslant 0.5~mg\cdot L^{-1}$, respectively. Finally, for μPAD detection of heavy metals, a colorimetric approach for image processing and quantification based on an iron-phenanthroline (Fe-phen) assay that has colored response with increasing concentration of iron was incorporated for the investigation of iron in water samples by Asano *et al.* [48]. The developed method allowed a direct analysis of tap and river water samples without pretreatment with a detection limit of 3.96 μM.

3.5. Detection of Other Food and Water Contaminants

Several methods have also been demonstrated for detecting other food and water contaminants using μPAD technology. Nie *et al.* [38] presented an electrochemical technique for ethanol detection in water for possible food quality control purposes. Electrochemical μPADs and a glucometer (Figure 6a) were used to amperometrically measure ethanol (LOD 0.1 mM) using ferricyanide as an electron-transfer mediator and alcohol dehydrogenase/β-NAD^+ as detecting components in the device. An electrochemical μPAD for halide detection in food supplement and water samples via cyclic voltammetry was also developed by Cuartero *et al.* [56]. The device utilizes silver elements as working and counter/reference electrodes as illustrated in Figure 6b. The oxidation of the silver foil working electrode is induced by an anodic potential scan resulting in a current that is related to the plating rate of the target halides in the sample as silver halides precipitate. This process is complemented by the reduction of the silver/silver halide element in the reference/counter electrode upon ion exchange movement of the Na^+ ion (halide counterion) through the permselective membrane to maintain the neutrality of charges in each paper compartment, and that leads to the release of halide ions into the solution. The two silver elements are regenerated to their previous states through the application of a backward potential sweep after the forward scan. The device was found capable of detecting bromide, iodide and chloride mixtures in food supplement, seawater, mineral water, tap water and river water samples with a detection limit of around 10^{-5} M of halide mixtures. Myers *et al.* [39] developed a multiplexed μPAD (called a saltPAD) that is capable of making an iodometric titration in a single printed card. Multiple reagents are stored on every compartment of each detection zone of the saltPAD and they are allowed to recombine and undergo surface-tension-enabled mixing upon introduction of the iodized salt sample solution for determination. During the iodometric titration process, triiodide is formed as excess iodide that reacts with iodate in the presence of acid. The triiodide is then titrated with thiosulfate that was previously stored in the saltPAD. Using starch as an indicator, the detection zone produces a blue color if the amount of triiodide exceeds the reducing capacity of the thiosulfate.

The indicator remains uncolored if the amount of triiodide is smaller than the reducing capacity of the thiosulfate. The detection limit of the device expressed as mg iodine/kg salt was 0.8 ppm.

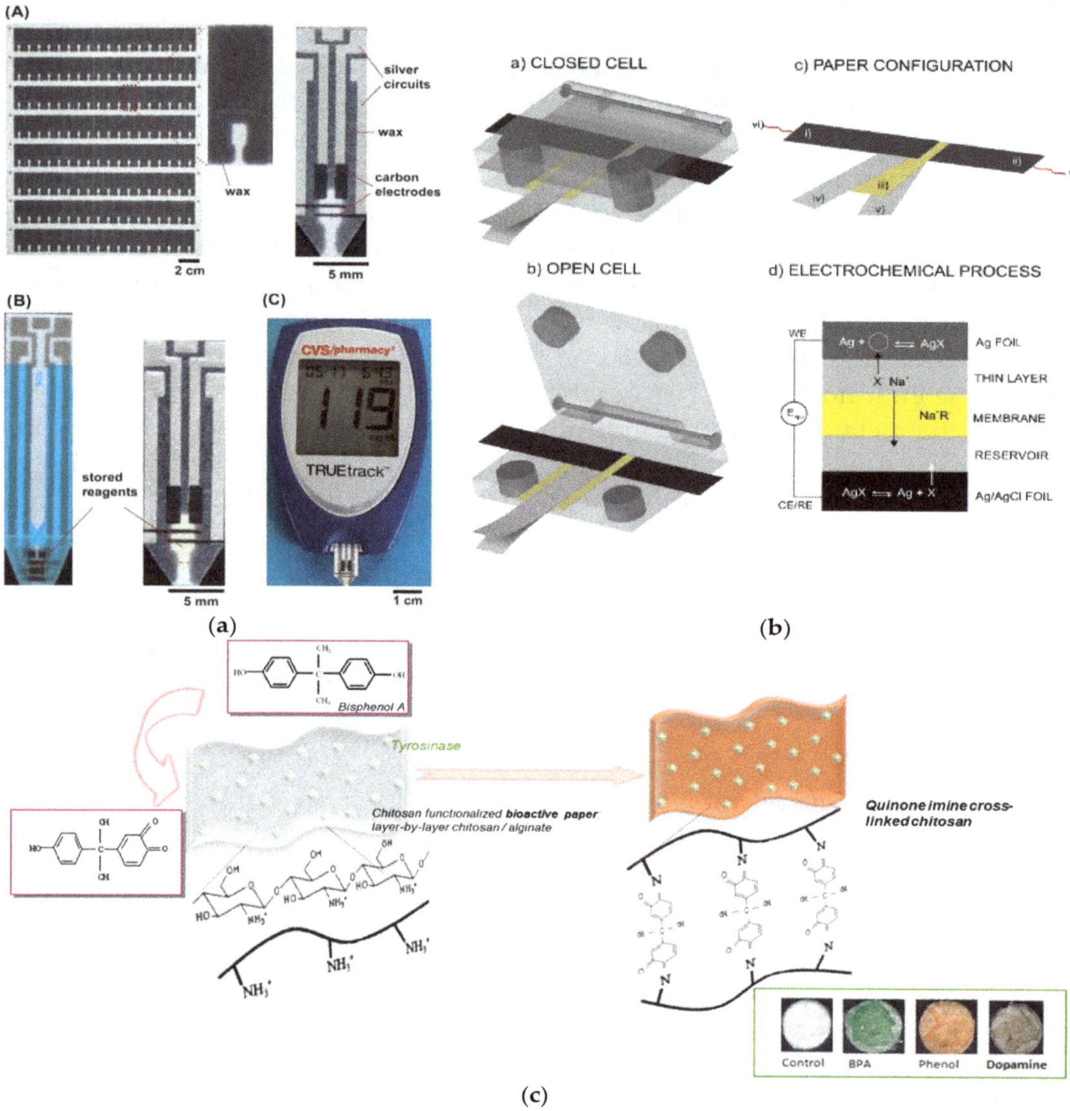

Figure 6. Detection methods for other food and water contaminants. (**a**) Components of the electrochemical detection system for ethanol using a glucometer as a readout device. Reprinted with permission from reference [38]. Copyright 2010 The Royal Society of Chemistry. (**b**) The configuration of the electrochemical cell for the analysis of halides utilizing silver components as electrodes on paper-assisted electrochemical detection. Reprinted with permission from reference [56]. Copyright 2015 American Chemical Society. (**c**) A representative paper-based colorimetric bioassay of BSA based on the enzymatically generated quinone from tyrosinase and chitosan interaction in the presence of the phenolic compound. Reprinted with permission from ref [59].

Cyanobacteria in drinking water pose a great threat to public health due to the cyanotoxins produced and released into water supplies. The most toxic of the cyanotoxins is microcystin-LR (MC-LR) [71,72]. Ge *et al.* [40] focused on the development of a method that specifically detects MC-LR in water using a gold-paper working electrode (Au-PWE) for electrochemical immunoassay. Differential pulse voltammetric measurements were performed by monitoring the oxidation process of thionine in the system for the quantification of MC-LR under the catalysis of HRP and peroxidase mimetics (Fe_3O_4). The sandwich immunoreaction produced a current proportional to the logarithm of MC-LR and gave a detection limit of 0.004 $\mu g \cdot mL^{-1}$. Phenolic compounds are generally produced as byproducts from industrial processes that present health risks to humans after consumption of contaminated food and water. For detection of phenolic compounds, Alkasir *et al.* [59] developed a paper sensor that produces different color responses for phenol (reddish-brown), bisphenol A (blue-green), dopamine (dark-brown), cathecol (orange), and m-cresol (orange) and p-cresol (orange) resulting from the specific binding of enzymatically generated quinone to chitosan immobilized in multiple layers on the paper. Figure 6c illustrates an example of the layer-by-layer paper-based bioassay for bisphenol A. The paper sensor was successfully applied to the analysis of tap and river water samples with a detection limit of 0.86 (±0.102) μg· L^{-1} for each of the phenolic compounds.

Finally, the only μPAD detection strategy based on electrochemiluminescence (ECL) detection for the specific analysis of food has been reported by Mani *et al.* [29]. The work described a device that specifically measures the genotoxic activity of a certain compound (benzo[a]-pyrene (B[a]P)) whose metabolite reacts with DNA and the responses are measured via ECL detection. The measurement essentially involves two steps, the first of which involves the conversion of the test compound B[a]P to a metabolite by a microsomal enzyme from rat liver microsomes. The second step is a DNA damage detection that involves the liberation of ECL light upon oxidation of the guanine in the damaged DNA by the (bis-2,2′-bipyridyl) ruthenium polyvinylpyridine ($[Ru(bpy)_2(PVP)_{10}]^{2+}$ or RuPVP) polymer of the electrochemical device. The technique was specifically tested on grilled chicken, and the detection limit was ~150 nM.

4. Conclusions and Future Directions

A review of microfluidic paper-based devices for food and water analysis has been presented. Table A1 (Appendix A) summarizes uses of microfluidic paper-based devices for detection of different pathogens, additives and contaminants in food and water that have been reported to date. μPADs in food and water safety and analysis represent a burgeoning technology that provides fast, economic, easy-to-use advantages and is highly applicable for point-of-need testing especially in resource limited environments. While the field of microfluidic paper-based sensors has expanded rapidly, food and water safety remains an area with many issues still to be addressed. One specific challenge in food analysis for example is the method of handling and pretreatment of the samples before μPAD detection. While fluid samples such as water and beverage usually do not require any pretreatment to the sample before introducing into the device for μPAD detection [22,28,38,48,49,51,53–56,59], food specimens could be in solid form, and therefore, a suitable pretreatment step is necessary for target sample collection before introducing into the μPAD for detection. In treating fruits, vegetables and meat samples for instance, most groups employ an extraction method to collect the target of interest [29,42], although an elution process [20,21], or boiling method [44], with the use of distilled water, followed by filtration are simple steps that are possibly performed to collect the target for μPAD detection. For pathogen collection, the swab sampling technique has also been performed which requires a significantly reduced enrichment times compared to the gold standard culture method before sample introduction and colorimetric paper-based detection [32]. While successful, the enzymatic assay systems point to the potential for exploring the use of specific inducers to enhance enzyme production as well as using selective enrichment media to inhibit the growth of competing microorganisms. Despite the current limitations on selectivity and sensitivity using paper as substrates for detection, the ability of μPADs to detect specific targets such as pathogenic bacteria, food additives and contaminants has been demonstrated in real food and water samples at levels that are vital to the safety and health of both animals and humans, therefore demonstrating its significant impact to the community for food and water safety and quality monitoring. Based on the number of references reporting the development of μPADs specifically directed to food and water safety and quality monitoring in the last six years, μPAD technology is still in its early stage and there are wide opportunities for developments and applications. Particularly exciting is the potential for application of μPADs for regular monitoring of food crops and drinking water sources, where, contamination is a risk from mining and industrial processes, and analytical measurements have traditionally been a cost limiting factor. From the detection of foodborne and waterborne infectious pathogens to different organic and inorganic analytes in general, μPADs offer the means to detect different targets using an inexpensive material like paper as their main substrate for qualitative as well as quantitative on-site food and water monitoring.

Acknowledgments: Lori Shayne Alamo Busa thanks Fatima Joy Cruz for her assistance in collecting some of the references, and the Ministry of Education, Culture, Sports, Science and Technology, Japan for the Ph.D. research scholarship. This research was partially supported by the Urakami Foundation for Food and Food Culture Promotion.

Author Contributions: Manabu Tokeshi conceived the structure of the review article; Lori Shayne Alamo Busa collected the references and wrote the paper; and Saeed Mohammadi, Masatoshi Maeki, Akihiko Ishida and Hirofumi Tani also contributed references and ideas for the review.

Abbreviations

The following abbreviations are used in this manuscript:

2,4-D	2,4-dichlorophenoxyacetic acid
Ach	acetylcholinesterase
AgNP	silver nanoparticle
AgNPl	silver nanoplate
ATP	adenosine triphosphate
B[a]P	benzo[a]pyrene
B-GAL	β-galactosidase
BPA	bisphenol A
CFU	colony-forming unit
CL	chemiluminescence
CMYK	cyan-magenta-yellow-key
CPRG	chlorophenol red β-galactopyranoside
DDV	dichlorvos
E. coli	*Escherichia coli*
ECL	electrochemiluminescence
FL	fluorescence
GO	graphene oxide
HRP	horseradish peroxidase
L. monocytogenes	*Listeria monocytogenes*
LOD	limit of detection
MCE	mixed cellulose esters
MC-LR	microcystin-LR
MIP	molecularly imprinted polymer
NEO	neomycin
PCP	pentachlorophenol
PDMS	poly(dimethylsiloxane)
PEC	photoelectrochemical detection
PIM	polymer inclusion membrane
Ppy	polypyrrole
PS-MS	paper spray mass spectrometry
Qdots	quantum dots
S. aureus	*Staphylococcus aureus*
S. enterica	*Salmonella enterica*
S. Typhimurium	*Salmonella* Typhimurium
SERS	surface-enhanced Raman spectroscopy
SWASV	square wave anodic stripping voltammetry
μPAD	microfluidic paper-based analytical device

Appendix A

Table A1. Summary of foodborne pathogens, toxins, pesticides and insecticides, heavy metals and food additives for food and water analyses on paper-based platforms.

Target	μPAD Wall Fabrication Method	Paper Substrate	Detection Method	Linear Detection Range	LOD	Real Sample Application	Reference
Pathogens							
E. coli O157:H7, *Salmonella* Typhimurium, *L. monocytogenes*	Wax printing	Whatman No. 1 filter paper	Colorimetric	-	10^6 CFU mL^{-1}, 10^4 CFU mL^{-1}, 10^8 CFU mL^{-1}	Bologna	[32]
Salmonella	Wax printing	Whatman No. 1 chromatography paper	CL	-	2.6×10^7 CFU mL^{-1}	-	[33]
S. Typhimurium	Photolithography	Chromatography paper	Optical (Mie scattering)	10^2–10^5 CFU mL^{-1}*	10^2 CFU mL^{-1}	-	[46]
S. aureus, *S. enterica*	Cutting by punching (PDMS/paper/glass hybrid)	Whatman chromatography paper	FL	10^4–10^6 CFU mL^{-1}, 42.2–675.0 CFU mL^{-1}	800.0 CFU mL^{-1}, 61.0 CFU mL^{-1}	-	[60]
E. coli	-	Millipore MCE membrane filter	Colorimetric and bioluminescence	-	4 CFU mL^{-1}	-	[57]
E. coli	Wax pencil drawing and PDMS screen printing	Whatman No. 1 filter paper	Colorimetric	-	57 CFU mL^{-1}	Drinking water	[30]
Pesticides and Herbicides							
2,4-D	-	Whatman No. 1 chromatography paper	CL	-	1.0 pM	Tap water, lake water	[49]
Paraoxon, Malathion	Wax printing	Whatman No. 1 filter paper	Colorimetric	1×10^{-8}–ca. 1×10^{-6} M	10 nM	-	[34]
Methyl-paraoxon, Chlorpyrifos-oxon	Polymer screen-printing	Whatman No. 4 filter paper	Colorimetric	0–0.1 μg· mL^{-1}, 0–60 ng· mL^{-1}	18 ng· mL^{-1}, 5.3 ng· mL^{-1}	For methyl-paraoxon: cabbage, dried green mussel	[42]
Dichlorvos	Cutting	Whatman 3MM Chr chromatography paper	CL	10 ng· mL^{-1}–1.0 μg· mL^{-1}	3.6 ng· mL^{-1}	Cucumber, tomato, cabbage	[20]
Dichlorvos	Cutting	Whatman 3MM Chr chromatography paper	CL	3.0 ng· mL^{-1}–1.0 μg· mL^{-1}	0.8 ng· mL^{-1}	Cabbage, tomato	[21]
Methomyl, Profenofos	Cutting	Canson paper	Colorimetric	-	6.16×10^{-4} mM, 0.27 mM	-	[58]
PCP	Wax screen-printing	Whatman No. 1 chromatography paper	PEC	0.01–100 ng· mL^{-1}	4 pg· mL^{-1}	-	[50]
Methyl viologen (paraquat)	Cutting	Whatman filter paper	FL	0.39 μmol· L^{-1}–3.89 μmol· L^{-1}	0.16 μmol· L^{-1}	-	[67]

Table A1. *Cont.*

Target	μPAD Wall Fabrication Method	Paper Substrate	Detection Method	Linear Detection Range	LOD	Real Sample Application	Reference
Food Additives							
Glucose	Cutting by punching	Whatman No. 1 filter paper	Electrochemical	1–5 mM	0.18 mM	Commercial soda beverages	[51]
Glucose, Fructose, Sucrose	Wax printing	Whatman No. 1 filter paper	Electrochemical	-	270 nM, 340 nM, 430 nM	Coca-Cola™, Orange Powerade™, Strawberry Lemonade Powerade™, Red Bull™, Vitamin Water™	[35]
Glucose	Paraffin stamping	Whatman grade 1 paper	PS-MS	1–500 μmol· L^{-1}	2.77 μmol· L^{-1}	Liquors	[43]
Sunset yellow, Lemon yellow	Cutting	Filter paper	SERS	-	10^{-5} M, 10^{-4} M	Grape juice, orange juice	[22]
Nitrite	Paraffin stamping	JProLab JP 40 filter paper	Colorimetric	0–100 μM	5.6 μM	Ham, sausage, preservative water	[44]
Nitrite	Alkylsilane assembling and UV-lithography	Whatman No. 1 filter paper	Colorimetric	0.156–2.50 mM	-	Processed red cubilose	[52]
Nitrite	Indelible ink contact stamping	Whatman No. 1 filter paper	Colorimetric	-	0.52 mg· L^{-1}	-	[45]
Nitrite, Nitrate	Inkjet printing	Whatman No. 1 and No.4 filter papers	Colorimetric	10–150 μM, 50–1000 μM	1.0 μM, 19 μM	Tap water, mineral water, pond water	[53]
Metals							
Pb(II)	Photolithography	Whatman No. 1 chromatography paper	Electrochemical	0–100 ppb	1.0 ppb	-	[47]
Hg(II), Cu(II), Cr(VI), Ni(II)	Wax printing	Whatman No. 1 paper	Colorimetric	-	~0.5–1 ppm	-	[31]
Pb(II), Cd(II)	Cutting	Whatman No. 1 filter paper	Electrochemical	10–100 ppb	2.0 ppb, 2.3 ppb	Carbonated electrolyte drinks	[54]
As(III)	Cutting	Whatman filter paper	Colorimetric	-	1.0 ppb	-	[23]
Pb(II), Cu(II)	Cutting	Whatman filter paper	Colorimetric	-	⩽10.0 ppb for both	-	[24]
Cu(II)	Cutting	Whatman No. 1 filter paper	Colorimetric	7.8–62.8 μM	7.8 nM or 0.5 μg· L^{-1}	Drinking water	[25]
Cu(II)	Wax printing	Whatman No. 1 filter paper	Colorimetric	0.5–200 ng· mL^{-1}	0.3 ng· mL^{-1}	Drinking water, ground water, tomato, rice	[36]
Cu(II)	Inkjet printing	Whatman No. 4 filter paper	Colorimetric	0.1–30.0 mg· L^{-1}	0.6 mg· L^{-1}	Hot tap water	[55]
Hg(II)	Wax screen printing	Whatman No. 1 filter paper	Colorimetric	5–75 ppm	0.12 ppm	Commercial bottled drinking water, tap water	[41]
Hg(II), Ag(I), NEO	Wax printing	Whatman No. 1 chromatography paper	FL	0–3 μM, 0–1.75 μM, 0–2 μM	121 nM, 47 nM, 153 nM	-	[37]

Table A1. *Cont.*

Target	µPAD Wall Fabrication Method	Paper Substrate	Detection Method	Linear Detection Range	LOD	Real Sample Application	Reference
Cu(II), Cr(VI), Ni(II)	Wax patterning	Whatman No. 1 chromatography paper	Colorimetric	-	$\geqslant$0.8 mg·L^{-1}, >0.5 mg·L^{-1}, $\geqslant$0.5 mg·L^{-1}	Tap water	[28]
Fe	Photolithography	Advantec No. 51B chromatography paper	Colorimetric	8.9–89 µM	3.96 µM	Tap water, river water	[48]
Others							
Ethanol	Wax printing	Whatman No. 1 chromatography paper	Electrochemical	0.1–3 mM	0.1 mM	Water	[38]
Phenol, Bisphenol A, Dopamine, Catechol, m-Cresol p-Cresol	Cutting by hole punching	Fisherbrand P5 filter paper	Colorimetric	1–400 µg·L^{-1}, 1–200 µg·L^{-1}, 1–300 µg·L^{-1}, 1–300 µg·L^{-1}, 1–500 µg·L^{-1}, 1–200 µg·L^{-1}	0.86 ($\pm$0.102) µg·L^{-1} for each of the phenolic compounds	Tap water, river water	[59]
Bromide, Iodide, Chloride	-	Whatman RC60 regenerated cellulose membrane filter	Electrochemical	$10^{-4.8}$–0.1 M for bromide and iodide, $10^{-4.8}$–0.6 M for chloride	10^{-5} M	Food supplement, seawater, mineral water, tap water, river water	[56]
Iodate	Wax printing	Ahlstrom 319 paper	Colorimetric	0.8–15 ppm iodine atoms from iodate	0.8 ppm iodine atoms	Iodized salt	[39]
MC-LR	Wax printing	Whatman No. 1 chromatography paper	Electrochemical	0.01–200 µg·mL^{-1}	0.004 µg·mL^{-1}	-	[40]
B[a]P	Wax patterning and screen printing	Whatman No. 1 filter paper	ECL	0.15–12.5 µM	~150 nM	Chicken skin	[29]

References

1. Hamburg, M.A. Advancing regulatory science. *Science* **2011**, *331*, 987. [CrossRef] [PubMed]
2. Escarpa, A. Lights and shadows on food microfluidics. *Lab Chip* **2014**, *14*, 3213–3224. [CrossRef] [PubMed]
3. Jéquier, E.; Constant, F. Water as an essential nutrient: The physiological basis of hydration. *Eur. J. Clin. Nutr.* **2010**, *64*, 115–123. [CrossRef] [PubMed]
4. Sasaki, Y.F.; Kawaguchi, S.; Kamaya, A.; Ohshita, M.; Kabasawa, K.; Iwama, K.; Taniguchi, K.; Tsuda, S. The comet assay with 8 mouse organs: Results with 39 currently used food additives. *Mutat. Res. Toxicol. Environ. Mutagen.* **2002**, *519*, 103–119. [CrossRef]
5. International Agency for Research in Cancer (IARC). 2-(2-Furyl)-3-(5-nitro-2-fyryl)acrylamide (AF-2). In *IARC Monographs on the Evaluation of the Carcinogenic Risk of Chemicals to Humans: Some Food Additives, Feed Additives and Naturally Occurring Substances*; World Health Organization: Lyon, France, 1983; Volume 31, p. 41.
6. Sajid, M.; Kawde, A.-N.; Daud, M. Designs, formats and applications of lateral flow assay: A literature review. *J. Saudi Chem. Soc.* **2014**, *19*, 689–705. [CrossRef]
7. Posthuma-Trumpie, G.A.; Korf, J.; van Amerongen, A. Lateral flow (immuno)assay: Its strengths, weaknesses, opportunities and threats. A literature survey. *Anal. Bioanal. Chem.* **2009**, *393*, 569–582. [CrossRef] [PubMed]
8. Whitesides, G.M. The origins and the future of microfluidics. *Nature* **2006**, *442*, 368–373. [CrossRef] [PubMed]
9. Manz, A.; Harrison, D.J.; Verpoorte, E.M.J.; Fettinger, J.C.; Paulus, A.; Lüdi, H.; Widmer, H.M. Planar chips technology for miniaturization and integration of separation techniques into monitoring systems. *J. Chromatogr. A* **1992**, *593*, 253–258. [CrossRef]
10. Ruano, J.M.; Benoit, V.; Aitchison, J.S.; Cooper, J.M. Flame hydrolysis deposition of glass on silicon for the integration of optical and microfluidic devices. *Anal. Chem.* **2000**, *72*, 1093–1097. [CrossRef] [PubMed]
11. Queste, S.; Salut, R.; Clatot, S.; Rauch, J.-Y.; Khan Malek, C.G. Manufacture of microfluidic glass chips by deep plasma etching, femtosecond laser ablation, and anodic bonding. *Microsyst. Technol.* **2010**, *16*, 1485–1493. [CrossRef]
12. Harris, N.R.; Hill, M.; Beeby, S.; Shen, Y.; White, N.M.; Hawkes, J.J.; Coakley, W.T. A silicon microfluidic ultrasonic separator. *Sens. Actuators B Chem.* **2003**, *95*, 425–434. [CrossRef]
13. Sanjoh, A.; Tsukihara, T. Spatiotemporal protein crystal growth studies using microfluidic silicon devices. *J. Cryst. Growth* **1999**, *196*, 691–702. [CrossRef]
14. McDonald, J.C.; Whitesides, G.M. Poly(dimethylsiloxane) as a material for fabricating microfluidic devices. *Acc. Chem. Res.* **2002**, *35*, 491–499. [CrossRef] [PubMed]
15. Leclerc, E.; Sakai, Y.; Fujii, T. Microfluidic PDMS (polydimethylsiloxane) bioreactor for large-scale culture of hepatocytes. *Biotechnol. Prog.* **2004**, *20*, 750–755. [CrossRef] [PubMed]
16. Pelton, R. Bioactive paper provides a low-cost platform for diagnostics. *TrAC Trends Anal. Chem.* **2009**, *28*, 925–942. [CrossRef]
17. Martinez, A.W.; Phillips, S.T.; Butte, M.J.; Whitesides, G.M. Patterned paper as a platform for inexpensive, low-volume, portable bioassays. *Angew. Chem. Int. Ed.* **2007**, *46*, 1318–1320. [CrossRef] [PubMed]
18. Mohammadi, S.; Maeki, M.; Mohamadi, R.M.; Ishida, A.; Tani, H.; Tokeshi, M. An instrument-free, screen-printed paper microfluidic device that enables bio and chemical sensing. *Analyst* **2015**, *140*, 6493–6499. [CrossRef] [PubMed]
19. Cate, D.M.; Adkins, J.A.; Mettakoonpitak, J.; Henry, C.S. Recent developments in paper-based microfluidic devices. *Anal. Chem.* **2015**, *87*, 19–41. [CrossRef] [PubMed]
20. Liu, W.; Kou, J.; Xing, H.; Li, B. Paper-based chromatographic chemiluminescence chip for the detection of dichlorvos in vegetables. *Biosens. Bioelectron.* **2014**, *52*, 76–81. [CrossRef] [PubMed]
21. Liu, W.; Guo, Y.; Luo, J.; Kou, J.; Zheng, H.; Li, B.; Zhang, Z. A molecularly imprinted polymer based a lab-on-paper chemiluminescence device for the detection of dichlorvos. *Spectrochim. Acta. A Mol. Biomol. Spectrosc.* **2015**, *141*, 51–57. [CrossRef] [PubMed]
22. Zhu, Y.; Zhang, L.; Yang, L. Designing of the functional paper-based surface-enhanced Raman spectroscopy substrates for colorants detection. *Mater. Res. Bull.* **2015**, *63*, 199–204. [CrossRef]
23. Nath, P.; Arun, R.K.; Chanda, N. A paper based microfluidic device for the detection of arsenic using a gold nanosensor. *RSC Adv.* **2014**, *4*, 59558–59561. [CrossRef]

24. Nath, P.; Arun, R.K.; Chanda, N. Smart gold nanosensor for easy sensing of lead and copper ions in solution and using paper strips. *RSC Adv.* **2015**, *5*, 69024–69031. [CrossRef]
25. Ratnarathorn, N.; Chailapakul, O.; Henry, C.S.; Dungchai, W. Simple silver nanoparticle colorimetric sensing for copper by paper-based devices. *Talanta* **2012**, *99*, 552–557. [CrossRef] [PubMed]
26. Hossain, S.M.Z.; Luckham, R.E.; Smith, A.M.; Lebert, J.M.; Davies, L.M.; Pelton, R.H.; Filipe, C.D.M.; Brennan, J.D. Development of a bioactive paper sensor for detection of neurotoxins using piezoelectric inkjet printing of sol-gel-derived bioinks. *Anal. Chem.* **2009**, *81*, 5474–5483. [CrossRef] [PubMed]
27. Hossain, S.M.Z.; Luckham, R.E.; McFadden, M.J.; Brennan, J.D. Reagentless bidirectional lateral flow bioactive paper sensors for detection of pesticides in beverage and food samples. *Anal. Chem.* **2009**, *81*, 9055–9064. [CrossRef] [PubMed]
28. Li, M.; Cao, R.; Nilghaz, A.; Guan, L.; Zhang, X.; Shen, W. "Periodic-table-style" paper device for monitoring heavy metals in water. *Anal. Chem.* **2015**, *87*, 2555–2559. [CrossRef] [PubMed]
29. Mani, V.; Kadimisetty, K.; Malla, S.; Joshi, A.A.; Rusling, J.F. Paper-based electrochemiluminescent screening for genotoxic activity in the environment. *Environ. Sci. Technol.* **2013**, *47*, 1937–1944. [CrossRef] [PubMed]
30. Ma, S.; Tang, Y.; Liu, J.; Wu, J. Visible paper chip immunoassay for rapid determination of bacteria in water distribution system. *Talanta* **2014**, *120*, 135–140. [CrossRef] [PubMed]
31. Hossain, S.M.Z.; Brennan, J.D. β-Galactosidase-based colorimetric paper sensor for determination of heavy metals. *Anal. Chem.* **2011**, *83*, 8772–8778. [CrossRef] [PubMed]
32. Jokerst, J.C.; Adkins, J.A.; Bisha, B.; Mentele, M.M.; Goodridge, L.D.; Henry, C.S. Development of a paper-based analytical device for colorimetric detection of select foodborne pathogens. *Anal. Chem.* **2012**, *84*, 2900–2907. [CrossRef] [PubMed]
33. Jin, S.-Q.; Guo, S.-M.; Zuo, P.; Ye, B.-C. A cost-effective Z-folding controlled liquid handling microfluidic paper analysis device for pathogen detection via ATP quantification. *Biosens. Bioelectron.* **2014**, *63*, 379–383. [CrossRef] [PubMed]
34. Sicard, C.; Glen, C.; Aubie, B.; Wallace, D.; Jahanshahi-Anbuhi, S.; Pennings, K.; Daigger, G.T.; Pelton, R.; Brennan, J.D.; Filipe, C.D.M. Tools for water quality monitoring and mapping using paper-based sensors and cell phones. *Water Res.* **2015**, *70*, 360–369. [CrossRef]
35. Adkins, J.A.; Henry, C.S. Electrochemical detection in paper-based analytical devices using microwire electrodes. *Anal. Chim. Acta* **2015**, *891*, 247–254. [CrossRef] [PubMed]
36. Chaiyo, S.; Siangproh, W.; Apilux, A.; Chailapakul, O. Highly selective and sensitive paper-based colorimetric sensor using thiosulfate catalytic etching of silver nanoplates for trace determination of copper ions. *Anal. Chim. Acta* **2015**, *866*, 75–83. [CrossRef] [PubMed]
37. Zhang, Y.; Zuo, P.; Ye, B.-C. A low-cost and simple paper-based microfluidic device for simultaneous multiplex determination of different types of chemical contaminants in food. *Biosens. Bioelectron.* **2015**, *68*, 14–19. [CrossRef] [PubMed]
38. Nie, Z.; Deiss, F.; Liu, X.; Akbulut, O.; Whitesides, G.M. Integration of paper-based microfluidic devices with commercial electrochemical readers. *Lab Chip* **2010**, *10*, 3163–3169. [CrossRef] [PubMed]
39. Myers, N.M.; Kernisan, E.N.; Lieberman, M. Lab on paper: Iodometric titration on a printed card. *Anal. Chem.* **2015**, *87*, 3764–3770. [CrossRef] [PubMed]
40. Ge, S.; Liu, W.; Ge, L.; Yan, M.; Yan, J.; Huang, J.; Yu, J. *In situ* assembly of porous Au-paper electrode and functionalization of magnetic silica nanoparticles with HRP via click chemistry for Microcystin-LR immunoassay. *Biosens. Bioelectron.* **2013**, *49*, 111–117. [CrossRef] [PubMed]
41. Apilux, A.; Siangproh, W.; Praphairaksit, N.; Chailapakul, O. Simple and rapid colorimetric detection of Hg(II) by a paper-based device using silver nanoplates. *Talanta* **2012**, *97*, 388–394. [CrossRef] [PubMed]
42. Nouanthavong, S.; Nacapricha, D.; Henry, C.; Sameenoi, Y. Pesticide analysis using nanoceria-coated paper-based devices as a detection platform. *Analyst* **2016**, *141*, 1837–1846. [CrossRef] [PubMed]
43. Colletes, T.C.; Garcia, P.T.; Campanha, R.B.; Abdelnur, P.V.; Romão, W.; Coltro, W.K.T.; Vaz, B.G. A new insert sample approach to paper spray mass spectrometry: Paper substrate with paraffin barriers. *Analyst* **2016**, *141*, 1707–1713. [CrossRef] [PubMed]
44. Cardoso, T.M.G.; Garcia, P.T.; Coltro, W.K.T. Colorimetric determination of nitrite in clinical, food and environmental samples using microfluidic devices stamped in paper platforms. *Anal. Methods* **2015**, *7*, 7311–7317. [CrossRef]

45. Lopez-Ruiz, N.; Curto, V.F.; Erenas, M.M.; Benito-Lopez, F.; Diamond, D.; Palma, A.J.; Capitan-Vallvey, L.F. Smartphone-based simultaneous pH and nitrite colorimetric determination for paper microfluidic devices. *Anal. Chem.* **2014**, *86*, 9554–9562. [CrossRef] [PubMed]
46. Park, T.S.; Li, W.; McCracken, K.E.; Yoon, J.-Y. Smartphone quantifies Salmonella from paper microfluidics. *Lab Chip* **2013**, *13*, 4832–4840. [CrossRef] [PubMed]
47. Nie, Z.; Nijhuis, C.A.; Gong, J.; Chen, X.; Kumachev, A.; Martinez, A.W.; Narovlyansky, M.; Whitesides, G.M. Electrochemical sensing in paper-based microfluidic devices. *Lab Chip* **2010**, *10*, 477–483. [CrossRef] [PubMed]
48. Asano, H.; Shiraishi, Y. Development of paper-based microfluidic analytical device for iron assay using photomask printed with 3D printer for fabrication of hydrophilic and hydrophobic zones on paper by photolithography. *Anal. Chim. Acta* **2015**, *883*, 55–60. [CrossRef] [PubMed]
49. Wang, S.; Ge, L.; Li, L.; Yan, M.; Ge, S.; Yu, J. Molecularly imprinted polymer grafted paper-based multi-disk micro-disk plate for chemiluminescence detection of pesticide. *Biosens. Bioelectron.* **2013**, *50*, 262–268. [CrossRef] [PubMed]
50. Sun, G.; Wang, P.; Ge, S.; Ge, L.; Yu, J.; Yan, M. Photoelectrochemical sensor for pentachlorophenol on microfluidic paper-based analytical device based on the molecular imprinting technique. *Biosens. Bioelectron.* **2014**, *56*, 97–103. [CrossRef] [PubMed]
51. Kuek Lawrence, C.S.; Tan, S.N.; Floresca, C.Z. A "green" cellulose paper based glucose amperometric biosensor. *Sens. Actuators B Chem.* **2014**, *193*, 536–541. [CrossRef]
52. He, Q.; Ma, C.; Hu, X.; Chen, H. Method for fabrication of paper-based microfluidic devices by alkylsilane self-assembling and UV/O3-patterning. *Anal. Chem.* **2013**, *85*, 1327–1331. [CrossRef] [PubMed]
53. Jayawardane, B.M.; Wei, S.; McKelvie, I.D.; Kolev, S.D. Microfluidic paper-based analytical device for the determination of nitrite and nitrate. *Anal. Chem.* **2014**, *86*, 7274–7279. [CrossRef] [PubMed]
54. Shi, J.; Tang, F.; Xing, H.; Zheng, H.; Lianhua, B.; Wei, W. Electrochemical detection of Pb and Cd in paper-based microfluidic devices. *J. Braz. Chem. Soc.* **2012**, *23*, 1124–1130. [CrossRef]
55. Jayawardane, B.M.; Coo, L.dlC.; Cattrall, R.W.; Kolev, S.D. The use of a polymer inclusion membrane in a paper-based sensor for the selective determination of Cu(II). *Anal. Chim. Acta* **2013**, *803*, 106–112. [CrossRef] [PubMed]
56. Cuartero, M.; Crespo, G.A.; Bakker, E. Paper-based thin-layer coulometric sensor for halide determination. *Anal. Chem.* **2015**, *87*, 1981–1990. [CrossRef] [PubMed]
57. Burnham, S.; Hu, J.; Anany, H.; Brovko, L.; Deiss, F.; Derda, R.; Griffiths, M.W. Towards rapid on-site phage-mediated detection of generic Escherichia coli in water using luminescent and visual readout. *Anal. Bioanal. Chem.* **2014**, *406*, 5685–5693. [CrossRef] [PubMed]
58. Badawy, M.E.I.; El-Aswad, A.F. Bioactive paper sensor based on the acetylcholinesterase for the rapid detection of organophosphate and carbamate pesticides. *Int. J. Anal. Chem.* **2014**, *2014*, 536823. [CrossRef] [PubMed]
59. Alkasir, R.S.J.; Ornatska, M.; Andreescu, S. Colorimetric paper bioassay for the detection of phenolic compounds. *Anal. Chem.* **2012**, *84*, 9729–9737. [CrossRef] [PubMed]
60. Zuo, P.; Li, X.; Dominguez, D.C.; Ye, B.-C. A PDMS/paper/glass hybrid microfluidic biochip integrated with aptamer-functionalized graphene oxide nano-biosensors for one-step multiplexed pathogen detection. *Lab Chip* **2013**, *13*, 3921–3928. [CrossRef] [PubMed]
61. Gilbert-López, B.; García-Reyes, J.F.; Molina-Díaz, A. Sample treatment and determination of pesticide residues in fatty vegetable matrices: A review. *Talanta* **2009**, *79*, 109–128. [CrossRef] [PubMed]
62. Wilson, C.; Tisdell, C. Why farmers continue to use pesticides despite environmental, health and sustainability costs. *Ecol. Econ.* **2001**, *39*, 449–462. [CrossRef]
63. Pundir, C.S.; Chauhan, N. Acetylcholinesterase inhibition-based biosensors for pesticide determination: A review. *Anal. Biochem.* **2012**, *429*, 19–31. [CrossRef] [PubMed]
64. Dai, M.; Copley, S.D. Genome shuffling improves degradation of the anthropogenic pesticide pentachlorophenol by sphingobium chlorophenolicum ATCC 39723. *Appl. Environ. Microbiol.* **2004**, *70*, 2391–2397. [CrossRef] [PubMed]
65. Fuentes, M.S.; Briceño, G.E.; Saez, J.M.; Benimeli, C.S.; Diez, M.C.; Amoroso, M.J. Enhanced removal of a pesticides mixture by single cultures and consortia of free and immobilized streptomyces strains. *Biomed Res. Int.* **2013**, *2013*, 392573. [CrossRef] [PubMed]

66. Nesakumar, N.; Ramachandra, B.L.; Sethuraman, S.; Krishnan, U.M.; Rayappan, J.B.B. Evaluation of inhibition efficiency for the detection of captan, 2,3,7,8-Tetrachlorodibenzodioxin, pentachlorophenol and carbosulfan in water: An electrochemical approach. *Bull. Environ. Contam. Toxicol.* **2016**, *96*, 217–223. [CrossRef] [PubMed]
67. Su, Y.; Ma, S.; Jiang, K.; Han, X. CdTe-paper-based Visual Sensor for Detecting Methyl Viologen. *Chin. J. Chem.* **2015**, *33*, 446–450. [CrossRef]
68. Liu, J.; Wang, H.; Manicke, N.E.; Lin, J.-M.; Cooks, R.G.; Ouyang, Z. Development, characterization, and application of paper spray ionization. *Anal. Chem.* **2010**, *82*, 2463–2471. [CrossRef] [PubMed]
69. Honikel, K.-O. The use and control of nitrate and nitrite for the processing of meat products. *Meat Sci.* **2008**, *78*, 68–76. [CrossRef] [PubMed]
70. Zhang, H.; Qi, S.; Dong, Y.; Chen, X.; Xu, Y.; Ma, Y.; Chen, X. A sensitive colorimetric method for the determination of nitrite in water supplies, meat and dairy products using ionic liquid-modified methyl red as a colour reagent. *Food Chem.* **2014**, *151*, 429–434. [CrossRef] [PubMed]
71. Kull, T.P.J.; Backlund, P.H.; Karlsson, K.M.; Meriluoto, J.A.O. Oxidation of the cyanobacterial hepatotoxin microcystin-LR by chlorine dioxide: Reaction kinetics, characterization, and toxicity of reaction products. *Environ. Sci. Technol.* **2004**, *38*, 6025–6031. [CrossRef] [PubMed]
72. Ding, Y.; Mutharasan, R. Highly sensitive and rapid detection of microcystin-LR in source and finished water samples using cantilever sensors. *Environ. Sci. Technol.* **2011**, *45*, 1490–1496. [CrossRef] [PubMed]

10

Magnetic Particle Plug-based Assays for Biomarker Analysis

Chayakom Phurimsak, Mark D. Tarn and Nicole Pamme *

Department of Chemistry, University of Hull, Cottingham Road, Hull, HU6 7RX, UK; chayakom.p@rmutsb.ac.th (C.P.); m.tarn@hull.ac.uk (M.D.T.)

* Correspondence: n.pamme@hull.ac.uk

Academic Editors: Manabu Tokeshi and Kiichi Sato

Abstract: Conventional immunoassays offer selective and quantitative detection of a number of biomarkers, but are laborious and time-consuming. Magnetic particle-based assays allow easy and rapid selection of analytes, but still suffer from the requirement of tedious multiple reaction and washing steps. Here, we demonstrate the trapping of functionalised magnetic particles within a microchannel for performing rapid immunoassays by flushing consecutive reagent and washing solutions over the trapped particle plug. Three main studies were performed to investigate the potential of the platform for quantitative analysis of biomarkers: (i) a streptavidin-biotin binding assay; (ii) a sandwich assay of the inflammation biomarker, C-reactive protein (CRP); and (iii) detection of the steroid hormone, progesterone (P4), towards a competitive assay. Quantitative analysis with low limits of detection was demonstrated with streptavidin-biotin, while the CRP and P4 assays exhibited the ability to detect clinically relevant analytes, and all assays were completed in only 15 min. These preliminary results show the great potential of the platform for performing rapid, low volume magnetic particle plug-based assays of a range of clinical biomarkers via an exceedingly simple technique.

Keywords: C-reactive protein (CRP); progesterone (P4); immunoassays; magnetic particles; magnetism; microfluidics; particle trapping

---

1. Introduction

Enzyme-linked immunosorbent assays (ELISA) are a powerful method of identification and quantification, utilising the specificity of labelled antibodies for their complementary antigens to give a signal (e.g., via fluorescence or chemiluminescence) dependent on the concentration of the latter [1,2]. However, while ELISA offers extremely low limits of detection and selectivity, the process is exceedingly slow, requiring multiple reagent and washing steps that are both laborious and time-consuming. The use of magnetic microparticles as solid supports has become incredibly popular for immunoassays and other applications thanks to their high surface-to-volume ratios, small sizes (0.1–100 μm), the range of functional groups that can be attached to the surfaces (e.g., antibodies, DNA, and chemical groups), and the ability to easily manipulate the particles via an applied magnetic field [3,4]. By employing antibody functionalised magnetic particles, immunoassay time frames can be greatly reduced, with permanent magnets used to enable the separation of antigens from the sample and speeding up the exchange of reaction and washing solutions. Even so, these magnetic particle-based assays still require multiple manual solution changes; hence, despite being faster than conventional ELISAs, they are still somewhat slow and require relatively large volumes of solutions.

The application of microfluidic devices [5–7], having channel networks with typical dimensions on the order of 1–100 s of micrometres, provides a number of advantages to immunoassays by reducing diffusion distances, reaction and washing time frames, as well as sample and reagent volumes [8–12].

Integration of magnetic particles with microfluidics thus combines the benefits of both [13–16], and has yielded great success for on-chip bioanalysis [12,16]. One of the easiest methods of performing techniques such as immunoassays is via the trapping of functionalised microparticles within the microchannel [17,18], before pumping solutions of sample, washing buffer and reagents over the particles. Trapping magnetic particles within a microchannel can easily be achieved by simply applying an external magnetic field that can be generated via a number of sources, including permanent magnets [19–32], integrated microelectromagnets [33–38], electromagnets [39–43], and externally magnetisable integrated microstructures [44–49].

Most often, this type of setup is used for the separation of target analytes from a sample; as the sample is pumped over the trapped magnetic particles the analytes bind to the functional groups on the particles, after which the particles are washed with buffer solution [13].

However, immunoassays are performed by consecutively flushing sample, reagent and washing solutions over antibody-coated particles, thereby enabling selection of a target analyte from a sample and its subsequent labelling (e.g., with a fluorescent tag) for detection. Such processes have been applied to assays for streptavidin-biotin [38,50], protein A [31,50], mouse IgG [26,29,51], parathyroid hormone [52], interleukin-5 [52], bovine serum albumin (BSA) [42], alkaline phosphatase [53], and glycine [50]. Modifications to this methodology have included the use of segmented flow, in which the consecutive reaction and washing solutions are contained within droplets that are pumped over the trapped particles [53], and the generation of a fluidised bed of magnetic particles to enhance mixing [29].

However, while modifications such as these can yield low detection limits with small sample volumes, it often comes at the cost of greater complexity. Further applications beyond immunoassays have included RNA isolation [54,55], DNA hybridisation [56–58] and separation [59], purification of polymerase chain reaction (PCR) products for gene synthesis [60], cell capture for DNA detection [61,62], reaction rate measurements [63], protein digestion [28,64–66], and the electrochemical detection of peroxide [48], among others.

Previously, we have demonstrated the trapping of plugs of magnetic particles in microchannels for performing simultaneous assays on particles featuring different surface functionalities [50]. In order to maintain a simple setup and user-friendliness, the apparatus consists only of a capillary placed between two permanent magnets and connected to a single syringe pump operating in withdrawal mode.

Functionalised magnetic particles are first pumped into the microchannel and trapped between the magnets, creating a particle plug that is consecutively exposed to reagent and washing solutions prior to detection of the target analyte using fluorescence (Figure 1). By placing multiple pairs of magnets upstream of each other, three different particle plugs (featuring glycine, protein A, and streptavidin surface groups) were generated for the simultaneous assays.

We have also demonstrated how diamagnetic repulsion forces can be employed for performing particle plug-based assays [67,68]. The proof-of-principle work thus far has involved only qualitative assays to test the platform. Here, we investigate the potential for using this simple platform for quantitative analysis towards its application in clinical diagnostics.

Three main approaches are described here: (i) the ability to generate a calibration curve and obtain a limit of detection for a streptavidin-biotin binding assay; (ii) the detection in a relevant concentration range of an inflammation and infection biomarker, C-reactive protein (CRP), via a sandwich immunoassay; and (iii) the detection of a clinically relevant steroid hormone, progesterone (P4), at multiple concentrations with a view to competitive assays.

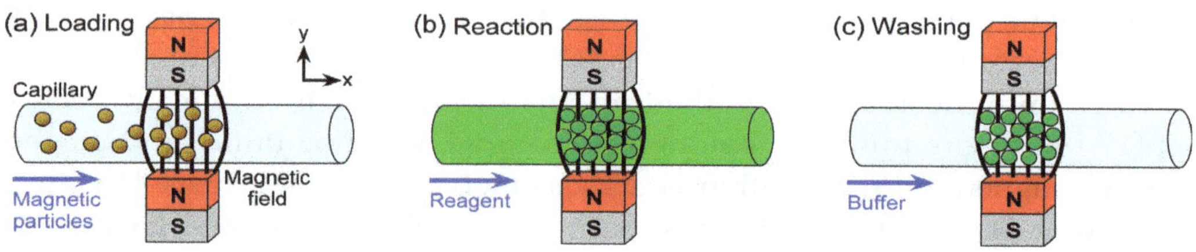

Figure 1. Principle of magnetic particle plug-based assays: (**a**) functionalised magnetic particles are introduced into a microchannel and trapped between two magnets, forming a plug; (**b**) a fluorescently labelled reagent or sample solution is flushed over the particle plug, with the reagent or target analyte binding to the particles; and (**c**) the microchannel is washed with buffer solution, allowing fluorescence detection of the trapped particle plug.

2. Materials and Methods

2.1. Reagents and Particles

Tris(hydroxymethyl)aminomethane (Tris), 2-(*N*-morpholino)ethanesulfonic acid (MES), *N*-(3-dimethylaminopropyl)-*N*′-ethylcarbodiimide hydrochloride (EDC), *N*-hydroxysuccinimide (NHS), and bovine serum albumin (BSA) were purchased from Sigma-Aldrich (Dorset, UK).

Superparamagnetic particles with a 2.8 μm diameter were purchased from Invitrogen (Paisley, UK) with two different surface functionalities: streptavidin (Dynabeads M-270 Streptavidin) and carboxylic acid (Dynabeads M-270 Carboxylic Acid). Biotin-4-fluorescein (λ_{ex} = 494 nm, λ_{em} = 524 nm) and phosphate buffered saline (PBS) tablets were also purchased from Invitrogen.

Recombinant human C-reactive protein (CRP) and primary CRP antibody (1° anti-CRP; biotinylated mouse anti-human C-reactive protein) were purchased from R&D Systems (Abington, UK). Secondary CRP antibody tagged with a fluorescent label (2° anti-CRP-FITC; polyclonal goat anti-human C-reactive protein conjugated to fluorescein isothiocyanate, λ_{ex} = 495 nm, λ_{em} = 521 nm) was purchased from Abcam (Cambridge, UK) in PBS solution at a stock concentration of 1 mg· mL^{-1}. Progesterone labelled with fluorescein isothiocyanate (P4-FITC, 1 mg· mL^{-1} stock solution) and progesterone antibody (anti-P4) were purchased from R&D Systems.

2.2. Preparation of Solutions

All solutions were prepared in double-filtered (0.05 μm) high purity water (18.2 MΩ· cm at 25 °C) via an ELGA Option 4 system that fed into an ELGA UHG PS system, both of which were from ELGA Process Water (Marlow, UK).

PBS solution (pH 7.45) was prepared by dissolving a tablet in 1000 mL water, and had BSA added to a concentration of 0.01% *w/v* in order to reduce non-specific binding of reagents and the sticking of magnetic particles to the capillary walls or to each other. Tris buffer (20 mM, pH 8) was prepared by dissolving tris(hydroxymethyl)aminomethane in water, with 0.1% *w/v* BSA added. MES buffer (pH 5) was prepared to a concentration of 25 mM in water.

Fluorescently labelled biotin (biotin-4-fluorescein) was dissolved in PBS solution to a stock concentration of 1 mg· mL^{-1} and protected from light by wrapping the container in aluminium foil. CRP antigen was reconstituted in Tris buffer to a concentration of 200 μg· mL^{-1}, as per the manufacturer's instructions, then diluted in PBS solution to concentrations of 1 μg· mL^{-1} and 10 μg· mL^{-1}. Primary CRP antibody (1° anti-CRP) was reconstituted in PBS solution to a concentration of 50 μg· mL^{-1}, as per the manufacturer's instructions, and then further diluted in PBS to 1 μg· mL^{-1}. Secondary CRP antibody (2° anti-CRP-FITC) was diluted in PBS to a concentration of 100 μg· mL^{-1}. Progesterone antibody (anti-P4) was dissolved in MES buffer (25 mM, pH 5) to a concentration of 1 μg· mL^{-1}, while fluorescently labelled progesterone (P4-FITC) was diluted in PBS solution to concentrations of 0.1–100 μg· mL^{-1}.

2.3. Preparation of Anti-CRP Functionalised Magnetic Particles

Immobilisation of biotinylated primary CRP antibodies (1° anti-CRP) onto streptavidin functionalised magnetic particles (Dynabeads M-270 Streptavidin) was achieved via the streptavidin-biotin interaction, as previously reported [69,70]. Briefly, 10 μL of stock particle suspension (6.5×10^8 particles$\cdot$ mL^{-1}) was added to a 1.5 mL microcentrifuge tube (VWR, Leicester, UK), followed by 200 μL of 1° anti-CRP solution at a concentration of 10 μg$\cdot$ mL^{-1}, and incubated for 15 min with slow tilt rotation in order for the biotinylated antibodies to bind to the streptavidin-coated particles. The particles were then washed three times using the following procedure.

The particles were pulled to the side of the tube via an external magnet and the supernatant removed using a pipette. PBS solution (1000 μL) was added to the tube, which was vortexed for 20 s to resuspend the particles. This washing process was repeated twice more, and the particles finally resuspended in PBS buffer solution.

2.4. Preparation of Anti-P4 Functionalised Magnetic Particles

Immobilisation of progesterone antibody (anti-P4) onto carboxylic acid functionalised magnetic particles (Dynabeads M-270 Carboxylic Acid) was achieved via amide bond formation between the carboxylic acid groups of the particles and the primary amine groups of the antibodies. The procedure was performed as per the manufacturer's instructions for a two-step coating procedure [71]. The first step of the process involved the "activation" of the magnetic particles with a carbodiimide (EDC) and *N*-hydroxysuccinimide (NHS).

One hundred microlitres of stock particle suspension (2×10^9 particles$\cdot$ mL^{-1}) was added to a microcentrifuge tube and washed twice, as described previously, with 100 μL of MES buffer (25 mM, pH 5). Immediately prior to use, a 50 mg$\cdot$ mL^{-1} solution of EDC was prepared in cold MES buffer, while a 50 mg$\cdot$ mL^{-1} solution of NHS was also prepared in MES buffer. The supernatant of the particle suspension was removed, and 50 μL of EDC solution and 50 μL of NHS solution were added to the magnetic particles. The suspension was mixed via a vortexer and allowed to incubate with slow tilt rotation at room temperature for 30 min.

Following incubation, the particles were washed twice with 100 μL of MES buffer. The supernatant was removed, and 60 μL of anti-P4 (1 μg$\cdot$ mL^{-1}) in MES buffer was added, followed by a further 40 μL of MES buffer. The mixture was incubated for 2 h at 20 °C, then the particles were washed four times with PBS solution (pH 7.45) and finally resuspended in PBS solution.

2.5. Instrumental Setup

Two rectangular neodymium-iron-boron (NdFeB) magnets ($4 \times 4 \times 6$ mm^3, Magnet Sales, Swindon, UK) were glued onto a glass microscope slide (6×2.5 cm^2) using Araldite Rapid epoxy resin (RS Components, Northants, UK), such that their opposing poles were facing and there was a 1 mm gap between them. A 10 cm long piece of fused silica capillary (150 μm ID (Inner Diameter), 363 μm OD (Outer Diameter), CM Scientific, Silsden, UK) had a section of its polyimide coating burned away with a lighter and wiped with a soft tissue to create a region for visualisation of trapped particles. The capillary was then placed between the pair of magnets (Figure 2a) and held in place using Blu-Tack (Bostick, UK). The two ends of the capillary were connected to Tygon tubing (254 μm ID, 762 μm OD, Cole-Parmer, London, UK), with one piece of tubing interfaced to a syringe on a syringe pump (PHD 22/2000, Harvard Apparatus, Kent, UK) and the other piece of tubing dipped into a microcentrifuge tube, acting as a reservoir, containing sample or buffer solution.

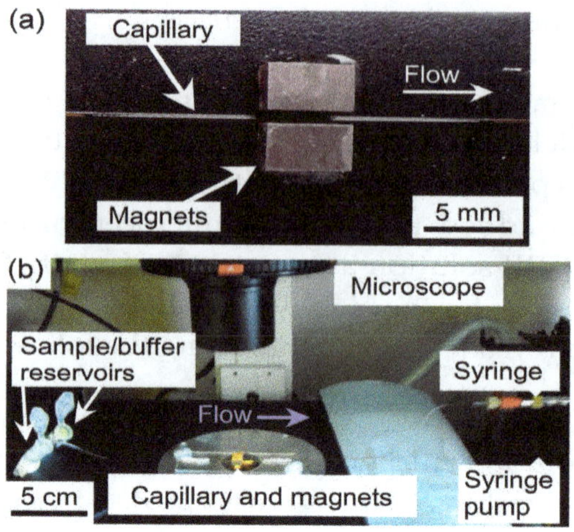

Figure 2. Setup of the microfluidic device: (**a**) photograph of a fused silica capillary located in the 1 mm gap between two 4 × 4 × 6 mm^3 NdFeB magnets that were fixed to a glass microscope slide; and (**b**) photograph of the glass microscope slide, holding the capillary and magnets, on the sample stage of an inverted fluorescence microscope. Samples, reagents and buffer solutions were introduced into the capillary from reservoirs via a syringe pump in withdrawal mode.

Solutions were drawn through the capillary from the sample/buffer reservoir via negative pressure from the syringe pump operating in withdrawal mode. The glass slide holding the magnets and capillary setup was situated on the sample stage of an inverted fluorescence microscope (TE-2000U, Nikon, Surrey, UK) (Figure 2b). Images were captured via a cooled CCD (charge-coupled device) camera (QImaging Retiga-EXL, Media Cybernetics, Buckinghamshire, UK) and Image-Pro Plus 6 software (Media Cybernetics, Buckinghamshire, UK). Such images were analysed using ImageJ software (US National Institutes of Health, Bethesda, MD, USA).

2.6. Experimental Procedures

2.6.1. Capillary-Based Particle Trapping and Reactions

Prior to performing an experiment, the capillary was cleaned and pre-treated by flushing consecutively with ethanol, water and PBS solution. Following this, the inlet tubing connected to the capillary was dipped into a suspension of magnetic particles in a microcentrifuge tube and negative pressure applied via the syringe pump to draw the particle suspension through the capillary. After a certain time frame, the syringe pump was stopped and the flow allowed to come to a halt (~30 s in order to prevent air from entering the system during solution exchange) before the inlet tubing was removed from the particle suspension vial and placed into a vial of PBS solution. PBS was then drawn through the capillary for several minutes to ensure that all particles within the capillary would reach the region between the two magnets to form a plug of particles. Characterisation of the particle plugs was performed at this stage by taking photographs via the microscope and CCD camera, and analysing the images with ImageJ software to determine the area of the plugs.

To perform a reaction on the particle plug, once the particle suspension had been introduced into the capillary, the microcentrifuge tube was exchanged for one containing a reagent solution, which was pumped through the capillary for several minutes such that it was allowed to wash over the particle plug, before the pump was again stopped. When the flow had stopped, the inlet tubing was placed in a vial of PBS solution, which was pumped over the particle plug to wash away any unbound material. The CRP assay, being a two-step sandwich assay, required a second reaction following the first. Finally, fluorescence images were taken of the particle plug and the fluorescence intensity of the plugs measured via ImageJ. Analysis was performed manually by drawing a small box inside the image of the particle plug, determining the maximum greyscale value (as a measure of fluorescence intensity), and deducting an average background intensity. This process was repeated for several regions inside the particle plug to provide average of the maximum fluorescence intensities.

2.6.2. Formation and Characterisation of Magnetic Particle Plugs

Prior to performing reactions, the formation of the particle plugs was characterised based on the applied flow rate and the particle concentration. A suspension of Dynabeads M-270 Carboxylic Acid particles in PBS buffer was pumped into the capillary for 90 s, then the solution swapped to PBS which was pumped through the capillary for a further 10 min. Images of the forming particle plug were collected every minute and experiments were repeated three times. Flow rates of 180–300 $\mu L \cdot h^{-1}$ (equivalent to linear velocities of 2.8–4.7 $mm \cdot s^{-1}$) and particle concentrations of 1×10^6 to 2×10^7 particles$\cdot mL^{-1}$ were studied.

2.6.3. Streptavidin-Biotin Assay

In order to test and optimise reactions on the setup, a streptavidin-biotin binding assay was investigated. Streptavidin functionalised magnetic particles (Dynabeads M-270 Streptavidin) in PBS solution (1×10^7 particles$\cdot mL^{-1}$) were pumped through the capillary for 2 min at a flow rate of 300 $\mu L \cdot h^{-1}$ (4.7 $mm \cdot s^{-1}$) to form a particle plug between the NdFeB magnets. The particle suspension vial was exchanged for one containing biotin-4-fluorescein solution, which was flushed over the trapped particle plug at 300 $\mu L \cdot h^{-1}$ for 3 min, then the sample vial was exchanged again for PBS solution. The PBS solution was drawn through the capillary for 3 min at 300 $\mu L \cdot h^{-1}$ to wash the particle plug, whose fluorescence intensity was then measured. The concentration of biotin-4-fluorescein was varied between 0.1–5 $\mu g \cdot mL^{-1}$. The effect of exposure time during the capture of fluorescence images was also studied using the streptavidin-biotin reaction for optimisation.

2.6.4. C-Reactive Protein (CRP) Assay

Magnetic particles featuring surface-bound primary CRP antibodies (1° anti-CRP), prepared as described in Section 2.3, in PBS solution (1×10^7 particles$\cdot mL^{-1}$) were introduced into the capillary at a flow rate of 300 $\mu L \cdot h^{-1}$ (4.7 $mm \cdot s^{-1}$) for 2 min in order to form the particle plug between the two magnets. The particle suspension tube was replaced with a tube containing CRP solution (1 or 10 $\mu g \cdot mL^{-1}$), which was pumped over the particle plug for 3 min at the same flow rate. Fluorescently tagged secondary antibody (2° anti-CRP-FITC) solution (100 $\mu g \cdot mL^{-1}$) was then flushed over the particle plug for 3 min at 300 $\mu L \cdot h^{-1}$, and the plug was finally washed with PBS solution for 5 min prior to fluorescence measurement of the particles.

2.6.5. Progesterone (P4) Assay

Magnetic particles functionalised with progesterone antibody (anti-P4), prepared as described in Section 2.4, in PBS solution (1×10^7 particles$\cdot mL^{-1}$) were pumped into the capillary at a flow rate of 300 $\mu L \cdot h^{-1}$ (4.7 $mm \cdot s^{-1}$) for 2 min for plug formation between the magnets. A solution of fluorescently labelled progesterone (P4-FITC), whose concentration was varied from 0.1–100 $\mu g \cdot mL^{-1}$, was subsequently flushed over the trapped plug at 300 $\mu L \cdot h^{-1}$ for 3 min, before washing the plug with PBS solution for 3 min to remove any unbound P4-FITC. Fluorescence analysis was then performed on the trapped magnetic particle plug.

3. Results and Discussion

3.1. Formation and Characterisation of Magnetic Particle Plugs

When particle suspensions were introduced into the fused silica capillary via negative pressure, they flowed freely through the tube until they approached the two magnets. At this point, they became trapped in the field between the magnets and began to form a plug of particles that grew larger as particles continued to be introduced into the 150 μm ID capillary. The particles remained stationary as they were trapped in the field, as opposed to the continuously recirculating plugs observed when diamagnetic particles are trapped in a magnetic fluid [67,68]. The location of particle plug formation was $x \approx 1.7$ mm from the centre ($x = 0$ mm) of the magnets (Figure 3a). In order to investigate this further, the magnetic field was modelled in FEMM 4.2 software (Figure 3b). The resultant simulations showed that the region of highest magnetic flux density (vectors not shown) was located between $x \approx -1.5$ mm and $x \approx 1.5$ mm (Figure 3c), hence the location of particle trapping was as predicted.

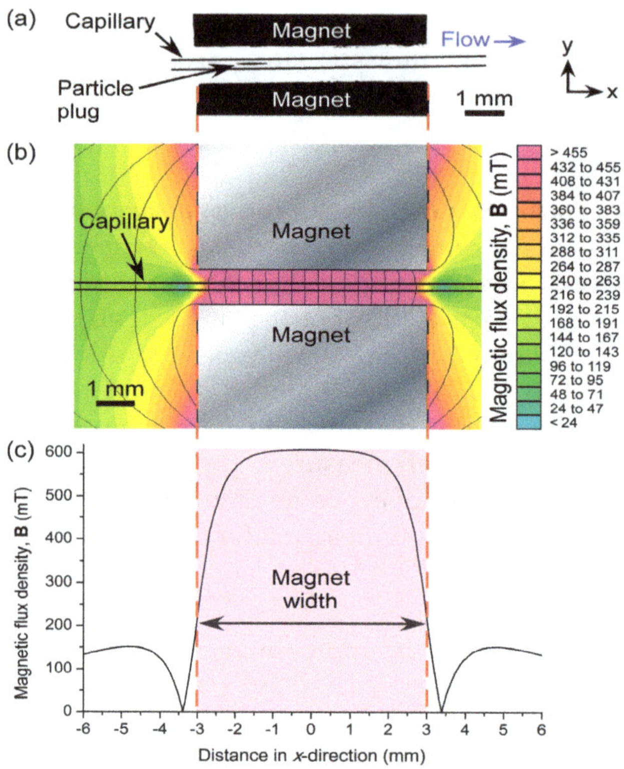

Figure 3. (**a**) Photograph of a plug of magnetic particles trapped between two NdFeB magnets in a capillary. (**b**) Simulation of the magnetic flux density (**B**) across the microfluidic channel, modelled using FEMM software. (**c**) Plot of the magnetic flux density along the length of the capillary (x-direction) between the two magnets.

As stated above, the particles were able to flow freely through the capillary. No evidence of sedimentation due to gravity was observed, with the theoretical forces on the particles due to gravity being calculated as 68 femtonewtons (fN), yielding a sedimentation velocity of 2.6 $\mu m \cdot s^{-1}$. This value was negligible compared to the minimal linear flow rate of 2.8 $mm \cdot s^{-1}$ (at a volumetric flow rate of 180 $\mu L \cdot h^{-1}$) in the capillary. Furthermore, inertial lift forces may also have helped to prevent particles from settling against the capillary wall while in flow. These observations also supported those from our previous work [50,67,68]. The sticking of particles to the capillary walls was minimal, with BSA added to solutions to prevent this from occurring, and was found not to interfere with experiments.

Prior to performing assays on trapped particles, two parameters affecting plug formation were investigated, namely the applied flow rate and the particle concentration, towards optimisation in terms of rapid formation of a plug deemed large enough for yielding suitable fluorescence signals. Carboxylic acid functionalised magnetic particles (2.8 μm) in PBS solution (pH 7.45) were employed for these tests.

3.1.1. Effect of Flow Rate

The applied flow rate is an important parameter since high flow rates would allow the collection of more particles in the trap in a shorter period of time. However, too high a flow rate could lead to particles escaping the trap when the hydrodynamic forces dominate the magnetic forces. Here, particle suspension (1×10^6 particles$\cdot$ mL^{-1}) was introduced into the capillary at a flow rate of 5 µL$\cdot$ min^{-1} for 90 s. The sample vial was then exchanged for one containing PBS solution, which was pumped through the capillary at flow rates of 180, 240 and 300 µL$\cdot$ h^{-1} (equivalent to linear velocities of 2.8, 3.8 and 4.7 mm$\cdot$ s^{-1}, respectively) in order to determine the effect of flow rate on plug formation. Images of the plugs were captured every minute for 10 min.

The build-up of the particle plugs can be seen in Figure 4a–c, which shows the size of the plugs at time frames of 1 min, 5 min, and 10 min after initialising the flow. The total area occupied by a particle plug (units of pixels^2, with one square pixel comprising approximately 5.6 µm^2) was measured using ImageJ and the results are plotted in Figure 4d, which clearly demonstrates that faster flow rates yielded larger plugs of trapped particles. Interestingly, however, it appeared that following the first minute of trapping, the plugs actually grew at similar rates at each of the three flow rates. This may have been due to the build-up of the plugs in three dimensions within the capillary while only a 2D image could be taken for analysis, a parameter that could be explored in future work. Importantly, even at the highest flow rate tested of 300 µL$\cdot$ h^{-1} (4.7 mm$\cdot$ s^{-1}), 100% trapping efficiency was achieved. Hence, this flow rate was employed in all subsequent experiments to yield the rapid formation of large plugs with no loss of particles.

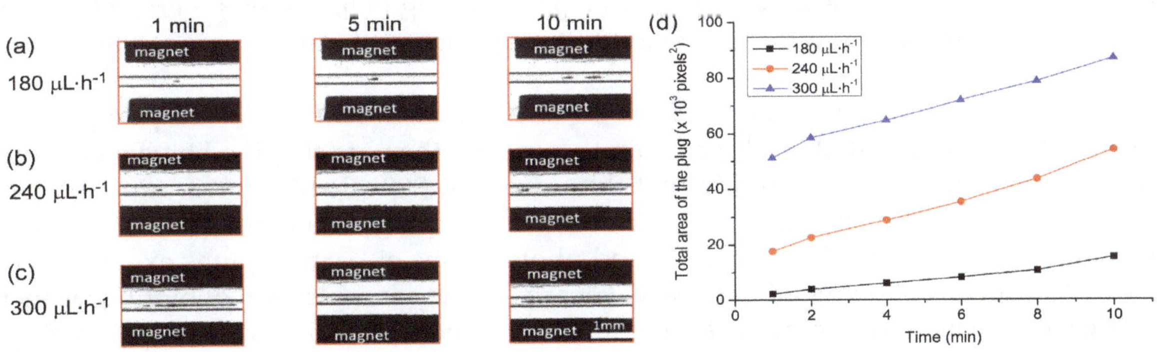

Figure 4. The effect of flow rate on magnetic particle plug formation. (**a–c**) Photographs of plug formation at time points of 1, 5 and 10 min for flow rates of: (**a**) 180 µL$\cdot$ h^{-1}; (**b**) 240 µL$\cdot$ h^{-1}; and (**c**) 300 µL$\cdot$ h^{-1}. (**d**) Plot of measured plug sizes over time at the three different flow rates. Each pixel was approximately equivalent to an area of 5.6 µm^2.

3.1.2. Effect of Particle Concentration

The effect of particle concentration on plug formation was investigated by pumping the magnetic particles through the capillary at concentrations of 5×10^6, 1×10^7 and 2×10^7 particles$\cdot$ mL^{-1} for 90 s at 300 µL$\cdot$ h^{-1}. PBS solution was then flushed through the capillary for 10 min at the same flow rate, with photographs taken of the growing particle plug every minute.

Images of the trapped particle plugs for each concentration can be seen in Figure 5a–c at 1, 5 and 10 min after starting the washing step. As expected, higher particle concentrations yielded larger plugs in shorter times. The total area of each plug was analysed using ImageJ software and the results are plotted in Figure 5d. Again, the rates of plug formation were actually largely quite similar in each case (from a 2D viewpoint at least), but nonetheless reaffirmed that the higher the particle concentration was, the larger the plug that was formed within a certain time frame. However, the plug formed at higher concentrations was a lot more spread out across the capillary (in the x-direction), hence a concentration of 1×10^7 particles$\cdot$ mL^{-1} was employed for subsequent experiments in order to form a fairly large plug with a better defined shape.

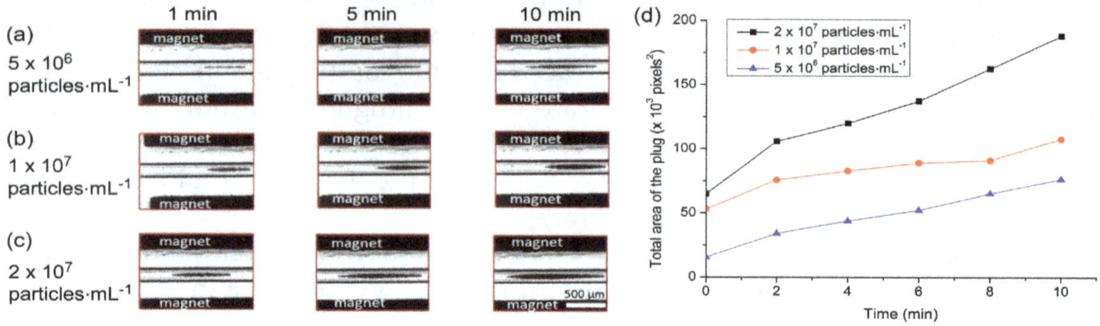

Figure 5. The effect of particle concentration on plug formation. (**a–c**) Photographs of plug formation at time points of 1, 5 and 10 min for particle concentrations of: (**a**) 5×10^6 particles$\cdot$ mL^{-1}; (**b**) 1×10^7 particles$\cdot$ mL^{-1}; and (**c**) 2×10^7 particles$\cdot$ mL^{-1}. (**d**) Plot of measured plug sizes over time at the three different particle concentrations. Each pixel was approximately equivalent to an area of 5.6 μm^2.

3.2. Streptavidin-Biotin Assay

In order to test the setup for performing reactions on particle plugs, proof-of-principle assays based on the streptavidin-biotin interaction were performed. While we have previously demonstrated a streptavidin-biotin binding assay on a magnetic particle plug [50], the tests were only qualitative. Here, we investigated the ability to produce calibration curves of biotin concentrations using the magnetic trapping platform.

Dynabeads M-270 Streptavidin particles (1×10^7 particles$\cdot$ mL^{-1}) were pumped into the capillary at a flow rate of 300 μL$\cdot$ h^{-1} for 2 min to generate the particle plug between the magnets. This was followed by a solution of biotin-4-fluorescein at 300 μL$\cdot$ h^{-1} for 3 min, and finally a solution of PBS for 3 min in order to wash the particle plug. A range of biotin-4-fluorescein concentrations (0.1–5 μg$\cdot$ mL^{-1}) were tested, and photographs of particle plugs exposed to each of these concentrations are shown in Figure 6a–e. Clearly, as the concentration of biotin was increased the fluorescence intensity of the particle plug also increased, indicating successful binding of the fluorescent biotin to the streptavidin-coated particles.

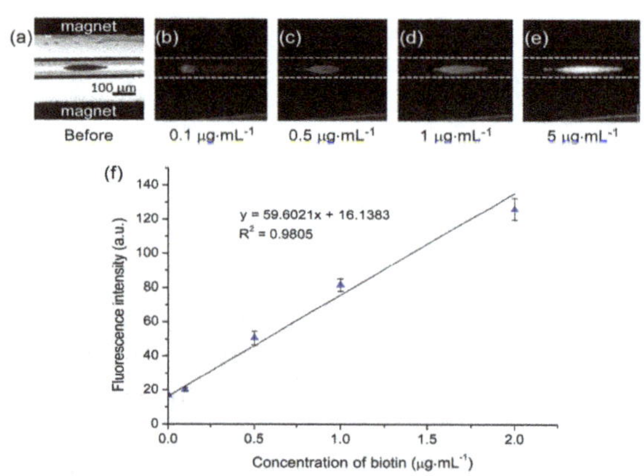

Figure 6. Streptavidin-biotin assays performed by flushing a solution of fluorescently labelled biotin over a trapped plug of streptavidin functionalised magnetic particles. (**a**) Bright-field image of the trapped particle plug. (**b–e**) Fluorescence images of streptavidin particle plugs exposed to varying concentrations of biotin: (**b**) 0.1 μg$\cdot$ mL^{-1}; (**c**) 0.5 μg$\cdot$ mL^{-1}; (**d**) 1 μg$\cdot$ mL^{-1}; and (**e**) 5 μg$\cdot$ mL^{-1}. (**f**) Calibration graph of particle plug fluorescence intensities exposed to a range of fluorescently labelled biotin concentrations.

In order to optimise the platform further for quantitative analysis, the effect of CCD camera exposure time was investigated alongside the ability to generate calibration curves. Multiple concentrations of biotin (0.1–5 μg· mL^{-1}) were flushed over particle plugs and analysed using a range of CCD camera exposure times (0.1–0.6 s). The resultant plots are shown in Figure S1a and demonstrate typical dose–response curves, with the fluorescence intensity first increasing sharply as the biotin concentration increased before reaching a plateau at ~2 μg· mL^{-1} as the number of streptavidin binding sites on the particles diminished. The polystyrene matrix of the magnetic particles exhibited auto-fluorescence, hence the non-zero fluorescence intensity even in the absence of biotin. Due to the plateau above 2 μg· mL^{-1}, the curves were re-plotted to show the linear responses between 0.1–2 μg· mL^{-1} (Figure S1b), demonstrating the suitability of the platform for quantitative analysis.

Clearly, as the CCD camera exposure time was increased the measured fluorescence intensities also increased, as expected since more of the fluorescence light was allowed to enter the CCD. However, while higher exposure times yielded greater intensities, they also exhibited poorer coefficients of determination (R^2) as demonstrated in Figure S1b. Based on this, the results obtained with the 0.3 s exposure time were selected as being the optimum, with the calibration curve (Figure 6f) yielding a limit of detection (LOD) of 40 ng· mL^{-1} and a limit of quantification (LOQ) of 134 ng· mL^{-1} for fluorescently labelled biotin.

These results demonstrated the potential of the platform for quantitative analysis with low limits of detection in a fast time frame (<10 min), while consuming only 15 μL of sample/reagent. Furthermore, some aspects could be optimised further, such as the reaction times, which would lead to shorter procedural times and lower reagent consumption, and the washing times. The detection method could also be improved to increase the coefficient of determination while reducing the limits of detection further. While these will be investigated further in later studies, the promising initial results prompted further studies with more clinically relevant biomarkers towards the use of the platform as a diagnostic tool.

3.3. C-Reactive Protein (CRP) Assay

The first clinically relevant biomarker tested using the magnetic plug platform was C-reactive protein (CRP) [72–75]. CRP is an acute phase reactant present in blood whose levels increase dramatically, up to 1000-fold, in response to inflammation, cell damage or tissue injury, hence its monitoring in a clinical setting for infections and inflammation. Normal levels of CRP in serum are considered to be 1–10 μg· mL^{-1}, with levels of 10–40 μg· mL^{-1} suggesting viral infection or mild inflammation while levels of 40–200 μg· mL^{-1} indicate active inflammation or bacterial infection. Chronic minor elevations in CRP levels may also be an indicator for cardiovascular disease (CVD), hence so-called high-sensitivity CRP (hs-CRP) testing is performed to monitor levels over time (<1μg· mL^{-1} = low CVD risk; 1–3 μg· mL^{-1} = medium risk, >3 μg· mL^{-1} = high risk) [76–78]. Due to its clinical relevance and the detection levels required, CRP was deemed to be an excellent choice for testing the ability to perform sandwich enzyme-linked immunosorbent (ELISA) assays using the magnetic particle plug platform.

Magnetic particles functionalised with 1° anti-CRP (1 × 10^7 particles· mL^{-1}) were introduced into the capillary at 300 μL· h^{-1} for 2 min for magnetic particle plug formation. This was followed by a solution of CRP for 3 min, allowing the CRP analyte to bind to the antibody-coated particles, before being flushed with a solution of 2° anti-FITC (100 μg· mL^{-1}) for 3 min that fluorescently labelled the captured CRP analyte. Finally, the particle plug was washed with PBS solution for 5 min and fluorescence images recorded for analysis.

Fluorescence images of particle plugs are shown in Figure 7a–c. Figure 7a demonstrates the auto-fluorescence of the particles prior to a reaction being performed, while Figure 7b,c show the effects of exposure to CRP concentrations of 1 μg· mL^{-1} and 10 μg· mL^{-1}, respectively, followed by reaction with the 2° anti-CRP-FITC (100 μg· mL^{-1}). The photographs clearly show an increase in fluorescence intensity with increasing CRP concentration, and the fluorescence intensity of the particle plugs are plotted in Figure 7d. While a full range of CRP standards was not tested, it is nonetheless clear that clinically relevant concentrations of CRP (>10 $\mu g \cdot mL^{-1}$ for inflammation and infection; 1–10 $\mu g \cdot mL^{-1}$ for CVD monitoring) could be distinguished from each other and from the unreacted particles. Furthermore, negative controls were performed to ensure that unspecific binding of reagents to the particles did not occur. Here, streptavidin-coated magnetic particles, having not undergone the 1° anti-CRP functionalisation step, were trapped in the capillary and flushed with CRP (10 $\mu g \cdot mL^{-1}$) and 2° anti-CRP-FITC (100 $\mu g \cdot mL^{-1}$). Image analysis showed no increase in fluorescence that confirmed a lack of unspecific binding, as previously demonstrated [69,70].

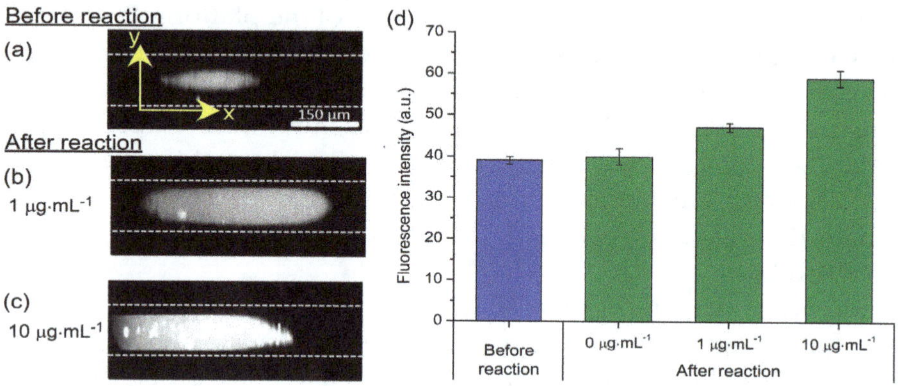

Figure 7. Results obtained via a magnetic particle plug-based sandwich assay for C-reactive protein (CRP). (**a**) Fluorescence image of a particle plug prior to the CRP assay, demonstrating the auto-fluorescence of the polystyrene-based particles. Magnetic particles were functionalised with primary CRP antibodies (1° anti-CRP). (**b**) Fluorescence exhibited by a particle plug after exposure to 1 $\mu g \cdot mL^{-1}$ CRP and subsequent labelling with fluorescently tagged secondary CRP antibody (2° anti-CRP-FITC; 100 $\mu g \cdot mL^{-1}$); and (**c**) after exposure to 10 $\mu g \cdot mL^{-1}$ CRP and labelling with 2° anti-CRP-FITC (100 $\mu g \cdot mL^{-1}$). (**d**) Plot of fluorescence intensities of the particle plugs at varying concentrations of CRP.

These results demonstrated the ability to perform two-step sandwich immunoassays for clinically relevant biomarkers. Further investigation will be required to determine whether a suitable calibration curve can be generated in the 1–10 $\mu g \cdot mL^{-1}$ region for hs-CRP testing, but its use for the determination of inflammation and infection, requiring less sensitivity, appears easily achievable. The 3 min analyte capture (CRP) and labelling (2° anti-CRP-FITC) steps at 300 $\mu L \cdot h^{-1}$ resulted in the consumption of 15 µL of both the sample and the relatively expensive labelling reagent. While already a low volume of each, this could be further reduced by optimising the reaction times. Furthermore, the total time of the magnetic plug-based was <15 min; far faster compared to conventional off-chip magnetic particle-based assays (50 min) and traditional ELISA testing (80 min) [69]. Future work will involve generation of a full calibration range for both conventional CRP and hs-CRP concentration ranges, and the analysis of real serum samples.

3.4. *Progesterone (P4) Assay*

Having established that the magnetic particle plug-based platform could be used for sandwich ELISAs, we next investigated the potential of the system towards achieving competitive ELISAs of clinically relevant biomarkers. Initial tests for this involved the detection of fluorescently labelled progesterone (P4-FITC). Progesterone (P4) is a steroid hormone that plays an important role in the menstrual cycle and pregnancy, being secreted to help prepare the uterus for pregnancy and, following conception, to ensure development of the embryo [79–82]. Thus, the monitoring of P4 can be used to determine the time at which fertility is highest, for the diagnosis of early pregnancy, to check for the risk or occurrence of miscarriage, and for the detection of adrenal or ovarian cancer.

Levels of progesterone are typically less than 1 ng· mL^{-1} pre-ovulation, increasing to 5–20 ng· mL^{-1} mid-menstruation cycle, 11.2–90.0 ng· mL^{-1} in the 1st trimester of pregnancy, 25.6–89.4 ng· mL^{-1} in the 2nd trimester, and 48–150 to ⩾300 ng· mL^{-1} in the 3rd trimester [82], with levels being present up to a maximum of 1 g· mL^{-1} in serum [83].

Preliminary tests were performed on the magnetic plug platform by first introducing anti-P4 functionalised magnetic particles (1×10^7 particles· mL^{-1}) into the capillary at 300 L· h^{-1} for 2 min to form the plug, before flushing the plug with P4-FITC for 3 min, and finally washing the plug for 3 min with PBS solution. A range of P4-FITC concentrations were tested, from 0.1–100 g· mL^{-1}, and the fluorescence intensity of the particle plug was measured at each concentration. Due to the proof-of-principle nature of this study, the levels of P4 tested covered the upper end of the P4 concentration range typically found during the 3rd trimester and the maximum level found in blood (1 g· mL^{-1}). Figure 8a–e shows fluorescence images at each of the concentrations. The fluorescence signals measured at each P4-FITC concentration are shown in Figure S2 in the Supplementary Material, demonstrating a typical dose–response curve with an initially rapid increase in signal intensity as the concentration increased, before reaching a plateau as the number of active sites on the magnetic particles was diminished. Plotting the fluorescence intensity against the logarithm of the P4-FITC concentration (background corrected) yielded a linear response over this wide calibration range (Figure 8f). However, the standard deviations of the results were quite large in this case, which may have been caused in part by the relatively low magnification employed for image capture (see the photographs in Figure 8), which would affect the signal intensity. This could be addressed by capturing images of the particle plugs at a higher magnification or by employing a different detection technique. Negative controls were also performed by flushing P4-FITC over a plug of carboxylic acid functionalised particles (*i.e.*, particles which had not had anti-P4 conjugated to them) for 1 h, which thereafter exhibited no fluorescence and so confirmed no issues with unspecific binding.

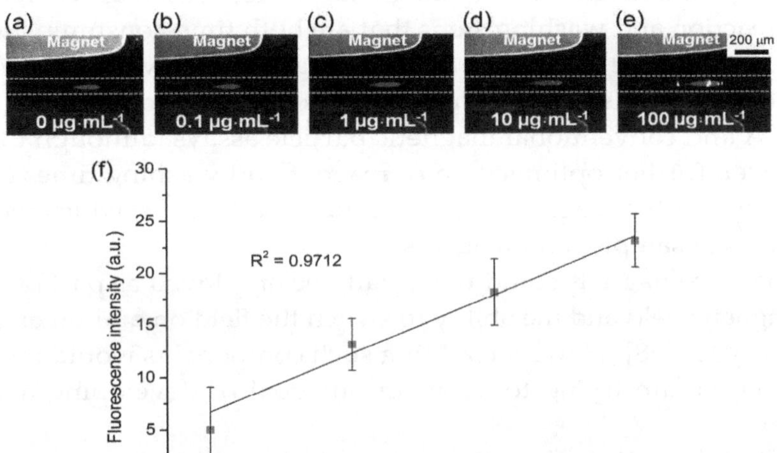

Figure 8. Results obtained for a progesterone (P4) assay, achieved by flushing P4-FITC over a trapped plug of anti-P4 functionalised magnetic particles. (**a–e**). Fluorescence images of particle plugs with increasing P4-FITC concentrations. (**f**) Plot of background-corrected particle plug fluorescence intensities at different concentrations (shown on a logarithmic scale) of P4-FITC.

These preliminary studies show the feasibility of performing competitive ELISAs for biomarkers, such as progesterone hormone, in 10 min and using only 10 µL of sample, though clearly further work is required in order to develop the method into a viable platform for clinical hormone analysis. The next steps towards this goal will be the introduction of unlabelled P4 at varying concentrations alongside the P4-FITC in order to perform actual competitive assays, while the limit of detection and linear range will be explored at the more clinically relevant levels of 1 ng· mL^{-1} to 1 μg· mL^{-1}. This would then lead to the testing of real serum samples.

4. Outlook

We have performed preliminary studies to establish the feasibility of applying the magnetic particle plug-based platform for clinically relevant bioassays. Characterisation of particle plug formation was performed in order to generate large particle plugs in a short timeframe, and three types of assay systems were investigated: streptavidin-biotin binding assays for evaluation of the platform for quantitative assays, C-reactive protein (CRP) assays for testing the ability to perform sandwich ELISAs of a biomarker in a clinically relevant concentration range, and fluorescently-labelled progesterone (P4-FITC) assays with a view to competitive ELISAs for hormone analysis. While optimisation is still required, these tests show great promise for quantitative analysis of a variety of biomarkers. In particular, the CRP assay, which can already be applied in a relevant concentration range, requires only a full calibration curve to be generated prior to analysis of real samples. The P4 analysis requires more work; while the assay mechanism operates as required and differences in concentration can be detected using the relatively high concentrations tested, the limits of detection need to be established and a calibration curve generated using fluorescently labelled and unlabelled P4 in a clinically relevant range before real serum analysis will be possible. In the case of both analyses, however, testing of robustness of the platform will also be required, including tests of inter-day and inter-chip variability.

The magnetic particle plug-based platform represents an extremely simple setup, requiring only a capillary, two NdFeB magnets, a syringe pump, and a detection system. It also brings several other advantages such as the speed with which assays can be performed. Each step of the process (particle loading, reagent addition, washing) took only 2–3 min each, meaning that total times for each assay was <15 min. Using other methods of performing CRP assays for comparison [69], typical ELISAs will take ~80 min and off-chip magnetic particle-based assays require ~50 min due to the multiple manual reaction and washing steps that are both time-consuming and labour-intensive. Hence, the on-chip platform here represents a far faster approach, while typically using only 15 μL of sample and 15 μL of the expensive labelling reagents compared to the hundreds of microliters of reagents used in ELISA and conventional magnetic particle assays (although ELISA only requires 5 μL of sample). However, further optimisation of reagent and washing times could lead to further reductions in time frames for the magnetic particle plug-based assays, which would in turn result in the use of lower volumes of samples and reagents.

Integrated microelectromagnets could potentially be employed as part of the system to enable finer control of the magnetic field and the ability to switch the field on-and-off as required, as has been demonstrated previously [33–38]. However, adding such components would increase the complexity of the setup; an aspect we are trying to avoid in our goal of developing a very simple, robust, user-friendly platform.

While the amount of time allowed for the reaction and washing steps is one method of reducing overall time frames, another would be to decrease the time required to switch between the different solutions (magnetic particle suspension, reagent solutions, washing buffer) being introduced into the microchannel. In the current system, the syringe pump was paused and the flow allowed to come to a stop before manually moving the inlet tubing from one sample or buffer reservoir to the next, with care taken not to allow introduction of air into the capillary during the exchange. These steps could be made far faster, and without the worry of introducing air bubbles, by employing a multi-port valve that allows simultaneous connection of each reservoir to the inlet tubing (e.g., the V-240 6-Way Selection Valve from IDEX Health & Science [84]). Alternatively, a moving array of microvial reservoirs at the capillary inlet could be employed, as has been successfully implemented for sample introduction in microfluidic capillary electrophoresis [85]. In addition, by writing a simple program for controlling such a valve or microarray system and the syringe pump, it would be very easy to automate the various steps of the assays. This would also help to enable multiplexed assays by easily allowing the generation of multiple particle plugs having different surface functionalities for various analytes, as we have demonstrated previously [50,68].

Furthermore, while current detection was achieved using a standard fluorescence microscope, which brings with it an associated bulk and expense, recent advances in miniaturised fluorescence detection systems could conceivably be applied to this platform to yield a far more compact and portable system [86–88]. While fluorescence detection was employed in the experiments described here, this could be replaced with a chemiluminescence setup by exchanging the fluorescent tags on the antibodies/antigens for a suitable enzyme, e.g., horseradish peroxidase (HRP), and the washing of the particle plug with a solution of a chemiluminescent substrate solution. This would also reduce the detection setup to a photomultiplier tube (PMT) without the need for a light source.

Clearly, there are a number of steps to be completed before a true analysis platform can be established, and the improvements suggested above would enable a faster, more sensitive, and more compact assay system that uses only small volumes of samples and reagents while requiring minimal manual steps. However, this represents a longer-term vision for the system. Nonetheless, the results described here have demonstrated the use of the miniaturised magnetic particle plug-based assay platform for the detection of several analytes at varying concentrations, showing great potential for fast, low volume sandwich ELISAs and competitive ELISAs for a range of clinically relevant biomarkers.

5. Conclusions

We have demonstrated a fast, low volume assay platform in which functionalised magnetic particles are introduced into a microchannel and trapped as a plug between two permanent magnets, allowing their subsequent exposure to consecutive reagent and washing solutions, followed by fluorescence analysis of the particle plug. The formation of the particle plug was characterised and the ability to perform quantitative analysis determined using a streptavidin-biotin binding assay (LOD = 40 ng$\cdot$ mL^{-1}). The capacity to detect clinically relevant biomarkers was explored using the inflammation marker, C-reactive protein (CRP), in a sandwich assay, and the steroid hormone, progesterone (P4), in a binding assay with a view to competitive ELISAs. Assays were achieved in less than 15 min, a significant reduction in time compared to conventional procedures, and used only 10–15 µL each of samples and reagents. This shows the potential of the platform for the rapid detection of a range of biomarkers, and future work will involve further optimisation of the setup and the procedure for the analysis of real samples.

Supplementary Materials
Figure S1: Optimisation of CCD camera exposure time. The fluorescence intensity of streptavidin functionalised particle plugs is plotted against the concentration of fluorescently labelled biotin at different CCD exposure times (0.1–0.6 s): (a) biotin concentrations of 0.1–5 µg$\cdot$ mL^{-1}, demonstrating a typical dose-response curve; and (b) linear range plotted for 0.1–2 g$\cdot$ mL^{-1} biotin concentration, Figure S2: Fluorescently labelled progesterone (P4-FITC) assay. Magnetic particles functionalised with anti-P4 were exposed to different concentrations of P4-FITC and the resultant fluorescence intensities were measured. The plot follows a typical dose–response curve.

Acknowledgments: Chayakom Phurimsak thanks the Royal Thai Government (Thailand) for financial support. The authors acknowledge Jessica A. Benton and Wilaiwan Phakthong for their assistance throughout the work.

Author Contributions: Chayakom Phurimsak and Nicole Pamme conceived and designed the experiments; Chayakom Phurimsak performed the experiments; Chayakom Phurimsak and Mark D. Tarn analysed the data; and Mark D. Tarn wrote the paper.

References

1. Wild, D. *The Immunoassay Handbook*, 3rd ed.; Elsevier B.V.: Amsterdam, The Netherlands, 2005.
2. Manz, A.; Dittrich, P.S.; Pamme, N.; Iossifidis, D. *Bioanalytical Chemistr*, 2nd ed.; Imperial College Press: London, UK, 2015.

3. Ruffert, C. Magnetic bead—Magic bullet. *Micromachines* **2016**, *7*, 21. [CrossRef]
4. Rios, A.; Zougagh, M.; Bouri, M. Magnetic (nano)materials as an useful tool for sample preparation in analytical methods. A review. *Anal. Methods* **2013**, *5*, 4558–4573. [CrossRef]
5. Tarn, M.D.; Pamme, N. Microfluidics. In *Elsevier Reference Module in Chemistry, Molecular Sciences and Chemical Engineering*; Reedijk, J., Ed.; Elsevier: Waltham, MA, USA, 2013.
6. Nge, P.N.; Rogers, C.I.; Woolley, A.T. Advances in microfluidic materials, functions, integration, and applications. *Chem. Rev.* **2013**, *113*, 2550–2583. [CrossRef] [PubMed]
7. Reyes, D.R.; Iossifidis, D.; Auroux, P.A.; Manz, A. Micro total analysis systems. 1. Introduction, theory, and technology. *Anal. Chem.* **2002**, *74*, 2623–2636. [CrossRef] [PubMed]
8. Lim, C.T.; Zhang, Y. Bead-based microfluidic immunoassays: The next generation. *Biosens. Bioelectron.* **2007**, *22*, 1197–1204. [CrossRef] [PubMed]
9. Lin, C.C.; Wang, J.H.; Wu, H.W.; Lee, G.B. Microfluidic immunoassays. *J. Lab. Autom.* **2010**, *15*, 253–274. [CrossRef]
10. Ng, A.H.C.; Uddayasankar, U.; Wheeler, A.R. Immunoassays in microfluidic systems. *Anal. Bioanal. Chem.* **2010**, *397*, 991–1007. [CrossRef] [PubMed]
11. Su, W.; Gao, X.; Jiang, L.; Qin, J. Microfluidic platform towards point-of-care diagnostics in infectious diseases. *J. Chromatogr. A* **2015**, *1377*, 13–26. [CrossRef] [PubMed]
12. Tarn, M.D.; Pamme, N. Microfluidic platforms for performing surface-based clinical assays. *Expert Rev. Mol. Diagn.* **2011**, *11*, 711–720. [CrossRef] [PubMed]
13. Pamme, N. Magnetism and microfluidics. *Lab Chip* **2006**, *6*, 24–38. [CrossRef] [PubMed]
14. Gijs, M.A.M. Magnetic bead handling on-chip: New opportunities for analytical applications. *Microfluid. Nanofluid.* **2004**, *1*, 22–40. [CrossRef]
15. Gijs, M.A.M.; Lacharme, F.; Lehmann, U. Microfluidic applications of magnetic particles for biological analysis and catalysis. *Chem. Rev.* **2010**, *110*, 1518–1563. [CrossRef] [PubMed]
16. Pamme, N. On-chip bioanalysis with magnetic particles. *Curr. Opin. Chem. Biol.* **2012**, *16*, 436–443. [CrossRef] [PubMed]
17. Nilsson, J.; Evander, M.; Hammarstrom, B.; Laurell, T. Review of cell and particle trapping in microfluidic systems. *Anal. Chim. Acta* **2009**, *649*, 141–157. [CrossRef] [PubMed]
18. Karimi, A.; Yazdi, S.; Ardekani, A.M. Hydrodynamic mechanisms of cell and particle trapping in microfluidics. *Biomicrofluidics* **2013**, *7*, 021501. [CrossRef] [PubMed]
19. Rashkovetsky, L.G.; Lyubarskaya, Y.V.; Foret, F.; Hughes, D.E.; Karger, B.L. Automated microanalysis using magnetic beads with commercial capillary electrophoretic instrumentation. *J. Chromatogr. A* **1997**, *781*, 197–204. [CrossRef]
20. Abonnenc, M.; Gassner, A.L.; Morandini, J.; Josserand, J.; Girault, H.H. Magnetic track array for efficient bead capture in microchannels. *Anal. Bioanal. Chem.* **2009**, *395*, 747–757. [CrossRef] [PubMed]
21. Brandl, M.; Mayer, M.; Hartmann, J.; Posnicek, T.; Fabian, C.; Falkenhagen, D. Theoretical analysis of ferromagnetic microparticles in streaming liquid under the influence of external magnetic forces. *J. Magn. Magn. Mater.* **2010**, *322*, 2454–2464. [CrossRef]
22. Chang, W.S.; Shang, H.; Perera, R.M.; Lok, S.M.; Sedlak, D.; Kuhn, R.J.; Lee, G.U. Rapid detection of dengue virus in serum using magnetic separation and fluorescence detection. *Analyst* **2008**, *133*, 233–240. [CrossRef] [PubMed]
23. Gassner, A.-L.; Abonnenc, M.; Chen, H.-X.; Morandini, J.; Josserand, J.; Rossier, J.S.; Busnel, J.-M.; Girault, H.H. Magnetic forces produced by rectangular permanent magnets in static microsystems. *Lab Chip* **2009**, *9*, 2356–2363. [CrossRef] [PubMed]
24. Krishnan, J.N.; Kim, C.; Park, H.J.; Kang, J.Y.; Kim, T.S.; Kim, S.K. Rapid microfluidic separation of magnetic beads through dielectrophoresis and magnetophoresis. *Electrophoresis* **2009**, *30*, 1457–1463. [CrossRef] [PubMed]
25. Smistrup, K.; Bu, M.Q.; Wolff, A.; Bruus, H.; Hansen, M.F. Theoretical analysis of a new, efficient microfluidic magnetic bead separator based on magnetic structures on multiple length scales. *Microfluid. Nanofluid.* **2008**, *4*, 565–573. [CrossRef]
26. Lacharme, F.; Vandevyver, C.; Gijs, M.A.M. Full on-chip nanoliter immunoassay by geometrical magnetic trapping of nanoparticle chains. *Anal. Chem.* **2008**, *80*, 2905–2910. [CrossRef] [PubMed]

27. Gassner, A.-L.; Morandini, J.; Josserand, J.; Girault, H.H. Ring magnets for magnetic beads trapping in a capillary. *Anal. Methods* **2011**, *3*, 614–621. [CrossRef]
28. Le Nel, A.; Minc, N.; Smadja, C.; Slovakova, M.; Bilkova, Z.; Peyrin, J.M.; Viovy, J.L.; Taverna, M. Controlled proteolysis of normal and pathological prion protein in a microfluidic chip. *Lab Chip* **2008**, *8*, 294–301. [CrossRef] [PubMed]
29. Tabnaoui, S.; Malaquin, L.; Descroix, S.; Viovy, J.L. Integrated microfluidic fluidized bed for sample preconcentration and immunoextraction. In Proceedings of the MicroTAS 2012 Conference, Okinawa, Japan, 28 October–1 November 2012; pp. 1408–1410.
30. Mohamadi, R.M.; Svobodova, Z.; Bilkova, Z.; Otto, M.; Taverna, M.; Descroix, S.; Viovy, J.-L. An integrated microfluidic chip for immunocapture, preconcentration and separation of beta-amyloid peptides. *Biomicrofluidics* **2015**, *9*, 054117. [CrossRef] [PubMed]
31. Degre, G.; Brunet, E.; Dodge, A.; Tabeling, P. Improving agglutination tests by working in microfluidic channels. *Lab Chip* **2005**, *5*, 691–694. [CrossRef] [PubMed]
32. Mohamadi, M.R.; Svobodova, Z.; Verpillot, R.; Esselmann, H.; Wiltfang, J.; Otto, M.; Taverna, M.; Bilkova, Z.; Viovy, J.-L. Microchip electrophoresis profiling of Aβ peptides in the cerebrospinal fluid of patients with Alzheimer's disease. *Anal. Chem.* **2010**, *82*, 7611–7617. [CrossRef] [PubMed]
33. Choi, J.W.; Liakopoulos, T.M.; Ahn, C.H. An on-chip magnetic bead separator using spiral electromagnets with semi-encapsulated permalloy. *Biosens. Bioelectron.* **2001**, *16*, 409–416. [CrossRef]
34. Ramadan, Q.; Samper, V.; Poenar, D.; Yu, C. On-chip micro-electromagnets for magnetic-based bio-molecules separation. *J. Magn. Magn. Mater.* **2004**, *281*, 150–172. [CrossRef]
35. Smistrup, K.; Hansen, O.; Bruus, H.; Hansen, M.F. Magnetic separation in microfluidic systems using microfabricated electromagnets-experiments and simulations. *J. Magn. Magn. Mater.* **2005**, *293*, 597–604. [CrossRef]
36. Smistrup, K.; Tang, P.T.; Hansen, O.; Hansen, M.F. Micro electromagnet for magnetic manipulation in lab-on-a-chip systems. *J. Magn. Magn. Mater.* **2006**, *300*, 418–426. [CrossRef]
37. Ramadan, Q.; Samper, V.; Poenar, D.; Yu, C. Magnetic-based microfluidic platform for biomolecular separation. *Biomed. Microdevices* **2006**, *8*, 151–158. [CrossRef] [PubMed]
38. Ramadan, Q.; Lau Ting, T. Reconfigurable translocation of microbeads using micro-engineered locally controlled magnetic fields. *J. Microelectromech. Syst.* **2011**, *20*, 1310–1323. [CrossRef]
39. Sinha, A.; Ganguly, R.; Puri, I.K. Magnetic separation from superparamagnetic particle suspensions. *J. Magn. Magn. Mater.* **2009**, *321*, 2251–2256. [CrossRef]
40. Smistrup, K.; Lund-Olesen, T.; Hansen, M.F.; Tang, P.T. Microfluidic magnetic separator using an array of soft magnetic elements. *J. Appl. Phys.* **2006**, *99*, 08P102. [CrossRef]
41. Teste, B.; Malloggi, F.; Gassner, A.-L.; Georgelin, T.; Siaugue, J.-M.; Varenne, A.; Girault, H.; Descroix, S. Magnetic core shell nanoparticles trapping in a microdevice generating high magnetic gradient. *Lab Chip* **2011**, *11*, 833–840. [CrossRef] [PubMed]
42. Moser, Y.; Lehnert, T.; Gijs, M.A.M. On-chip immuno-agglutination assay with analyte capture by dynamic manipulation of superparamagnetic beads. *Lab Chip* **2009**, *9*, 3261–3267. [CrossRef] [PubMed]
43. Rida, A.; Gijs, M.A.M. Manipulation of self-assembled structures of magnetic beads for microfluidic mixing and assaying. *Anal. Chem.* **2004**, *76*, 6239–6246. [CrossRef] [PubMed]
44. Chen, H.T.; Kaminski, M.D.; Caviness, P.L.; Liu, X.Q.; Dhar, P.; Torno, M.; Rosengart, A.J. Magnetic separation of micro-spheres from viscous biological fluids. *Phys. Med. Biol.* **2007**, *52*, 1185–1196. [CrossRef] [PubMed]
45. Deng, T.; Prentiss, M.; Whitesides, G.M. Fabrication of magnetic microfiltration systems using soft lithography. *Appl. Phys. Lett.* **2002**, *80*, 461–463. [CrossRef]
46. Deng, T.; Whitesides, G.M.; Radhakrishnan, M.; Zabow, G.; Prentiss, M. Manipulation of magnetic microbeads in suspension using micromagnetic systems fabricated with soft lithography. *Appl. Phys. Lett.* **2001**, *78*, 1775–1777. [CrossRef]
47. Nawarathna, D.; Norouzi, N.; McLane, J.; Sharma, H.; Sharac, N.; Grant, T.; Chen, A.; Strayer, S.; Ragan, R.; Khine, M. Shrink-induced sorting using integrated nanoscale magnetic traps. *Appl. Phys. Lett.* **2013**, *102*, 063504. [CrossRef] [PubMed]
48. Armbrecht, L.; Dincer, C.; Kling, A.; Horak, J.; Kieninger, J.; Urban, G. Self-assembled magnetic bead chains for sensitivity enhancement of microfluidic electrochemical biosensor platforms. *Lab Chip* **2015**, *15*, 4314–4321. [CrossRef] [PubMed]

49. Saliba, A.-E.; Saias, L.; Psychari, E.; Minc, N.; Simon, D.; Bidard, F.-C.; Mathiot, C.; Pierga, J.-Y.; Fraisier, V.; Salamero, J.; *et al.* Microfluidic sorting and multimodal typing of cancer cells in self-assembled magnetic arrays. *Proc. Natl. Acad. Sci. USA* **2010**, *107*, 14524–14529. [CrossRef] [PubMed]
50. Bronzeau, S.; Pamme, N. Simultaneous bioassays in a microfluidic channel on plugs of different magnetic particles. *Anal. Chim. Acta* **2008**, *609*, 105–112. [CrossRef] [PubMed]
51. Choi, J.W.; Oh, K.W.; Thomas, J.H.; Heineman, W.R.; Halsall, H.B.; Nevin, J.H.; Helmicki, A.J.; Henderson, H.T.; Ahn, C.H. An integrated microfluidic biochemical detection system for protein analysis with magnetic bead-based sampling capabilities. *Lab Chip* **2002**, *2*, 27–30. [CrossRef] [PubMed]
52. Hayes, M.A.; Polson, N.A.; Phayre, A.N.; Garcia, A.A. Flow-based microimmunoassay. *Anal. Chem.* **2001**, *73*, 5896–5902. [CrossRef] [PubMed]
53. Teste, B.; Ali-Cherif, A.; Viovy, J.L.; Malaquin, L. A low cost and high throughput magnetic bead-based immuno-agglutination assay in confined droplets. *Lab Chip* **2013**, *13*, 2344–2349. [CrossRef] [PubMed]
54. Lien, K.Y.; Lin, J.L.; Liu, C.Y.; Lei, H.Y.; Lee, G.B. Purification and enrichment of virus samples utilizing magnetic beads on a microfluidic system. *Lab Chip* **2007**, *7*, 868–875. [CrossRef] [PubMed]
55. Jiang, G.F.; Harrison, D.J. mRNA isolation in a microfluidic device for eventual integration of cDNA library construction. *Analyst* **2000**, *125*, 2176–2179. [CrossRef] [PubMed]
56. Fan, Z.H.; Mangru, S.; Granzow, R.; Heaney, P.; Ho, W.; Dong, Q.P.; Kumar, R. Dynamic DNA hybridization on a chip using paramagnetic beads. *Anal. Chem.* **1999**, *71*, 4851–4859. [CrossRef] [PubMed]
57. Smistrup, K.; Kjeldsen, B.G.; Reimers, J.L.; Dufva, M.; Petersen, J.; Hansen, M.F. On-chip magnetic bead microarray using hydrodynamic focusing in a passive magnetic separator. *Lab Chip* **2005**, *5*, 1315–1319. [CrossRef] [PubMed]
58. Lund-Olesen, T.; Dufva, M.; Hansen, M.F. Capture of DNA in microfluidic channel using magnetic beads: Increasing capture efficiency with integrated microfluidic mixer. *J. Magn. Magn. Mater.* **2007**, *311*, 396–400. [CrossRef]
59. Doyle, P.S.; Bibette, J.; Bancaud, A.; Viovy, J.L. Self-assembled magnetic matrices for DNA separation chips. *Science* **2002**, *295*, 2237–2237. [CrossRef] [PubMed]
60. Huang, M.C.; Ye, H.; Kuan, Y.K.; Li, M.H.; Ying, J.Y. Integrated two-step gene synthesis in a microfluidic device. *Lab Chip* **2009**, *9*, 276–285. [CrossRef] [PubMed]
61. Liu, R.H.; Yang, J.N.; Lenigk, R.; Bonanno, J.; Grodzinski, P. Self-contained, fully integrated biochip for sample preparation, polymerase chain reaction amplification, and DNA microarray detection. *Anal. Chem.* **2004**, *76*, 1824–1831. [CrossRef] [PubMed]
62. Furdui, V.I.; Harrison, D.J. Immunomagnetic T cell capture from blood for PCR analysis using microfluidic systems. *Lab Chip* **2004**, *4*, 614–618. [CrossRef] [PubMed]
63. Caulum, M.M.; Henry, C.S. Measuring reaction rates on single particles in a microfluidic device. *Lab Chip* **2008**, *8*, 865–867. [CrossRef] [PubMed]
64. Liu, J.Y.; Lin, S.; Qi, D.W.; Deng, C.H.; Yang, P.Y.; Zhang, X.M. On-chip enzymatic microreactor using trypsin-immobilized superparamagnetic nanoparticles for highly efficient proteolysis. *J. Chromatogr. A* **2007**, *1176*, 169–177. [CrossRef] [PubMed]
65. Slovakova, M.; Minc, N.; Bilkova, Z.; Smadja, C.; Faigle, W.; Futterer, C.; Taverna, M.; Viovy, J.L. Use of self assembled magnetic beads for on-chip protein digestion. *Lab Chip* **2005**, *5*, 935–942. [CrossRef] [PubMed]
66. Bilkova, Z.; Slovakova, M.; Minc, N.; Futterer, C.; Cecal, R.; Horak, D.; Benes, M.; le Potier, I.; Krenkova, J.; Przybylski, M.; *et al.* Functionalized magnetic micro-and nanoparticles: Optimization and application to μ-chip tryptic digestion. *Electrophoresis* **2006**, *27*, 1811–1824. [CrossRef] [PubMed]
67. Peyman, S.A.; Kwan, E.Y.; Margarson, O.; Iles, A.; Pamme, N. Diamagnetic repulsion—A versatile tool for label-free particle handling in microfluidic devices. *J. Chromatogr. A* **2009**, *1216*, 9055–9062. [CrossRef] [PubMed]
68. Tarn, M.D.; Peyman, S.A.; Pamme, N. Simultaneous trapping of magnetic and diamagnetic particle plugs for separations and bioassays. *RSC Adv.* **2013**, *3*, 7209–7214. [CrossRef]
69. Phurimsak, C.; Tarn, M.D.; Peyman, S.A.; Greenman, J.; Pamme, N. On-chip determination of C-reactive protein using magnetic particles in continuous flow. *Anal. Chem.* **2014**, *86*, 10552–10559. [CrossRef] [PubMed]
70. Phurimsak, C.; Yildirim, E.; Tarn, M.D.; Trietsch, S.J.; Hankemeier, T.; Pamme, N.; Vulto, P. Phaseguide assisted liquid lamination for magnetic particle-based assays. *Lab Chip* **2014**, *14*, 2334–2343. [CrossRef] [PubMed]

71. Dynabeads®M-270 Carboxylic Acid. 2012. Available online: https://tools.thermofisher.com/content/sfs/manuals/dynabeads_m270carboxylicacid_man.pdf (accessed on 26 April 2012).
72. Black, S.; Kushner, I.; Samols, D. C-reactive protein. *J. Biol. Chem.* **2004**, *279*, 48487–48490. [CrossRef] [PubMed]
73. Clyne, B.; Olshaker, J.S. The C-reactive protein. *J. Emerg. Med.* **1999**, *17*, 1019–1025. [CrossRef]
74. Agassandian, M.; Shurin, G.V.; Ma, Y.; Shurin, M.R. C-reactive protein and lung diseases. *Int. J. Biochem. Cell Biol.* **2014**, *53*, 77–88. [CrossRef] [PubMed]
75. Vashist, S.K.; Venkatesh, A.G.; Marion Schneider, E.; Beaudoin, C.; Luppa, P.B.; Luong, J.H.T. Bioanalytical advances in assays for C-reactive protein. *Biotechnol. Adv.* **2016**, *34*, 272–290. [CrossRef] [PubMed]
76. Greenland, P.; Alpert, J.S.; Beller, G.A.; Benjamin, E.J.; Budoff, M.J.; Fayad, Z.A.; Foster, E.; Hlatky, M.A.; Hodgson, J.M.; Kushner, F.G.; *et al.* 2010 ACCF/AHA Guideline for Assessment of Cardiovascular Risk in Asymptomatic Adults: Executive Summary. *Circulation* **2010**, *122*, 2748–2764. [CrossRef] [PubMed]
77. Lagrand, W.K.; Visser, C.A.; Hermens, W.T.; Niessen, H.W.M.; Verheugt, F.W.A.; Wolbink, G.J.; Hack, C.E. C-reactive protein as a cardiovascular risk factor—More than an epiphenomenon? *Circulation* **1999**, *100*, 96–102. [CrossRef] [PubMed]
78. MedlinePlus. C-Reactive Protein. Available online: http://www.nlm.nih.gov/medlineplus/ency/article/003356.htm (accessed on 13 February 2016).
79. Arévalo, F.J.; Messina, G.A.; Molina, P.G.; Zón, M.A.; Raba, J.; Fernández, H. Determination of progesterone (P4) from bovine serum samples using a microfluidic immunosensor system. *Talanta* **2010**, *80*, 1986–1992. [CrossRef] [PubMed]
80. Monerris, M.J.; Arévalo, F.J.; Fernández, H.; Zon, M.A.; Molina, P.G. Integrated electrochemical immunosensor with gold nanoparticles for the determination of progesterone. *Sens. Actuators B* **2012**, *166–167*, 586–592. [CrossRef]
81. Ehrentreich-Förster, E.; Scheller, F.W.; Bier, F.F. Detection of progesterone in whole blood samples. *Biosens. Bioelectron.* **2003**, *18*, 375–380. [CrossRef]
82. MedlinePlus. Serum progesterone. Available online: https://www.nlm.nih.gov/medlineplus/ency/article/003714.htm (accessed on 13 February 2016).
83. Sanghavi, B.J.; Moore, J.A.; Chávez, J.L.; Hagen, J.A.; Kelley-Loughnane, N.; Chou, C.-F.; Swami, N.S. Aptamer-functionalized nanoparticles for surface immobilization-free electrochemical detection of cortisol in a microfluidic device. *Biosens. Bioelectron.* **2016**, *78*, 244–252. [CrossRef] [PubMed]
84. IDEX Health & Science. V-240-Selection Valve, 6 Position-7 Port, .040 Black. Available online: https://www.idex-hs.com/valves/flow-regulating-valves/selection-valves/selection-valve-6-position-7-port-040-black.html (accessed on 13 February 2016).
85. He, Q.-H.; Fang, Q.; Du, W.-B.; Huang, Y.-Z.; Fang, Z.-L. An automated electrokinetic continuous sample introduction system for microfluidic chip-based capillary electrophoresis. *Analyst* **2005**, *130*, 1052–1058. [CrossRef] [PubMed]
86. Ahrberg, C.D.; Ilic, B.R.; Manz, A.; Neuzil, P. Handheld real-time PCR device. *Lab Chip* **2016**, *16*, 586–592. [CrossRef] [PubMed]
87. Novak, L.; Neuzil, P.; Pipper, J.; Zhang, Y.; Lee, S. An integrated fluorescence detection system for lab-on-a-chip applications. *Lab Chip* **2007**, *7*, 27–29. [CrossRef] [PubMed]
88. Zhu, H.; Isikman, S.O.; Mudanyali, O.; Greenbaum, A.; Ozcan, A. Optical imaging techniques for point-of-care diagnostics. *Lab Chip* **2013**, *13*, 51–67. [CrossRef] [PubMed]

11

High-Pressure Acceleration of Nanoliter Droplets in the Gas Phase in a Microchannel

Yutaka Kazoe [1,†], Ippei Yamashiro [2,†], Kazuma Mawatari [2] and Takehiko Kitamori [2,*]

1 Department of Hemolysis and Apheresis, Graduate School of Medicine, The University of Tokyo, 7-3-1 Hongo, Bunkyo, Tokyo 113-8656, Japan; kazoe@icl.t.u-tokyo.ac.jp
2 Deparment of Applied Chemistry, Graduate School of Engineering, The University of Tokyo, 7-3-1 Hongo, Bunkyo, Tokyo 113-8656, Japan; yamashiro@icl.t.u-tokyo.ac.jp (I.Y.); kmawatari@icl.t.u-tokyo.ac.jp (K.M.)
* Correspondence: kitamori@icl.t.u-tokyo.ac.jp
† These authors contributed equally to this work.

Academic Editors: Manabu Tokeshi and Kiichi Sato

Abstract: Microfluidics has been used to perform various chemical operations for pL–nL volumes of samples, such as mixing, reaction and separation, by exploiting diffusion, viscous forces, and surface tension, which are dominant in spaces with dimensions on the micrometer scale. To further develop this field, we previously developed a novel microfluidic device, termed a microdroplet collider, which exploits spatially and temporally localized kinetic energy. This device accelerates a microdroplet in the gas phase along a microchannel until it collides with a target. We demonstrated 6000-fold faster mixing compared to mixing by diffusion; however, the droplet acceleration was not optimized, because the experiments were conducted for only one droplet size and at pressures in the 10–100 kPa range. In this study, we investigated the acceleration of a microdroplet using a high-pressure (MPa) control system, in order to achieve higher acceleration and kinetic energy. The motion of the nL droplet was observed using a high-speed complementary metal oxide semiconductor (CMOS) camera. A maximum droplet velocity of ~5 m/s was achieved at a pressure of 1–2 MPa. Despite the higher fluid resistance, longer droplets yielded higher acceleration and kinetic energy, because droplet splitting was a determining factor in the acceleration and using a longer droplet helped prevent it. The results provide design guidelines for achieving higher kinetic energies in the microdroplet collider for various microfluidic applications.

Keywords: microfluidics; microchannel; droplet; gas phase

---

1. Introduction

Microfluidics has enabled the fabrication of miniaturized chemical systems, known as lab-on-a-chip and micro-total analysis systems (μTAS), for chemical analysis, medical diagnosis, and chemical synthesis [1,2]. By exploiting diffusion, viscous forces, and surface tension, which are dominant in small spaces due to the short diffusion distance and increased surface-to-volume ratio, effective and fast micro-unit operations (MUOs) such as mixing, reaction, and separation have been developed [3–6]. Based on these principles, integrated chemical systems have exhibited superior performances, i.e., shorter processing times (from days or hours to minutes or seconds) and smaller sample/reagent volumes (down to μL), compared to conventional bulk chemical systems.

Our group has developed a new microfluidic device, termed a microdroplet collider, which exploits spatially and temporally localized kinetic energy in small spaces [7]. A microdroplet in the gas phase is formed in a microchannel, accelerated, and made to collide with a target. Acceleration to a velocity of ~1 m/s was demonstrated, which is a more than 100 times faster velocity (i.e., 10,000 times higher kinetic energy) than the values achievable by conventional microfluidic transport of a droplet in an oil phase [4,8]. The inelastic and minimally deformable collisions exploited by using the confined spaces with dimensions on the micrometer scale achieved highly efficient energy transfer compared to collisions between droplets in free space, which involve energy dissipation by deformation mechanisms such as bouncing, coalescence, disruption, and fragmentation [9,10]. Compared to mixing by diffusion of similar-sized droplets, the microdroplet collider achieved 6000-fold faster mixing. The higher the droplet acceleration, the higher the kinetic energy and efficiency of the chemical operations performed.

However, the droplet acceleration has not been optimized since the experiments were conducted for only one droplet size and at pressures of 10–100 kPa. Previously, we found that the microdroplet splits at a high velocity, due to wetting of the droplet tail on the channel wall [11], suggesting an upper limit to the droplet acceleration.

Recently, we have developed a high-pressure (MPa) control system [12], which may allow higher droplet acceleration. In the present study, we investigated the acceleration of microdroplets in the gas phase in a microchannel at a pressure in the order of 1 MPa. We adjusted the droplet length by varying the channel design, and applied pressures in the order of 1 MPa for higher acceleration. The motion of the accelerated droplets was observed using a high-speed complementary metal oxide semiconductor (CMOS) camera. Based on our results, we discuss the effects of droplet size and pressure on the acceleration and kinetic energy of the droplets.

2. Experimental Section

Figure 1 illustrates schematics the experimental setup and microfluidic process used for droplet formation and acceleration. The microchip for the microdroplet collider was connected to a high-pressure (MPa) control system for pressure-driven flow control [12]. The control system was equipped with an inverted microscope, a 2× objective lens, and a high-speed CMOS camera (FASTCAM, Photron, Tokyo, Japan) with a pixel size of 20 μm.

The microchip consisted of a droplet launcher with a width of 70 μm and a depth of 30 μm, a Laplace valve ($40^W \times 10\ \mu m^D$), and an acceleration microchannel ($70^W \times 30\ \mu m^D$). We prepared two microchips with droplet launcher lengths, $L = 1$ and 2 mm. Microchannels with two different depths were fabricated on a glass substrate by two-step photolithographic wet etching, as reported previously [13]. The substrate was thermally bonded with another glass substrate having inlet and outlet holes for gas and liquid injection. The channel wall was modified with an amorphous fluoropolymer (INT-332VE, NI material) to make the surface hydrophobic. The static contact angle measured by a contact angle meter was $\theta = 117°$. Due to the wet etching, the cross-sectional shape of the channel was rounded-rectangular, as illustrated in Figure 1a. We approximated the cross-sectional area of the channel, A, and the wetted perimeter, P, using the equations $A \approx (W - 2D)D + \pi D^2/2$ and $P \approx 2W - 2D + \pi D$, respectively, where W is the channel width and D is the channel depth. Based on these approximations, the hydraulic diameter of the channel, D_h, was 39 μm, as given by $D_h = 4A/P$.

The microfluidic process used for droplet formation and acceleration (Figure 1b) resembled that used in our previous reports [7,11]. A key feature of the process is the Laplace valve [14], which utilizes the Laplace pressure given by the Young-Laplace equation, $P_{LP} = -4\gamma\cos\theta/D_h$, where γ is the surface tension. The Laplace pressure of the valve was calculated to be $P_{LP} = 8.4$ kPa, using the surface tension at the water–air interface, $\gamma = 72.3$ mN/m. First, water was injected into the droplet launcher at a pressure below the Laplace pressure ($P_{IN} < P_{LP}$). Then, air was injected into the channel to form the microdroplet ($P_{IN} < P_{LP}$). Finally, a pressure higher than the Laplace pressure was applied to accelerate the droplet in the acceleration channel ($P_{IN} > P_{LP}$). The volumes of the microdroplet were 1.7 and 3.4 nL, as estimated using the size and length of the droplet launcher.

In this study, the droplet motion was captured using a high-speed CMOS camera at frame rates of 13,333 to 25,000 Hz. The droplet position in the images captured was determined with a spatial resolution of 10 μm, which corresponds to the image pixel size, given by pixel size/magnification.

Then, the droplet velocity and droplet length were determined at a measurement position 8 mm away from the launching point.

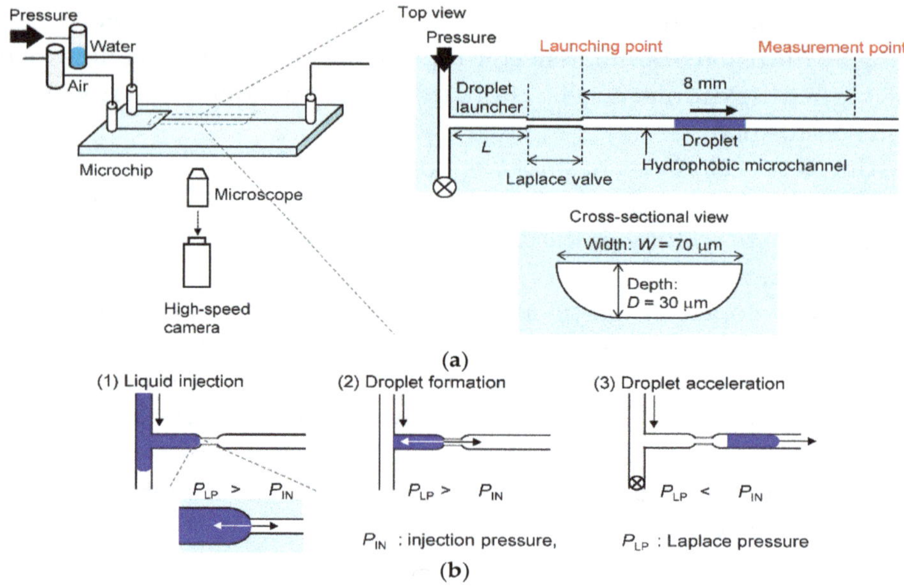

Figure 1. Schematics of (**a**) experimental setup for acceleration of a microdroplet in gas phase in a microchannel and (**b**) microfluidic process for droplet formation and acceleration. The microfluidic device consists of a droplet launcher, a Laplace valve and an acceleration microchannel. The channel wall is modified hydrophobically. During the process, (1) liquid is injected into the droplet launcher, (2) a droplet is formed by air flow, and (3) the droplet is accelerated in the microchannel.

3. Results and Discussion

3.1. Acceleration of Microdroplet under MPa-Order Pressure

Figure 2 shows the motion of an accelerated droplet in a microchannel with $L = 2.00$ mm at an applied pressure of 1800 kPa. A video of the accelerated droplet is presented in the Supplementary Material (Video S1). Within 1.5 ms, the droplet moves through the channel, begins to split from its tail, and disappears. Since the volume of the droplet is only in the order of nL, the influence of water evaporation on the droplet splitting and disappearance must be considered. In order to verify the influence, we estimated the time it would take for the nL droplet to evaporate, based on the evaporation rate of a water droplet on a solid surface reported in a previous study [15]. Assuming that evaporation occurred at the air–water interfacial area in the cross-section of the channel, the time for the evaporation of the nL droplet was estimated to be 10 s. This time scale is much larger than that for the droplet splitting and disappearance in 1 ms as shown in Figure 2. Therefore, the effect of evaporation is considered to be negligible.

Next, the velocity, length, and kinetic energy of the droplet were determined from the captured images. In case of the droplet splitting, we evaluated the length of the first droplet, because it collides with a target in the process. The kinetic energy of the droplet was estimated from $1/2\rho A L_D U^2$, where ρ is the density (998 kg/m^3 for water), L_D is the droplet length, and U is the droplet velocity. We used the cross-sectional area of channel A as the cross-sectional area of the droplet, assuming that the cross-section of the channel is filled entirely by the droplet. However, it is still unclear whether the droplet fills a sharp corner of the wet-etched channel (Figure 1a), where the channel wall is modified hydrophobically. Since the channel cross-section seems to be almost entirely filled by the droplet (see Figure 2), which was confirmed in a previous study [11], the dry area in the corner is considered to be small compared to the total area, even if the channel cross-section is not completely filled.

Figure 2. Images of a droplet in a microchannel with $L = 2.00$ mm at an applied pressure of 1800 kPa. $t = 0$ is defined as the time at which the droplet front passes the point 4 mm from the launching point.

As shown in Figure 3a, when the droplet front passes the point 4 mm from the launching point ($t = 0$), the droplet velocity and length are 3.5 m/s and 1.8 mm, respectively. The velocity increases when the droplet splits, reaching 7.5 m/s just before disappearance, because the fluid resistance decreases in proportion to the droplet length. The fluctuation in the droplet velocity is due to the instability of the wetting, caused in turn by the variation in the dynamic contact angle with splitting, as reported in our previous study [11]. During droplet splitting, the droplet length initially decreases gradually, and no large droplet fragments are observed ($t = 0$ ms and $t = 0.56$ ms in Figure 2). After the droplet passes the 7.5 mm mark, it becomes unstable and rapidly splits with large fragments ($t = 1.04$ ms and $t = 1.44$ ms in Figure 2). This rapid splitting with large fragments occurs when the droplet length falls below a certain value depending on the launcher length and pressure, and often shows variability between experiments conducted under nominally identical conditions. In the rapid splitting state, in addition to the instability of the wetting, Rayleigh instabilities are considered to be a dominant factor in the splitting, because splitting is accompanied by fluctuation of the droplet width, and the droplet splits once the droplet width becomes too small, as shown in Figure 2 and Video S1.

Since the droplet length decreases with splitting, the kinetic energy cannot be at maximum when the droplet velocity reaches a maximum. In the case of the channel with $L = 2.00$ mm at an applied pressure of 1800 kPa, the maximum kinetic energy is attained at a 7.4-mm distance from the launching point, as shown in Figure 3b. At the maximum kinetic energy, the droplet velocity and length were 4.8 m/s and 1.5 mm, respectively.

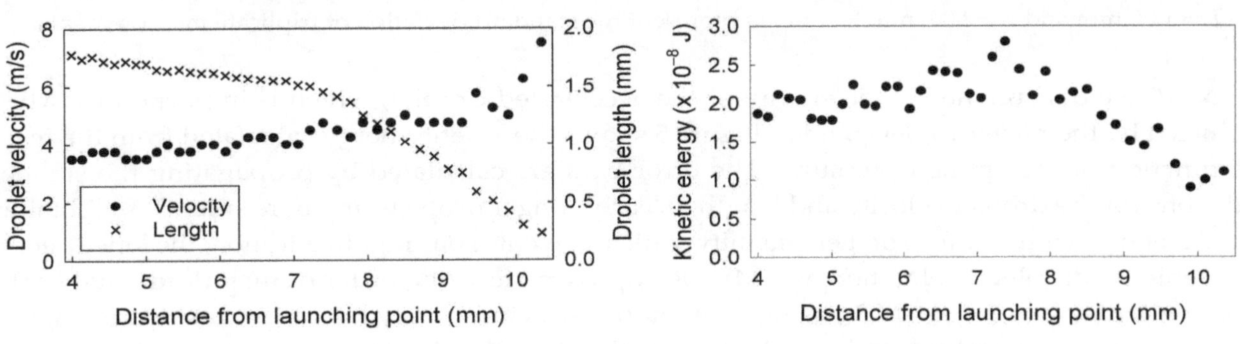

(a) (b)

Figure 3. (**a**) Velocity, length, and (**b**) kinetic energy of accelerated droplet as functions of distance from the launching point in a microchannel with $L = 2.00$ mm at an applied pressure of 1800 kPa.

3.2. Relationship between Droplet Acceleration and Applied Pressure

Figure 4 shows the droplet velocity and length at a 8-mm distance from the launching point, as a function of the applied pressure. The error bars represent the standard deviation of triplicate measurements. As shown in Figure 4a, for both channels (L = 1 and 2 mm), the droplet velocity is around 1 m/s at a pressure below 200 kPa, and increases linearly with increasing pressure above 200 kPa. These results suggest that a longer droplet is more stable without splitting and suitable for higher acceleration. As shown in Figure 4b, the droplet in the L = 1 mm channel does not split at 200 kPa, but rather starts to split at higher pressures, and disappears when the pressure exceeds 400 kPa. This is because the droplet velocity increases with increasing pressure as shown in Figure 4a, which leads to the droplet splitting as a result of wetting and Rayleigh instabilities, triggered by an increase in wall friction. The maximum velocity with keeping the initial length is 1.2 m/s, and that with splitting is 1.8 m/s. In the case of the L = 2 mm channel, the droplet does not split at 600 kPa, but rather at a higher pressure, and disappears at pressures exceeding 1800 kPa. The maximum velocity with keeping the initial length is 2.1 m/s, and that with splitting is 4.3 m/s. Thus, the maximum velocity achieved in the L = 2 mm channel was roughly double that of the L = 1 mm channel. The opposite trend is observed in the case of the fluid resistance, which increases in proportion to the droplet length: compared to the L = 1 mm channel, the droplet in the L = 2 mm channel has twice the fluid resistance, and requires twice the pressure to get a similar velocity. This is because droplet splitting, rather than fluid resistance, is the determining factor in droplet acceleration. Therefore, a design that maintains the droplet stability is most important for high acceleration.

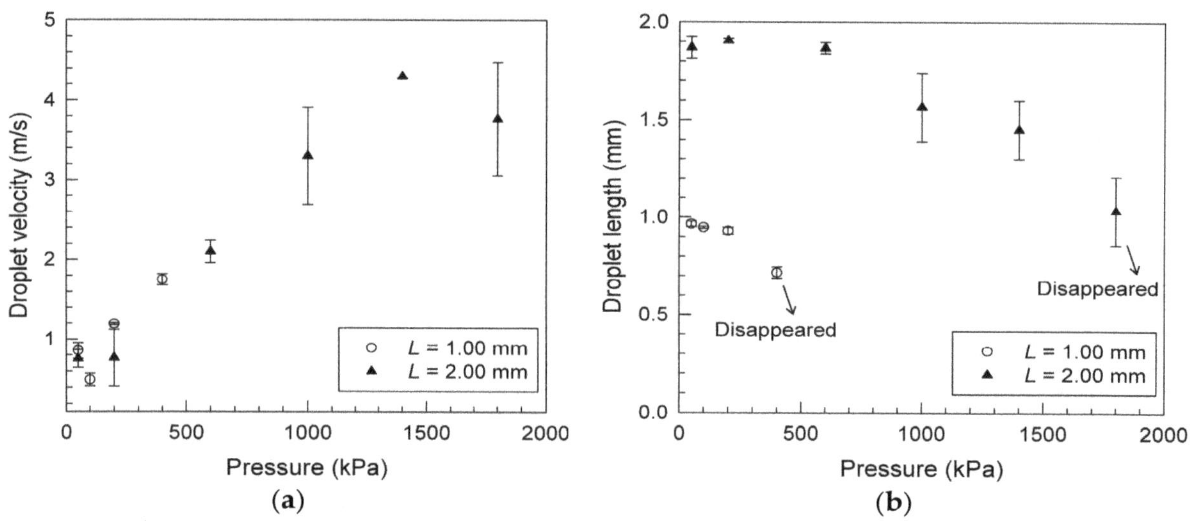

Figure 4. (**a**) Velocity and (**b**) length of droplet as function of applied pressure in microchannels of L = 1.00 mm and L = 2.00 mm. Error bars represent the standard deviation of triplicate measurements.

Next, we discuss the kinetic energy of the accelerated droplet, which is important for MUOs conducted by the microdroplet collider. Figure 5 shows the kinetic energy calculated from the results as a function of the applied pressure. The error bars are calculated by propagating the standard deviations for the droplet velocity and length. Clearly, longer droplets are more suitable for obtaining higher kinetic energy because of their stability, which prevents splitting. In addition, the kinetic energy achievable by droplet acceleration at 1 MPa is approximately one order of magnitude higher than the value achievable at 10^{-1} MPa, as reported in our previous study [11]. The knowledge obtained from these results provides a design guideline for the microdroplet collider. For example, in order to get the highest possible kinetic energy, multistep acceleration with droplet splitting at a pressure of 1 MPa was effective because of the higher droplet velocity. On the other hand, if a well-controlled droplet volume is required for a given application, acceleration without droplet splitting at 10^{-1} MPa would be more suitable. Thus, this study will contribute greatly to applications using the microdroplet collider, such as enhanced mixing/reactions, injection into continuous flows, and chemical injection into cells and tissues, by exploiting spatially and temporally localized kinetic energy.

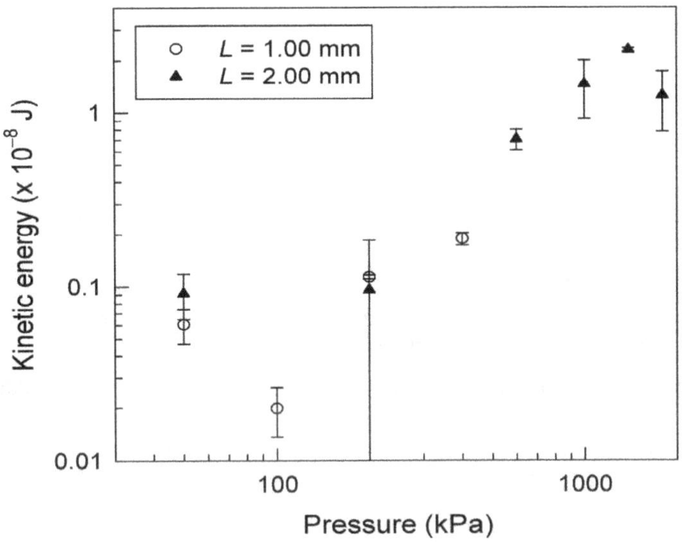

Figure 5. Kinetic energy of the droplet as function of applied pressure in microchannels of $L = 1.00$ mm and $L = 2.00$ mm. Error bars are calculated by propagating the standard deviations for the droplet velocity and length.

Finally, we estimated the level of improvement that can be expected through higher acceleration of the droplet in the case of mixing. In our previous work [11], we observed the mixing process of two droplets by a high-speed CMOS camera combined with an image intensifier. The results suggested that during the mixing, after an accelerated droplet collides with a target droplet, the cusp of the accelerated droplet penetrates the target droplet, maintaining its velocity owing to the effect of inertia. Since the inertial force, with a Reynolds number of 10^1–10^2, is dominant, it is thought that convection, which is proportional to the fluid velocity, is a dominant factor in the mixing. By using a pressure in the order of MPa, we achieved a 10-fold increase in droplet velocity relative to our previous study [11]; thus, a 10-fold increase in the speed of mixing can be expected.

4. Conclusions

We investigated the acceleration of microdroplets with volumes in the order of nL in the gas phase in microchannels using a high-pressure (MPa) control system. We characterized the droplet motion by using a high-speed CMOS camera to determine the droplet velocity and length. A maximum droplet velocity of ~5 m/s was achieved at a high pressure of 1–2 MPa. The velocities achieved by the microdroplet collider, which are comparable to those achieved by inkjet nozzles [16,17], are 100 times faster than the microfluidic transport of droplets in gas phase reported in a previous study [18], and 10,000 times faster than the velocities obtained when simply sliding a droplet on an inclined surface [19,20]. Despite the higher fluid resistance, longer droplets were more suitable for higher acceleration because droplet splitting, which limits the droplet acceleration, could be prevented by using longer droplets. These results suggested that multistep acceleration with droplet splitting at a pressure of 10^0 MPa was effective for obtaining the highest kinetic energy. On the other hand, if a given application requires well-controlled droplet volumes, acceleration without droplet splitting at 10^{-1} MPa pressure would be more suitable. The understanding and insight gained through this study will help to establish design guidelines for the microdroplet collider for various microfluidic applications.

Acknowledgments: This work was supported by a Grant-in-Aid for Challenging Exploratory Research from the Japan Society for the Promotion of Science (JSPS).

Author Contributions: T.K., K.M. and Y.K. conceived and designed the experiments; I.Y. and Y.K. performed the experiments and analyzed the data. Y.K. wrote the paper.

References

1. Janasek, D.; Franzke, J.; Manz, A. Scaling and the design of miniaturized chemical-analysis systems. *Nature* **2006**, *442*, 374–380. [CrossRef] [PubMed]
2. Mawatari, K.; Kazoe, Y.; Aota, A.; Tsukahara, T.; Sato, K.; Kitamori, T. Microflow systems for chemical synthesis and analysis: Approaches to full integration of chemical process. *J. Flow Chem.* **2011**, *1*, 3–12. [CrossRef]
3. Tokeshi, M.; Minagawa, T.; Uchiyama, K.; Hibara, A.; Sato, K.; Hisamoto, H.; Kitamori, T. Continuous-flow chemical processing on a microchip by combining microunit operations and a multiphase flow network. *Anal. Chem.* **2002**, *74*, 1565–1571. [CrossRef] [PubMed]
4. Teh, S.-Y.; Lin, R.; Hung, L.-H.; Lee, A.P. Droplet microfluidics. *Lab Chip* **2008**, *8*, 198–220. [CrossRef] [PubMed]
5. Sun, J.; Wang, W.; He, F.; Chen, Z.-H.; Xie, R.; Ju, X.-J.; Liu, Z.; Chu, L.-Y. On-chip thermo-triggered coalescence of controllable Pickering emulsion droplet pairs. *RSC Adv.* **2016**, *6*, 64182–64192. [CrossRef]
6. Song, H.; Tice, J.D.; Ismagilov, F. A microfluidic system for controlling reaction networks in time. *Angew. Chem. Int. Ed.* **2003**, *47*, 767–772. [CrossRef] [PubMed]
7. Takahashi, K.; Mawatari, K.; Sugii, Y.; Hibara, A.; Kitamori, T. Development of a microdroplet collider; the liquid-liquid system utilizing the spatial-temporal localized energy. *Microfluid. Nanofluid.* **2010**, *9*, 945–953. [CrossRef]
8. Chen, D.; Du, W.; Liu, Y.; Liu, W.; Kuznetsov, A.; Mendez, F.E.; Philipson, L.H.; Ismagilov, R.F. The chemistrode: A droplet-based microfluidic device for stimulation and recording with high temporal, spatial, and chemical resolution. *PNAS* **2008**, *105*, 16843–16868. [CrossRef] [PubMed]
9. Abbott, C.E. A survey of waterdrop interaction experiments. *Rev. Geophys. Space Phys.* **1977**, *15*, 363–374. [CrossRef]
10. Yamada, T.; Sakai, K. Observation of collision and oscillation ofmicrodroplets with extremely large shear deformation. *Phys. Fluids* **2012**, *24*, 022103. [CrossRef]
11. Takahashi, K.; Sugii, Y.; Mawatari, K.; Kitamori, T. Experimental investigation of droplet acceleration and collision in the gas phase in a microchannel. *Lab Chip* **2011**, *11*, 3098–3105. [CrossRef] [PubMed]
12. Ishibashi, R.; Mawatari, K.; Takahashi, K.; Kitamori, T. Development of a pressure-driven injection system for precisely time controlled attoliter sample injection into extended nanochannels. *J. Chromatogr. A* **2012**, *1228*, 51–56. [CrossRef] [PubMed]
13. Hibara, A.; Iwayama, S.; Matsuoka, S.; Ueno, M.; Kikutani, Y.; Tokeshi, M.; Kitamori, T. Surface modification method of microchannels for gas-liquid two-phase flow in microchips. *Anal. Chem.* **2005**, *77*, 943–947. [CrossRef] [PubMed]
14. Takei, G.; Nonogi, M.; Hibara, A.; Kitamori, T.; Kim, H.-B. Tuning microchannel wettability and fabrication of multi-step Laplace valves. *Lab Chip* **2007**, *7*, 596–602. [CrossRef] [PubMed]
15. Birdi, K.S.; Vu, D.T.; Winter, A. A study of the evaporation rates of small water drops placed on a solid surface. *J. Phys. Chem.* **1989**, *93*, 3702–3703. [CrossRef]
16. Son, Y.; Kim, C.; Yang, D.H.; Ahn, D.J. Spreading of an inkjet droplet on a solid surface with a controlled contact angle at low Weber and Reynolds numbers. *Langmuir* **2008**, *24*, 2900–2907. [CrossRef] [PubMed]
17. Shin, P.; Sung, J.; Lee, M.H. Control of droplet formation for low viscosity fluid by double waveforms applied to a piezoelectric inkjet nozzle. *Microelectron. Reliab.* **2011**, *51*, 797–804. [CrossRef]
18. Günter, A.; Jhunjhunwala, M.; Thalmann, M.; Schmidt, M.A.; Jensen, K.F. Micromixing of miscible liquids in segmented gas-liquid flow. *Langmuir* **2005**, *21*, 1547–1555. [CrossRef] [PubMed]
19. Podgorski, T.; Flesselles, J.-M.; Limat, L. Corners, cusps, and pearls in running drops. *Phys. Rev. Lett.* **2001**, *87*, 036102. [CrossRef] [PubMed]
20. Kim, H.-Y.; Lee, H.J.; Kang, B.H. Sliding of liquid drops down an inclined solid surface. *J. Colloid Interface Sci.* **2002**, *247*, 372–380. [CrossRef] [PubMed]

12

Light-Addressable Potentiometric Sensor as a Sensing Element in Plug-based Microfluidic Devices

Ko-Ichiro Miyamoto [1,*], Takuya Sato [1], Minami Abe [1], Torsten Wagner [2], Michael J. Schöning [2,3] and Tatsuo Yoshinobu [1,4]

1 Department of Electronic Engineering, Tohoku University, 6-6-05 Aza-Aoba, Aramaki, Aoba-ku, Sendai 980-8579, Japan; t.sato.tohoku@gmail.com (T.S.); minami.abe.1190@gmail.com (M.A.); nov@ecei.tohoku.ac.jp (T.Y.)

2 Institute of Nano- and Biotechnologies, Aachen University of Applied Sciences, Heinrich-Mußmann-Str. 1, Jülich 52428, Germany; torsten.wagner@fh-aachen.de (T.W.); schoening@fh-aachen.de (M.J.S.)

3 Peter-Grünberg Institute (PGI-8), Research Centre Jülich, Jülich 52425, Germany

4 Department of Biomedical Engineering, Tohoku University, 6-6-05 Aza-Aoba, Aramaki, Aoba-ku, Sendai 980-8579, Japan

* Correspondence: k-miya@ecei.tohoku.ac.jp

Academic Editors: Manabu Tokeshi and Kiichi Sato

Abstract: A plug-based microfluidic system based on the principle of the light-addressable potentiometric sensor (LAPS) is proposed. The LAPS is a semiconductor-based chemical sensor, which has a free addressability of the measurement point on the sensing surface. By combining a microfluidic device and LAPS, ion sensing can be performed anywhere inside the microfluidic channel. In this study, the sample solution to be measured was introduced into the channel in a form of a plug with a volume in the range of microliters. Taking advantage of the light-addressability, the position of the plug could be monitored and pneumatically controlled. With the developed system, the pH value of a plug with a volume down to 400 nL could be measured. As an example of plug-based operation, two plugs were merged in the channel, and the pH change was detected by differential measurement.

Keywords: chemical sensor; plug-based microfluidic device; light-addressable potentiometric sensor

---

1. Introduction

Microfluidic devices are advantageous in various aspects for biological or clinical assays where it is necessary to handle small-volume samples. Common processes such as filtering, mixing, separating, heating, cooling, and sensing of the final products after a series of steps are realized by microfluidic systems constructed on a small substrate known as Micro-TAS or Lab-on-a-chip. They are expected to reduce the cost by minimizing the amount of required samples and reagents.

In pursuit of the volume advantage of microfluidic devices, it is reasonable to divide the sample solution into small volumes and manipulate them in the form of plugs or droplets [1–4]. Consumption of reagents can be saved by reducing the volume, while a plug or a droplet can act as an independent reaction chamber by itself. To realize such a system, a sensing element is required that can probe a plug or a droplet inside the channel.

In this study, we propose a plug-based microfluidic device based on a light-addressable potentiometric sensor (LAPS [5]), which is a chemical sensor based on the field effect in a semiconductor. In our previous papers, we applied a LAPS for measurement of continuously flowing samples in a microfluidic device [6,7], which featured (1) a complete flat sensor surface due to a simple structure consisting of silicon, insulator and electrolyte; (2) a flexible definition of measurement areas by illumination; and (3) the label-free detection and visualization of chemical species. It was also possible to visualize the spatial distribution of ions inside the flow [8]. In the case of continuously flowing samples, however, the consumption of the solution was still large, which spoiled the main advantage of microfluidic devices. To solve this problem, a plug-based version of a microfluidic device combined with LAPS is developed, which can (1) control the position of a plug in the channel; (2) mix two plugs for reaction; and (3) detect the change by differential measurement.

2. Experiments

A microfluidic channel on the sensing surface of the LAPS is depicted in Figure 1a. It consists of a LAPS sensor plate, polydimethylsiloxane (PDMS) film and a glass cover with an Ag/AgCl electrode.

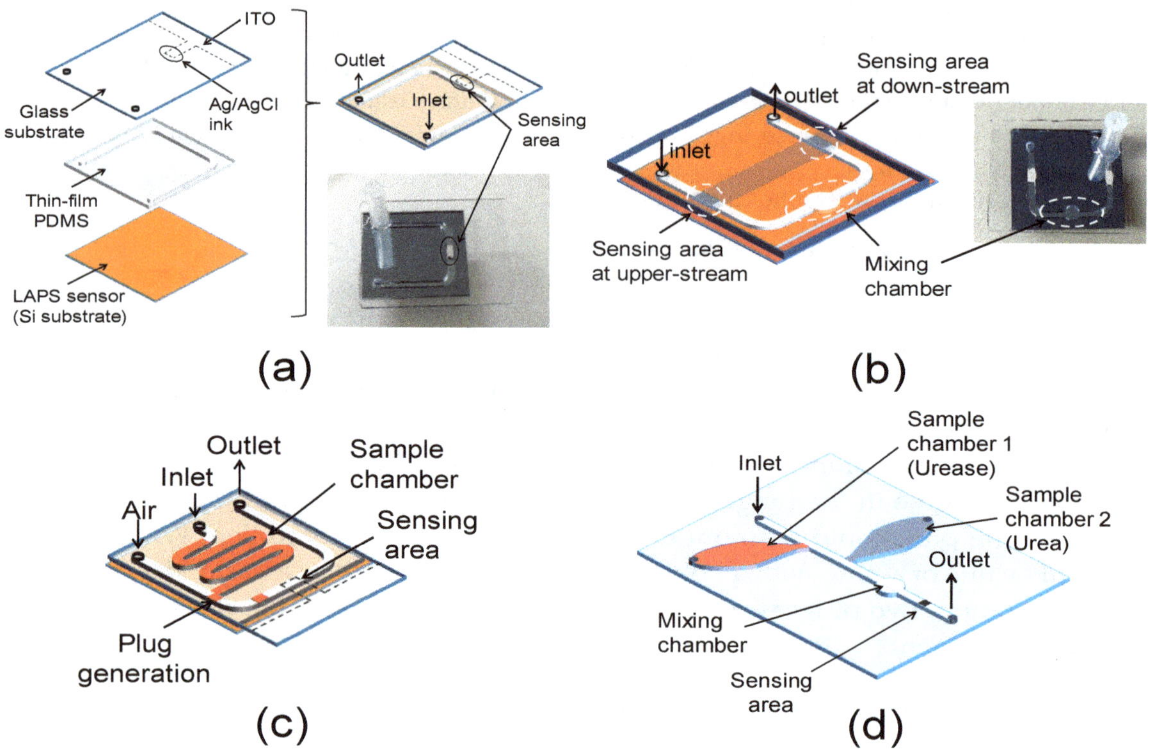

Figure 1. (**a**) Test structure of the microfluidic device combined with LAPS; (**b**) Channel design with a chamber for merging and differential measurement; (**c**) Channel design to generate plugs on chip; (**d**) Test structure with two sample chambers, one merging chamber, and one sensing area.

LAPS: The sensor substrate was n-type silicon with a size of 18 mm × 18 mm and a thickness of 200 μm, insulated with a thermal oxide layer and a silicon nitride layer deposited by LP-CVD [6–8].

PDMS film: A 500-μm-thick PDMS film to define the channel pattern was prepared by casting the PDMS (Silpot 184, Dow Corning Toray Co., Ltd., Tokyo, Japan) in a polished aluminum mold, which was pressed and cured by heating. The width and the height of the U-shaped channel defined by the PDMS film were 1 mm and 500 μm, respectively.

Glass cover and assembly: To allow optical observation of a plug in the channel, a glass cover was used as the ceiling of the channel, on which an Ag/AgCl electrode was prepared by the following process. Firstly, the electrode pattern was defined by a masking tape on an indium tin oxide (ITO)-coated glass plate, and unnecessary ITO was removed by etching with HCl after application of the mixture of zinc powder and glycerin (3 g/mL). Then, the patterned PDMS thin film was bonded onto the glass cover and Ag/AgCl ink for the reference electrode (No. 011464, ALS Co., Ltd., Tokyo, Japan) was painted to cover the ITO pattern inside the channel. The width, the length and the thickness of the Ag/AgCl pattern were 1 mm, 2 mm, and about 70 μm, respectively. Finally, the other side of the PDMS film was bonded onto the sensor surface after opening the inlet and outlet by an ultrasonic drill. Prior to each bonding process, the surfaces were treated by O_2 plasma under a pressure of 0.4 mbar with a power of 10 W for 1 min using a plasma cleaner (Zepto, Diener electronic GmbH, Ebhausen, Germany). The bonding was finalized by heating at 95 °C for 10 min. If an optical observation of the plug is not necessary, the whole area of the ceiling may be covered with Ag/AgCl so that the measurement can be done at any point inside the channel. In this case, the measured points are defined by light beams that induce a photocurrent dependent on the local value of pH.

Figure 1b shows another channel pattern with a wider portion proposed by Itoh et al. [9], in which two successive plugs can be merged. In this structure, the first plug remains inside until the second plug arrives, which is merged with the first plug and leaves the chamber together. Two measurement areas are defined by Ag/AgCl electrodes in the upstream and the downstream of the chamber, which allow differential measurement of the pH change. In this study, this configuration was applied to a measurement of enzymatic reactions, where the two plugs contained an enzyme and its substrate, respectively.

Figure 1c shows a channel pattern with a sample chamber. The chamber acts as a dispenser, which divides the sample solution into a series of plugs. The sample solution in the sample chamber was pushed into the main channel by air from a syringe pump, and another continuous air flow in the main channel repeatedly split the injected sample every time the main channel was occluded. Figure 1d shows an integrated structure of two sample chambers, one merging chamber and one measurement point in the downstream, which was examined in this study.

Measurement setup: The measurement setup used in this study is depicted in Figure 2. A droplet of sample solution is supplied at the inlet using a micro-syringe (SGE analytical science, Victoria, Australia) and sucked into the channel as a plug by a peristaltic pump (AC-2120, ATTO Corp., Tokyo, Japan) when the vent valve (PM-0815W, Takasago Electric Inc., Nagoya, Japan) in the downstream is closed. The position of the plug is controlled by opening/closing the vent valve. The line velocity of the plug was 0.78 mm/s by aspiration of the pump. The sensing area of the LAPS is addressed by illumination of a modulated light from a red-colored LED guided by an optical fiber (ϕ = 1 mm). The photocurrent signal is amplified by a transimpedance amplifier (10^6 V/A) and recorded after 16-bit AD conversion at a sampling frequency of 100 kHz and a sampling number of 10^4. The whole measurement process is controlled by a homemade software developed with LabVIEW (National Instruments, Austin, TX, USA).

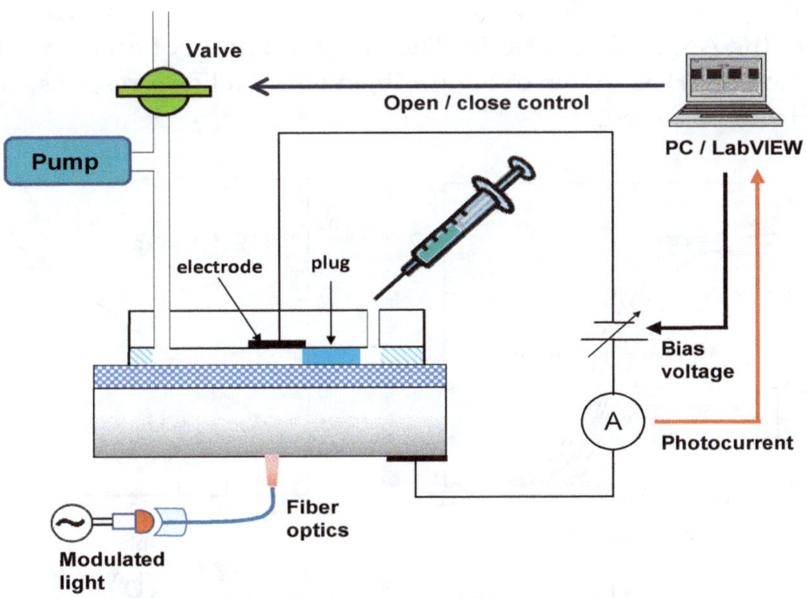

Figure 2. Schematic view of the measurement system.

3. Results and Discussion

3.1. Control of the Plug Motion and LAPS Measurement

After applying a droplet at the inlet and closing the vent valve, the plug moves through the channel at a constant velocity. A bias voltage of −0.5 V is applied to the Ag/AgCl electrode with respect to the silicon substrate and the arrival of the plug at the electrode position is detected by monitoring the photocurrent signal as shown in Figure 3. The increase of the photocurrent was due to the arrival of the plug, which connected the Ag/AgCl electrode and the illuminated point of the sensor. When the photocurrent exceeds a predefined threshold, the vent valve is automatically opened so that the plug stays at the electrode position during the LAPS measurement, in which the photocurrent (I) is recorded as a function of the bias voltage (V). When the I-V curve is obtained, the vent valve is closed again and the plug is led to the outlet.

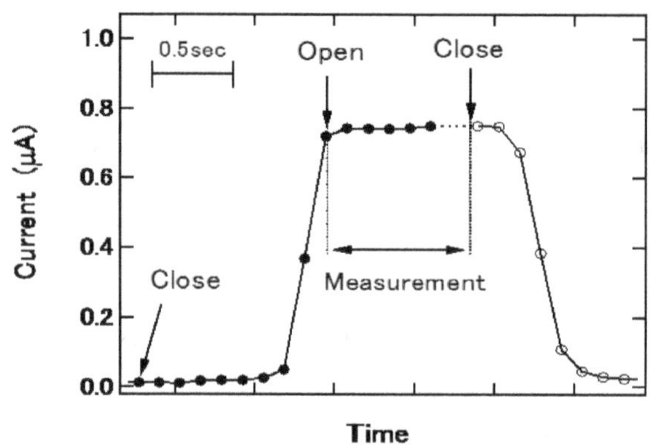

Figure 3. Detection of the plug by the photocurrent signal.

Figure 4a shows an example of I-V curves obtained with the channel in Figure 1a for plugs of pH 4 to pH 10, each with a length of 2 mm and a volume of 1 μL. The bias voltage was swept in the range of −1.0 V to +1.0 V with a step width of 10 mV. In this series of experiments, different amounts of NaCl were added to each pH buffer (Titrisol, Merck, KGaA, Darmstadt, Germany), so that the total chloride concentration of the plug was always 0.05 M, in order to avoid the influence of the chloride sensitivity of the Ag/AgCl electrode. The inflection point of each I-V curve was calculated and plotted as a function of pH as shown in Figure 4b, in which a linear response with a pH sensitivity of 45.9 ± 1.0 mV/pH was observed.

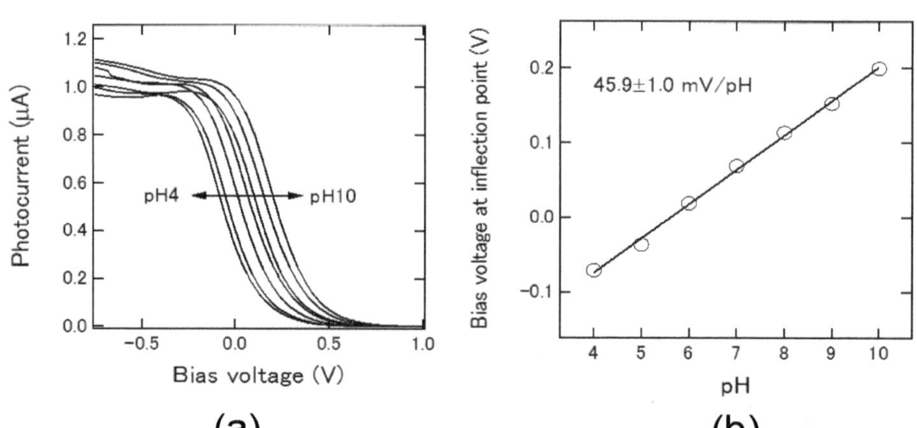

Figure 4. (**a**) I-V curves measured for plugs with different pH values; (**b**) Inflection points of I-V curves plotted as a function of pH.

In order to further reduce the volume of a plug, the height of the channel was reduced from 500 μm to 200 μm, while keeping the same width of 1 mm. A stable plug with a length of 2 mm and a volume of 400 nL could be reproducibly formed. The amplitude of the obtained photocurrent signal was almost the same as in the case of a 1 μL plug, as the contacting area of the plug with the sensing surface and the intensity of illumination were the same. The shift of the I-V curve (data not shown) was again linearly dependent on the pH value, with a pH sensitivity of 47.5 ± 1.9 mV/pH.

3.2. Merging Plugs and Differential Measurement

In many assays, a test reagent is added to the sample to cause some reactions to be detected. The addition of enzyme to its substrate, the addition of antibody to antigen, or vice versa are commonly used in such tests. To mimic such a situation, two plugs were merged inside the microfluidic channel using the structure proposed in Reference [9] and the pH change was detected using the channel pattern in Figure 1b. The measurement areas in the upstream and the downstream of the chamber were simultaneously monitored by using two optical fibers illuminating these areas with two light beams modulated at different frequencies of 15 kHz and 14.9 kHz, respectively. By extracting each frequency component from the obtained photocurrent signal, the pH values at both positions could be independently determined [10]. Firstly, a solution with the total chloride concentration of 0.154 M was prepared by adding NaCl to 1 mM HCl solution. Then, 1 μL of this solution was introduced into the channel as the first plug, which stayed inside the merging chamber. The second plug, 1 μL of 0.154 M NaCl solution, was introduced into the channel and delivered to the first measurement area in the upstream, where the I-V curve in Figure 5a was obtained. The horizontal position of this curve was shifted by −76.8 mV with respect to that of a pH 7 buffer measured at the same position, and considering that the pH sensitivity at this position determined by a preliminary experiment was 51.5 ± 0.9 mV/pH, the pH value of the second plug was calculated to be 5.5. The second plug then entered the merging chamber and the merged plug of 2 μL was delivered to the measurement area in the downstream. The I-V curve for the merged plug is shown in Figure 5b, which is shifted by −145.9 mV with respect to that of a pH 7 buffer measured at the same position. Using the pH sensitivity value of 43.7 ± 1.1 mV/pH at this position, the pH value of the merged plug was calculated to be 3.7. This value was slightly higher than 3.3, calculated for 1 mM HCl after dilution to double. The result in this section confirms that the proposed device can be used for detecting a pH change induced by the addition of a plug of reagent to a plug of sample, which is expected to be applicable to various types of enzymatic and immuno-assays.

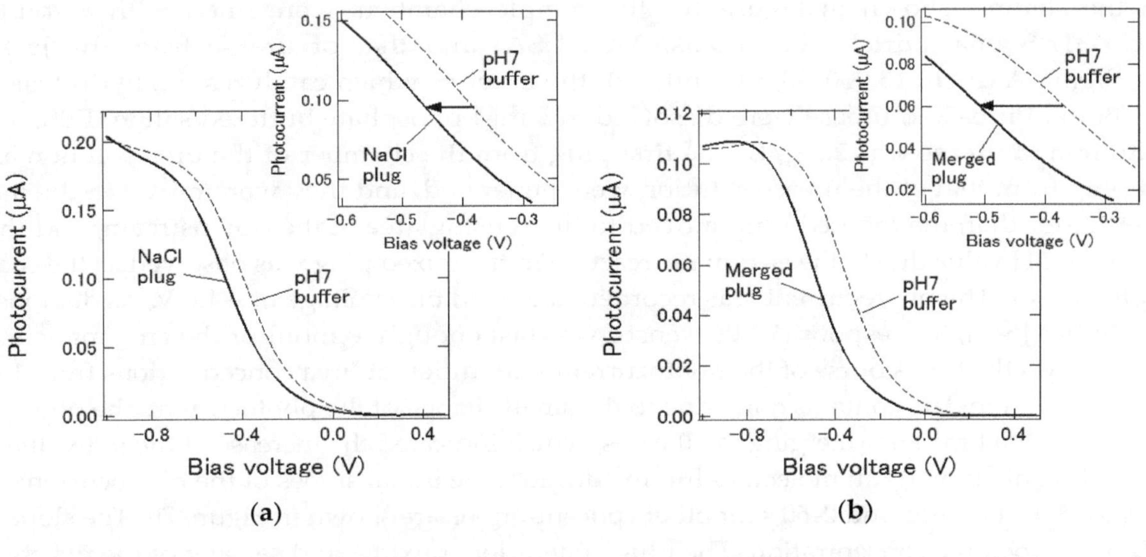

Figure 5. (**a**) I-V curves for the second plug measured in the upstream before merging; (**b**) I-V curves for the merged plug measured in the downstream.

3.3. Plug Generation in the Channel

In the channels shown in Figure 1a,b, the plug was manually prepared before measurement; however, for practical use, it is desired that the plug can be generated on a chip. Thus, plug generation using the channel structure in Figure 1c was examined. Typical air flow in the main channel and the injection speed of the sample were 1.67 μL/s and 3.0 μL/s, respectively. The channel was treated with a fluorine coating agent (FS-1060-2.0, Fluoro Technology Co., Ltd., Aichi, Japan) to make the surface hydrophobic. Figure 6 shows the effect of the treatment on the reproducibility of plug generation. The variation of the plug volume was suppressed by the fluorine coating.

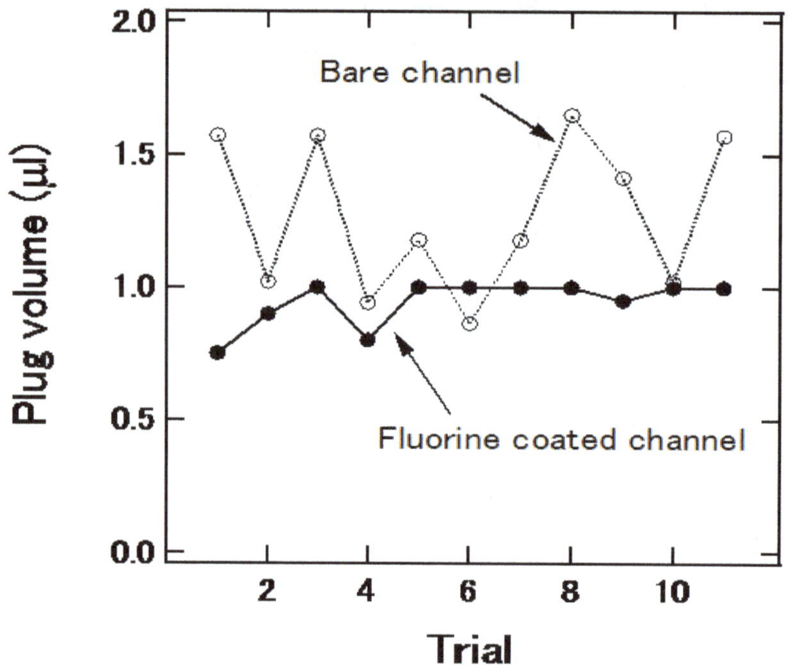

Figure 6. Effect of fluorine treatment of channel on the variation of plug volume.

3.4. Mixing of Two Plugs for Reaction

In the channel shown in Figure 1d, the sample chambers were filled with a solution of urea (U0631, Sigma-Aldrich, St. Louis, MO, USA) and that of urease from the Jack bean (U1500, Sigma-Aldrich, 15,000–50,000 units/g), the latter of which catalyzes the hydrolysis of the former. Both of urea and urease were dissolved in 5 mM phosphate buffered saline (PBS), and the concentration of urease was 2.8 g/L. The first plug from the chamber of the urea solution and the second plug from that of the urease solution were generated, and they were mixed in the merging chamber. After that, the merged plug moved to the sensing area at the downstream, and then the change of the pH value due to the enzymatic reaction in the mixed plug was observed as the change of the photocurrent. The photocurrent was recorded at a fixed bias voltage of −1.0 V. As demonstrated in past studies [6–8], the response of the sensor was fast enough to monitor the enzymatic reaction. Figure 7a shows the time-course of the photocurrent with different urea concentrations from 1 mM to 8 mM. Although the data contains noise due to the small change of the photocurrent, the photocurrent increased with the time after merging in all cases, which indicated the increase of the pH value due to the production of ammonium molecules [6]. In addition, the initial slopes of the photocurrent change in 0–25 s for 8 mM of urea and 0–50 s for other concentrations are shown in Figure 7b. The slope varied depending on the urea concentration. The plug generation, mixing, and sensing of the pH change of the plug on the chip were successfully demonstrated.

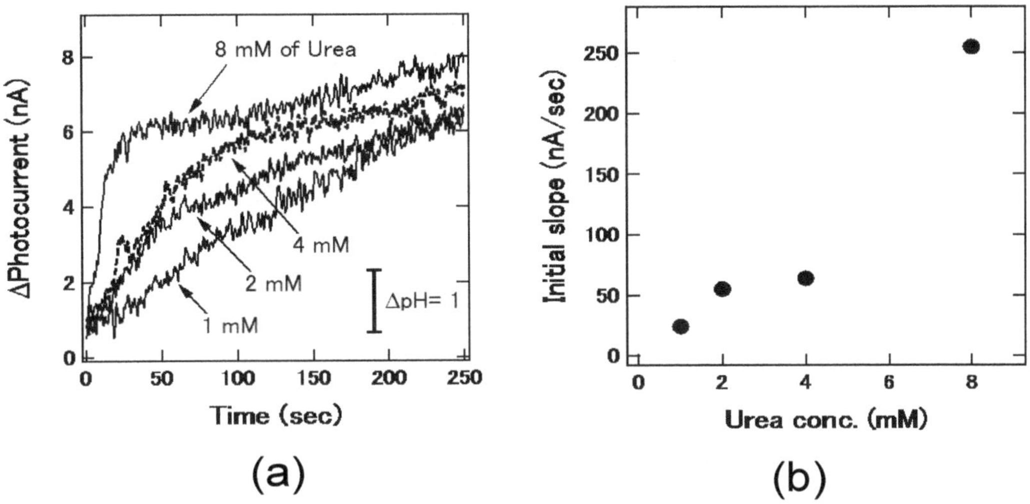

Figure 7. (**a**) Temporal change of photocurrent after merging a plug of urea solution and that of urease solution. The photocurrent response varied depending on concentrations of urea; (**b**) Initial slope of photocurrent change as a function of the urea concentration.

4. Conclusions

A plug-based microfluidic device was developed on the basis of a LAPS. The motion of the plug was pneumatically controlled so that the plug stayed at the electrode position during the LAPS measurement. A linear shift of the I-V curve depending on the pH value was observed for a plug with a volume down to 400 nL. Two successive plugs were merged inside the channel and the pH values before and after merging were successfully measured in the upstream and the downstream of the merging chamber. In addition, plug generation in the channel and detection of enzymatic reaction were also examined. A pH change caused by hydrolysis of urea catalyzed by urease was observed. Based on the light-addressability of LAPS, many detection points can be defined within the channel, and, therefore, the system is expected to be applicable to various kinds of assays where a small volume of samples are tested with reagents.

Acknowledgments: This work was supported by JSPS Grant-in-Aid for Scientific Research (B) (contract No. 24310098). A part of this research was carried out at the Machine Shop Division of Fundamental Technology Center, Research Institute of Electrical Communication, Tohoku University.

Author Contributions: T. Y. and M. J. S. conceived and designed the experiments; T. S. and M. A. performed the experiments; K. M. and T. W. analyzed the data and wrote the paper.

References

1. Atencia, J.; Beebe, D.J. Controlled microfluidic interfaces. *Nature* **2005**, *437*, 648–655. [CrossRef] [PubMed]
2. Garstecki, P.; Fuerstman, M.J.; Stone, H.A.; Whitesides, G.M. Formation of droplets and bubbles in a microfluidic T-junction-scaling and mechanism of break-up. *Lab Chip* **2006**, *6*, 437–446. [CrossRef] [PubMed]
3. Song, H.; Chen, D.L.; Ismagilov, R.F. Reactions in droplets in microfluidic channels. *Angew. Chem.* **2006**, *45*, 7336–7356. [CrossRef] [PubMed]
4. Huebner, A.; Sharma, S.; Srisa-Art, M.; Hollfelder, F.; Edel, J.B.; deMello, A.J. Microdroplets: A sea of applications? *Lab Chip* **2008**, *8*, 1244–1254. [CrossRef] [PubMed]
5. Hafeman, D.G.; Parce, J.W.; McConnell, H.M. Light-addressable potentiometric sensor for biochemical systems. *Science* **1988**, *240*, 1182–1185. [CrossRef] [PubMed]
6. Miyamoto, K.; Yoshida, M.; Sakai, T.; Matsuzaka, A.; Wagner, T.; Kanoh, S.; Yoshinobu, T.; Schöning, M.J. Differential setup of light-addressable potentiometric sensor with an enzyme reactor in a flow channel. *Jpn. J. Appl. Phys.* **2011**, *50*, 04dl08. [CrossRef]

7. Miyamoto, K.; Hirayama, Y.; Wagner, T.; Schöning, M.J.; Yoshinobu, T. Visualization of enzymatic reaction in a microfluidic channel using chemical imaging sensor. *Electrochim. Acta* **2013**, *113*, 768–772. [CrossRef]
8. Miyamoto, K.; Itabashi, A.; Wagner, T.; Schöning, M.J.; Yoshinobu, T. High-speed chemical imaging inside a microfluidic channel. *Sens. Actuators B Chem.* **2014**, *194*, 521–527. [CrossRef]
9. Itoh, D.; Sassa, F.; Nishi, T.; Kani, Y.; Murata, M.; Suzuki, H. Droplet-based microfluidic sensing system for rapid fish freshness determination. *Sens. Actuators B Chem.* **2012**, *171–172*, 619–626. [CrossRef]
10. Zhang, Q.; Wang, P.; Parak, W.J.; George, M.; Zhang, G. A novel design of multi-light LAPS based on digital compensation of frequency domain. *Sens. Actuators B Chem.* **2001**, *73*, 152–156.

13

High-Resolution Microfluidic Paper-Based Analytical Devices for Sub-Microliter Sample Analysis

Keisuke Tenda, Riki Ota, Kentaro Yamada, Terence G. Henares, Koji Suzuki and Daniel Citterio *

Department of Applied Chemistry, Keio University, 3-14-1 Hiyoshi, Kohoku-ku, Yokohama 223-8522, Kanagawa, Japan; kei-54ppp@keio.jp (K.T.); ricky@keio.jp (R.O.); ymkn.z3@keio.jp (K.Y.); tghenares@gmail.com (T.G.H.); suzuki@applc.keio.ac.jp (K.S.)

* Correspondence: citterio@applc.keio.ac.jp

Academic Editors: Manabu Tokeshi and Kiichi Sato

Abstract: This work demonstrates the fabrication of microfluidic paper-based analytical devices (μPADs) suitable for the analysis of sub-microliter sample volumes. The wax-printing approach widely used for the patterning of paper substrates has been adapted to obtain high-resolution microfluidic structures patterned in filter paper. This has been achieved by replacing the hot plate heating method conventionally used to melt printed wax features into paper by simple hot lamination. This patterning technique, in combination with the consideration of device geometry and the influence of cellulose fiber direction in filter paper, led to a model μPAD design with four microfluidic channels that can be filled with as low as 0.5 μL of liquid. Finally, the application to a colorimetric model assay targeting total protein concentrations is shown. Calibration curves for human serum albumin (HSA) were recorded from sub-microliter samples (0.8 μL), with tolerance against ±0.1 μL variations in the applied liquid volume.

Keywords: μPAD; wax printing; inkjet printing; colorimetry; protein assay

---

1. Introduction

In recent years, microfluidic paper-based analytical devices, commonly referred to as μPADs, have gained a lot of attention as potential alternative tools for various analytical tasks. The attractive features of μPADs are to a large extent related to the use of paper as the substrate material and include low cost, easy disposability, as well as external power-free sample transport driven by capillary forces. In addition, μPADs are generally easy to fabricate, user-friendly, and offer simple readouts of analytical assay results. The continuously growing interest in this type of paper-based devices is demonstrated by the fact that more than 100 research articles including the term "μPADs" have been published during the 2014–2015 period alone [1]. The advantageous characteristics described above have led to the application of μPADs especially for biomedical and environmental analysis, as well as for the monitoring of food and beverage contamination [2–6].

One disadvantage of μPADs in general, however, is that they mostly require larger sample volumes compared to conventional microfluidic devices based on glass or polymeric substrates [7–9]. This is due to the mostly larger dimensions of microfluidic structures patterned on paper substrates and the fact that μPADs are generally open systems prone to evaporation of sample fluids during the slow wicking process in paper microfluidic channels. Although there are many types of practical applications where this is not an issue, it will pose restrictions on the use of μPADs with samples that are only available in very small quantities, such as tear fluid [10] or blood plasma obtained from a finger prick blood sample passed through a blood cell retaining filter [11]. Sample volumes in the sub-microliter order are not sufficient to wet out the microfluidic structure of μPADs and to reach the zone where signal creation takes place.

Since the first report on photolithographically patterned μPADs by the Whitesides group in 2007 [12], a large number of more low cost, simple, and rapid alternative patterning methods for paper substrates has been developed. Major approaches include wax printing, inkjet printing, flexographic printing, knife cutting, laser cutting, and stamping [2]. Microfluidic channels with widths down to the order of 300 μm have been achieved. Among all of these methods, wax printing has evolved into the most widely used patterning technique, because of its ease and flexibility of computer-based pattern design (Adobe Illustrator, Microsoft PowerPoint, *etc.*), small number of process steps (printing followed by heating), and suitability for mass production at low cost [13,14]. The method relies on the printing of a solid wax-based ink in a pattern outlining the desired microfluidic structure, followed by a heating step wherein the wax is melted to penetrate into the paper, finally resulting in the formation of water impermeable hydrophobic barriers throughout the thickness of the paper.

However, fabricating highly resolved microfluidic structures meeting the requirements for sub-microliter sample volume assays by this technique remains challenging. One main issue is that the printed wax does not only vertically penetrate into the paper substrate to form the desired hydrophobic barriers, but at the same time also undergoes horizontal diffusion, leading to an inevitable blurring of the originally sharp printed structures [13,15]. This drawback prevents the applicability of the simplest paper patterning method for the fabrication of μPADs applicable to assays with very small sample volumes. Although a paper-based device obtained by a craft punch technique suitable for the handling of sub-microliter sample volumes has been reported, no sample liquid transport in a microfluidic channel is involved, and samples have to be applied directly to the detection zone [16].

The goal of the present work is the fabrication of μPADs enabling the distribution of a sub-microliter sample from a single inlet into multiple detection zones. Focus is set on a hot laminator as the post-print heating method, since one cause of horizontal wax diffusion lies in the use of hot plates or ovens conventionally chosen to melt the printed wax into the paper substrate. If a laminator is used in combination with lamination films, the evaporation of sample fluid can be prevented at the same time. Therefore, this approach is expected to eliminate the two major causes for the requirement of larger sample volumes with μPAD assays compared to conventional microfluidics mentioned above. Finally, the closed space produced by lamination of the entire device contributes to a more controlled sample uptake by the μPAD, increasing the tolerance in the amount of applied sample volume by the end user. This is important when working with minuscule sample volumes, where precise pipetting is more challenging. Although the combination of wax printing with a hot laminator has been used before [15,17], high-resolution μPADs for the analysis of sub-microliter samples, to the best of our knowledge, have not been investigated.

2. Materials and Methods

2.1. Reagents and Equipment

All reagents were used as received without further purification. Citric acid and tetrabromophenol blue (TBPB) were purchased from Sigma-Aldrich (St. Louis, MO, USA). The phospholipid 1,2-dioleoyl-*sn*-glycero-3-phosphocholine was obtained from TCI (Tokyo, Japan), and ethanol (for HPLC 99.5%) from Kanto Chemical (Tokyo, Japan) and *N*-(2-hydroxyethyl)-1-piperazine ethanesulfonic acid (HEPES) was purchased from Rikaken (Aichi, Japan). Sodium hydrogen phosphate, sodium chloride, potassium chloride, calcium chloride, magnesium chloride hexahydrate, and human serum albumin (HSA) were purchased from Wako Pure Chemical Industries (Osaka, Japan). Whatman No.1 filter paper (460 mm × 570 mm) was purchased from GE Healthcare Life Sciences (Buckinghamshire, UK). A Xerox ColorQube 8570 wax printer (Xerox, Norwalk, CT, USA) was used to print microfluidic structures designed in Adobe Illustrator CC software. The colorimetric assay reagent was deposited onto the paper devices using an iP2700 inkjet printer (Canon, Tokyo, Japan), where the original ink cartridges have been cut open and cleaned, and the sponges removed. Hot lamination films (thickness 150 μm) were obtained from Jointex (Tokyo, Japan). A Silhouette Cameo electronic knife blade cutting device

(Silhouette, Lehi, UT, USA) in double cutting mode was used to cut sample inlet holes into the top lamination film layer. Hot lamination was performed on a QHE325 laminator (Meiko Shokai, Tokyo, Japan). The instrument used only allows for the optimization of substrate thickness and feeding speed. In all experiments, substrate thickness was set to the maximum available value (the "250 μm" position) and the "fast" feeding speed (experimentally measured as 39.4 ± 0.4 cm/min; n = 8). According to information provided by the instrument manufacturer, the maximum thickness setting corresponds to maximum heating temperature, but temperature values are not provided. The hot plate used in comparison experiments (Nissin NHS-450ND) was obtained from Nissinrika (Tokyo, Japan). A 9000F MARK II color scanner (Canon, Tokyo, Japan) was used to acquire colorimetric data, and the color intensity was measured by the Image J color image analysis software (NIH, Bethesda, MD, USA). A DVM2500 digital microscope (Leica, Wetzlar, Germany) was used for measuring the widths of hydrophobic barriers and microfluidic channels.

2.2. Device Fabrication

In general, identical patterns of wax were printed in alignment on both sides of the filter paper substrate cut into A4 size before use. An exception are the experiments described in Figure S1 of the Supplementary Materials, where patterns were only printed on one side of the paper. After printing, the wax-modified paper substrate was placed between the two layers of the laminate film, with 0.85 mm diameter holes cut out of the top layer for the sample inlet if required. The aligned devices were passed through the hot laminator twice in order to melt the printed wax into the paper. Only for comparison experiments discussed below Figure 1, wax was melted into paper substrates by hot lamination without laminate film coverage on the top side. In this case, the wax-modified side of the paper was covered with aluminum foil to prevent the wax from contaminating the rollers of the hot laminator, while the back side of the paper was in contact with laminate film. Alternatively, the wax was melted into the paper by placing it on the hot plate at 150 °C for 180 s. A solution of 0.5 wt % of acid yellow was used to visualize the microfluidic structures in all cases.

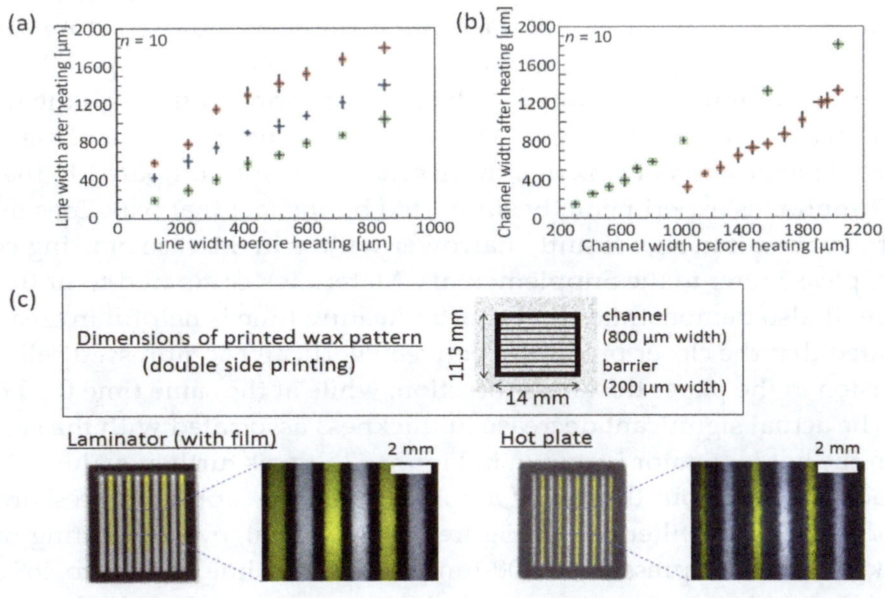

Figure 1. Comparison of resolution achieved for melting of printed wax into paper substrates by hot plate or hot lamination. (**a**) Wax line widths observed directly after printing and after heating with hot plate (red circles), hot laminator without top side lamination (blue diamonds), and hot laminator with full lamination (green squares) (mean value ± 1σ); (**b**) Microfluidic channel widths after heating by hot plate (red circles) or hot laminator with full lamination (green squares) (mean value ± 1σ); (**c**) Dimensions and photographs of 10 parallel microfluidic channels after lamination or hot plate (150 °C for 15 s) treatment. Channels visualized by application of colored aqueous solution.

For devices used in HSA assays, the colorimetric reagent TBPB was dissolved in 95% (v/v) aqueous ethanol at a concentration of 3.3 mM and mixed with citric acid buffer (pH 4.00, 0.10 M) in a 1:1 ratio. This reagent ink was filled into the black ink cartridge of the inkjet printer and deposited before hot lamination with the printer in black-and-white mode (R, G, and B values set to 0) in 20 printing cycles as a rectangle with a height of 1.4 mm covering all of the detection areas.

2.3. Human Serum Albumin Assay

HSA standard solutions of various concentrations (0, 2, 4, 6, 7, 8, 9, 10 mg/mL) were prepared in a HEPES buffer (pH 7.40, 10 mM) with a background of 150 mM NaCl, 20 mM KCl, 1.0 mM $CaCl_2$, 0.6 mM $MgCl_2$, and 0.36 mM phospholipid, simulating the composition of human tear fluid [18,19]. The standard solutions (0.7–0.9 μL) were pipetted onto the sample inlet port of the μPADs. After drying, the devices were scanned, and the color intensities measured. Throughout this work, the mean red (R) intensities obtained from the four detection areas were used as the quantitative signal.

3. Results and Discussion

3.1. Heating Method of Melting Printed Wax into Paper Substrates

At first, it was experimentally demonstrated in terms of the sharpness and resolution of patterned features that hot lamination offers advantages over a hot plate in melting printed wax into filter paper substrates. For this purpose, black wax lines (100–900 μm wide, in increments of 100 μm) were printed on the top side of the filter paper substrate. Similarly, two parallel lines (400 μm wide) were printed with various gaps (laminator: 300–2000 μm, hot plate: 1000–2000 μm) for the comparison of formed hydrophilic channel widths. As previously mentioned, horizontal wax diffusion inevitably reduces the resolution of printed patterns. However, Figure 1a demonstrates that the line widths after heating by a hot laminator are smaller than those obtained by hot plate treatment, and are even further reduced by full lamination (film thickness of 150 μm). Accordingly, decreased horizontal diffusion by hot lamination compared to plate heating also results in wider hydrophilic channels for patterns printed with equal line distances (Figure 1b). Patterning of microfluidic structures by hot lamination results in features with visibly higher resolution, as demonstrated by the 10 parallel microchannels in Figure 1c. The better resolution is attributed to the pressure applied through the hot rollers during the heating step from both sides of the paper, which results in not only shorter heating times but also an enhanced vertical penetration of wax over horizontal diffusion compared to the hot plate-based method. This assumption is experimentally supported by the fact that wax lines melted into paper by hot plate treatment result in significantly narrower widths in the case of using compressed filter paper (for details, please refer to the Supplementary Materials), compared to untreated filter paper (Figure S1). Figure S1 also demonstrates that shorter heating time is helpful in creating thinner wax lines. It is postulated that the closer proximity between vertically compressed cellulose fibers leads to dominant diffusion in the paper thickness direction, while at the same time the horizontal flow of wax is reduced. The actual significant decrease in thickness associated with the compression of the paper substrate in the hot laminator is shown in Figure S2. It was further evaluated whether similar effects could be achieved without the use of a hot laminator by applying pressure during the hot plate treatment of wax-printed filter paper. Figure S1 shows that, even if heating time is shortened to 30 s under 22 kg of uniform pressure, a 300-μm printed wax line widens to 869 $\pm$ 32 μm (n = 5). This value is much larger than in the case of the full lamination-based method (405 $\pm$ 47 μm; n = 10) (Figure 1a). Shortening the heating time further and placing heavier weights on the hot plate would be detrimental to experimental simplicity. It can be concluded that the conventional hot plate treatment cannot compete with the presented hot lamination method in terms of achievable resolution.

The horizontal error bars in Figure 1a,b, indicating the standard deviations in line and channel widths before heating, are an indication of the achievable precision of the wax printing method itself and of variations in line widths caused by the inherent roughness of the filter paper used.

3.2. Device Design and Optimization

In order to achieve high-resolution μPADs, which can be used with sample volumes of no more than 1.0 μL, a variety of factors had to be considered. Besides the more obvious need to create the narrowest possible reliable hydrophobic barriers and channel widths for decreasing the overall size of the system, factors such as the influence of cellulose fiber direction and overall device geometry on sample wicking were evaluated before deciding on an appropriate design model. Throughout this work, microfluidic structures were created by printing identical patterns on both sides of the filter paper. This is in contrast to previous reports on wax-printed μPADs, where patterns are printed on one surface of the substrate and allowed to penetrate throughout the thickness of the paper during the post-print heating process. The double-sided patterning approach allows the printing of narrower line features, since wax penetrates into the paper from two directions, reducing the amount of ink required on each surface of the paper. It should be noted that the printing of identical patterns in full alignment on opposite paper surfaces calls for careful handling. First, if paper substrates are not precut, size-reproducible cutting is required, but can be readily achieved by using a paper cutting device. Second, the absolute position of features printed on the two paper surfaces needs to be fine-tuned in the graphic software, since actually printed patterns subtly go out of alignment upon printing on both sides of a paper sheet. Shifts need to be applied particularly perpendicular to the paper feeding direction. Thus, a trial-and-error approach is initially inevitable to perfectly match wax patterns on both sides of the paper substrates, since the required shifts are dependent on the position along the paper feeding direction. Finally, attention must be paid to reversible paper feeding into the printer, which is best achieved by relying on the manual feed tray, where the paper can be manually placed and fixed.

3.2.1. Influence of Cellulose Fiber Direction

During the general paper fabrication process, cellulose fibers become dominantly aligned parallel to a specific direction, known as the machine direction (MD), in contrast to the perpendicular cross direction (CD) [20]. In the context of μPADs and in particular with narrow flow channels, this fiber direction is expected to affect the sample wicking process. To realize μPADs for the analysis of sub-microliter samples, the influence of different fiber directions has to be taken into account. For this reason, the effects of fiber directions were experimentally evaluated by measuring the maximal sample flow distance of a colored aqueous solution in differently orientated microfluidic channels (Figure 2a). The corresponding results are shown in Figure 2b.

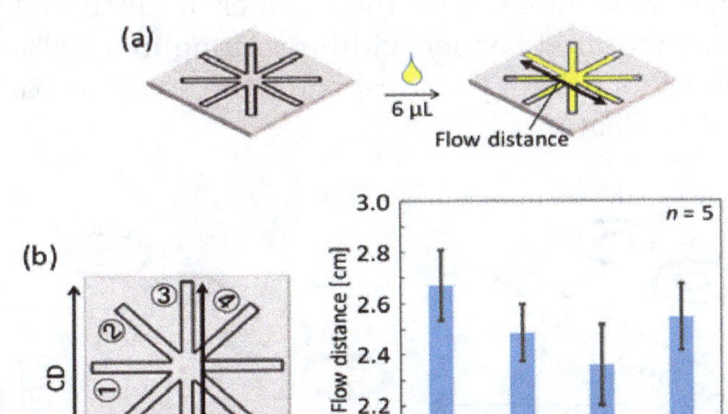

Figure 2. (**a**) Schematic representation of the evaluation of the influence of cellulose fiber direction on sample wicking in patterned filter paper (channel width: 553 ± 31 μm (n = 20) after lamination). The flow distances are measured as indicated by the arrow; (**b**) Quantitative results averaged for 5 independently fabricated devices (mean value ± 1σ). Circled numbers indicate the respective flow direction.

Figure 2b clearly demonstrates that significantly longer flow distances are achieved in microfluidic channels aligned to the machine direction (① in Figure 2b) with parallel orientation of the cellulose fibers, compared to the cross direction (③ in Figure 2b), where flow occurs perpendicular to the cellulose fiber orientation. Channels arranged in a 45° angle to MD or CD (② and ④ in Figure 2b) showed flow distances approximately between the two extremes. The minor differences observed for the two 45° oriented channels (② and ④), as well as the relatively large error bars, can be attributed to the fact that, in spite of the paper-machining-induced general trend of fiber orientation, random local variations do exist. These results confirm that the cellulose fiber direction in the paper substrate does affect sample flow distance. To further investigate the influence of fiber direction on sample fluid transport, the flow velocity of dye solutions has been measured in single channels of different orientation. As shown in Figure S3, fiber orientation does influence the wicking speed. Observed flow speeds decrease in the order of MD-oriented (③ in Figure S3b), 45° rotated (② in Figure S3b), and CD-oriented (① in Figure S3b) microfluidic channels. These effects have to be taken into account when designing a μPAD for small sample volumes, where the machine direction is the more suitable orientation for efficient sample liquid transport. As a consequence, all microfluidic channels described in the following sections have been patterned along MD, with the exception of circular shapes, where the cellulose fiber orientation is irrelevant.

3.2.2. Optimization of Flow Channel Barrier Width

One important factor determining the achievable spatial resolution of the microfluidic structure on a μPAD is the thickness of the hydrophobic barriers that is required to prevent the leaking of the transported sample liquid. Thinner barriers allow for higher densities of microfluidic channels, which contributes to an overall size reduction of devices and can therefore work with smaller sample volumes. The narrowest printed wax line widths resulting in reliable fluid barriers were experimentally determined. Leaking tests were performed similarly to a previously described method, as schematically outlined in Figure 3 [15]. Two concentric circles of wax with 1.5-mm and 4.0-mm diameters were printed on both sides of the filter paper and melted into the paper by hot lamination. While the line width of the outer circle was kept constant at 500 μm, the value of the inner circle was varied between 200–350 μm in increments of 50 μm.

A barrier is regarded as being functional, if no leaking of liquid from the inner circle occurs. The performance of a hydrophobic barrier depends on the applied sample volume per area. For this reason, the current experiment was performed with samples of larger volume per area compared to those encountered in a final device. Table 1 lists the number of functional barriers obtained among 20 repetitions. The minimally required barrier width resulting in a 100% success rate (20/20) was determined as 300 μm (value set for printing), corresponding to an actual barrier width of 467 ± 33 μm after passing through the hot laminator.

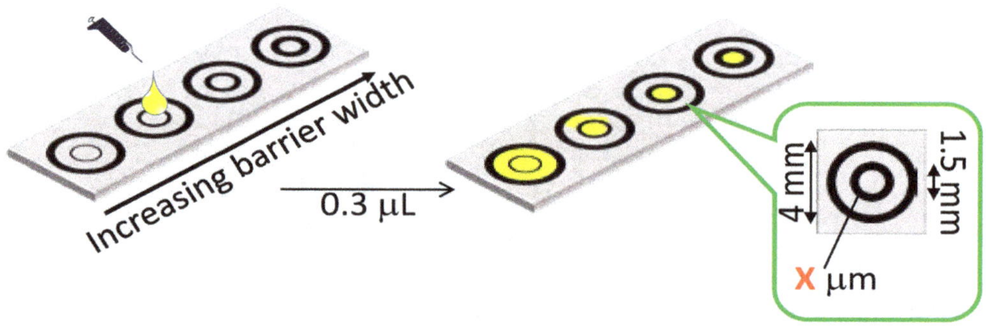

Figure 3. Schematic of the experimental method used to evaluate the minimally required wax barrier width. The indicated dimensions (x = 200–350 μm) refer to the values set for wax printing and do not represent the actual dimensions obtained after hot lamination.

Table 1. Evaluation of minimally required printed wax barrier width (images converted to grayscale and contrast adjusted for improved visibility).

Printer Set Wax Barrier Width (μm)	200	250	300	350
Number of functional barriers (n = 20)	13/20	12/20	20/20	20/20
Actual pictures	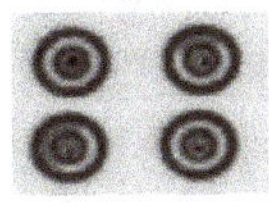	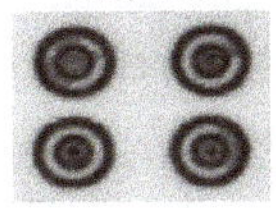	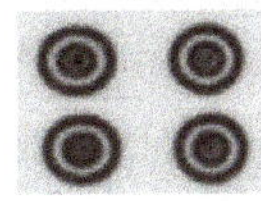	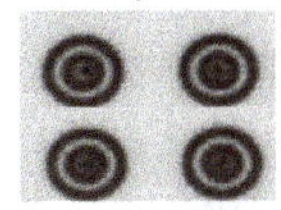

3.2.3. Optimization of Flow Channel Width

The width and length of the microfluidic channels in a μPAD directly influence the minimal amount of sample liquid required to fully wet the device. While the channel lengths can be adapted to the needs arising from specific assays to be performed on a μPAD, their minimal width is determined by the requirement to achieve reliable fluid transport. To decide on the narrowest channel width for unhindered wicking of the sample, liquid flow with varying channel widths was experimentally evaluated as shown in Figure 4. Based on the results described above, flow channels were aligned in parallel to the cellulose fiber direction of the paper (MD) for this experiment and printed with 300-μm barrier line widths and channel widths of 300–500 μm on both sides of the filter paper.

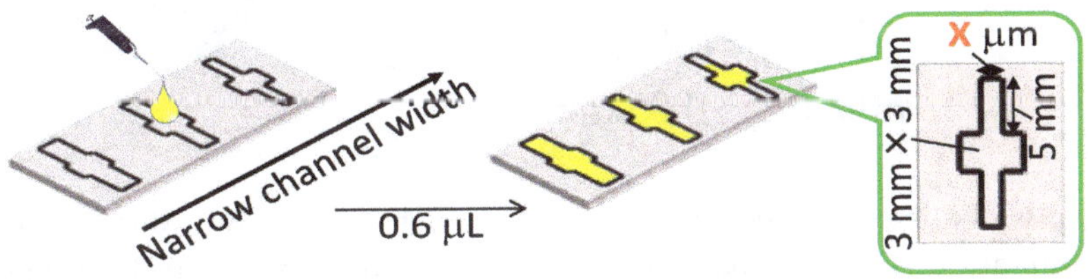

Figure 4. Schematic of the experimental method used to evaluate the minimally required width of microfluidic channels aligned to the cellulose fiber direction in the filter paper. The indicated dimensions refer to the values set for wax printing and do not represent the actual dimensions obtained after hot lamination.

A sample volume of 0.6 μL was selected to guarantee a liquid volume sufficient for complete wetting of the microfluidic paper structure with the widest channels evaluated. Table 2 summarizes the results of 20 independent trials for each channel width. As in the case of optimizing the hydrophobic barrier width, a flow channel was only regarded as reliable in the case of a 100% success rate (20/20). According to Table 2, 400 μm of printed channel width is the narrowest channel value to obtain unhindered sample transport. This printer set width results in actual microfluidic flow channels of 228 ± 30 μm after lamination, which is currently the narrowest value reported by any printing method for μPADs.

Table 2. Evaluation of narrowest possible printed microfluidic channel width (images converted to grayscale and contrast adjusted for improved visibility).

Printer Set Channel Width (μm)	300	400	500
Number of functional channels (n = 20)	4/20	20/20	20/20
Actual pictures	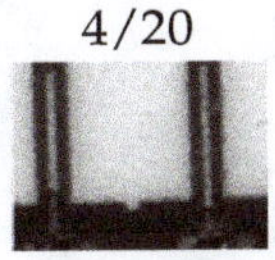	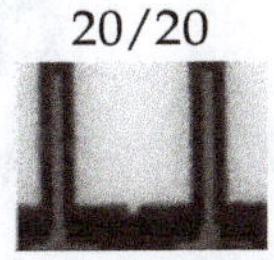	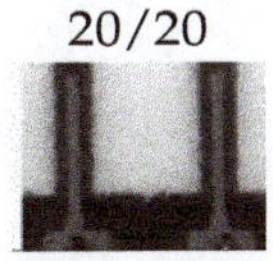

3.2.4. Optimized μPAD Design

Based on the experimental results described above, the following conditions were considered to be important to obtain a μPAD, in which a single sub-microliter sample is reliably and reproducibly distributed into multiple detection zones through a network of microfluidic channels: (a) double-sided wax printing of aligned identical patterns; (b) a symmetric pattern design with all microfluidic channels aligned parallel to the cellulose fiber direction of the paper; and (c) identical flow distances from the sample inlet to all detection zones. One possible device design meeting these conditions is shown in Figure 5. It consists of four detection zones located at the end of microfluidic flow channels and one single sample inlet. The size of the detection zones has been selected to allow for the acquisition of colorimetric signals by conventional desktop scanners without having to refer to microscopic techniques.

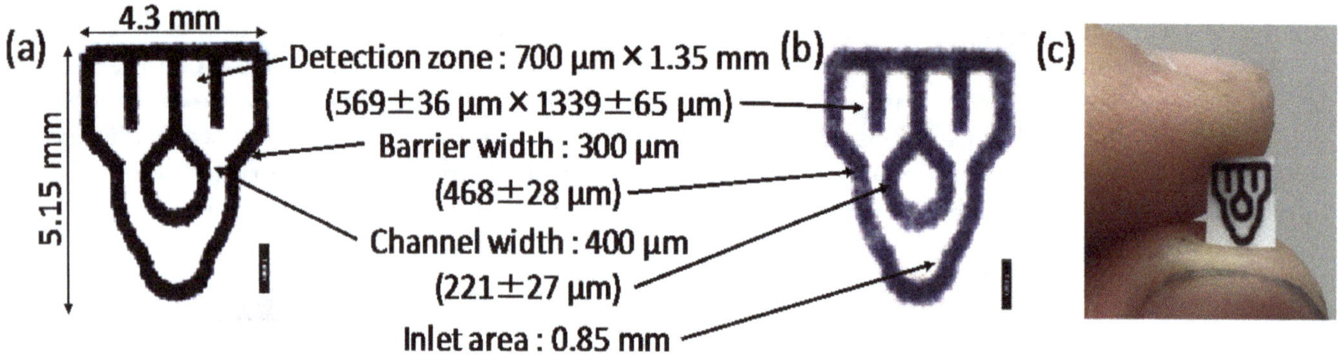

Figure 5. Micrographs of a wax-printed microfluidic paper-based analytical device (μPAD): (**a**) before and (**b**) after hot lamination. The scale bars correspond to a length of 1 mm. Dimensions are indicated before and after hot lamination (values in parentheses). (**c**) Corresponding photograph.

The minimum amount of aqueous sample liquid required to fully wet all of the four detection zones of the μPAD was evaluated by applying 0.3–0.6 μL of colored aqueous solution. Table 3 lists the results of five independent trials for each respective liquid volume. The results demonstrate that a volume of 0.5 μL is sufficient to reliably fill all of the four detection zones. The experiment also reveals that lower sample volumes lead to incomplete wetting of detection zones with significant differences in flow distance. An optimized symmetric pattern design with identical lengths of flow channels cannot completely eliminate the influence of inherent spatial differences in a filter paper substrate in terms of cellulose fiber orientation and density. The present required liquid volume of 0.5 μL is to the best of our knowledge the lowest reported for a paper-based device with a single sample inlet and a patterned microfluidic structure terminating in multiple detection zones.

Table 3. Evaluation of the minimal sample volume required to completely fill the model μPAD with multiple detection zones; the scale bars correspond to a length of 1 mm (images converted to grayscale and contrast adjusted for improved visibility).

Sample Volume (μL)	**0.3**	**0.4**	**0.5**	**0.6**
Success rate (n = 5)	0/5	1/5	5/5	5/5
Actual pictures				

3.3. Human Serum Albumin (HSA) Assay

3.3.1. Colorimetric Assay with Sub-Microliter Sample Volume

The colorimetric quantification of total protein content, with human serum albumin (HSA) as a representative protein, was used to demonstrate the analytical performance of an optimized μPAD design in a simple model assay. The detection relies on the pH-sensitive indicator tetrabromophenol blue (TBPB), which upon interaction with proteins at acidic pH undergoes a color change from yellow to blue [21]. A sample matrix simulating the composition of human tear fluid was selected, since it represents a type of sample typically only available at small volumes. The identical colorimetric reagent was deposited onto all of the four detection zones of the μPAD before lamination using a conventional desktop inkjet printer. In order to eliminate any issues caused by a potential misalignment of printed reagent ink, TBPB solution was printed in a continuous rectangular shape covering all detection zones. The chosen arrangement enables the simultaneous acquisition of quadruple data points from a single application of sample liquid. However, in the present model assay, it mainly serves the purpose of demonstrating the small interchannel variations obtained with a μPAD design optimized for small sample volumes.

Figure 6a shows a calibration curve obtained for HSA after the single application of 0.8 μL of sample liquid. A separate μPAD was used for each single HSA concentration. The linear range of 0–10 mg/mL corresponds to typical total protein concentrations found in human tear fluid [22]. Figure 6b shows a series of micrographs of the corresponding μPADs. It should be noted, however, that the colorimetric data analysis has been performed on standard color scanned images without the use of a microscope.

The mean of the relative standard deviations for the colorimetric signal intensity in the four detection zones was calculated as 2.40% (Table 4). This value indicates a reasonably low interchannel variation in sample flow passing through the parallel microfluidic channels.

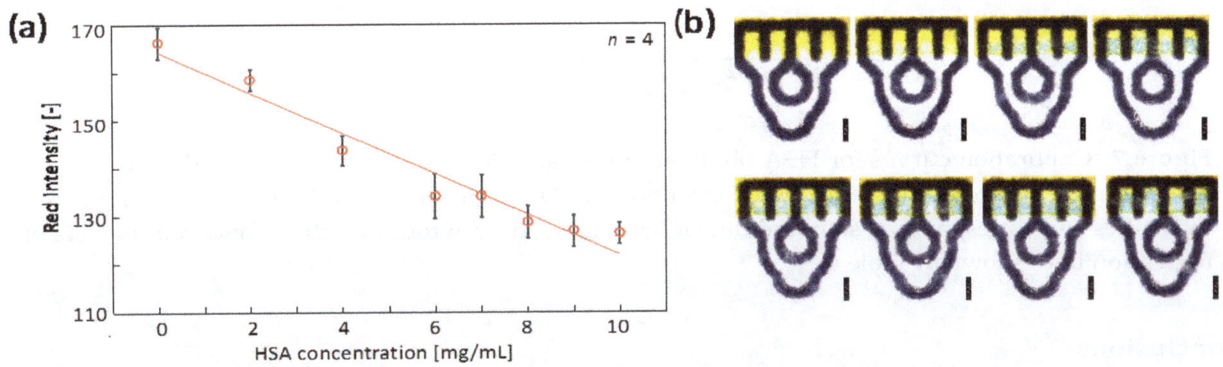

Figure 6. Colorimetric human serum albumin (HSA) analysis through the application of 0.8 μL of sample liquid to an optimized μPAD. (**a**) Calibration curve with data plots and error bars representing mean red intensities and corresponding standard deviations extracted from the four detection zones (parameters of regression line shown in Table 4); (**b**) Micrographs of μPADs after application of 0, 2, 4, 6, 7, 8, 9, and 10 mg/mL HSA (from upper left to lower right). The scale bars correspond to a length of 1 mm (brightness and contrast adjusted for improved visibility).

Table 4. Slope and y-intercept values of linear HSA response curves (0–10 mg/mL) recorded by application of variable sample volumes.

Sample Volume (μL)	Slope	*y*-Intercept	R^2	Mean of Relative Standard Deviations
0.7	−4.21	166	0.979	2.08%
0.8	−4.22	164	0.956	2.40%
0.9	−4.01	163	0.931	1.78%
Average	−4.15 ± 0.12	164 ± 1.5	-	-

3.3.2. Sample Volume Variation Tolerance of the Colorimetric Signal

Colorimetric signals obtained in assays performed on µPADs are generally dependent on the volume of the applied sample liquid. The colorimetric signal is determined by the absolute amount of analyte transported through the cellulose fiber network to the colorimetric detection zone, rather than the concentration of the sample alone. When working with sub-microliter samples, the risk of pipetting-related variations in sample volume inevitably increases. On the other hand, it has been anticipated that the closed space created by full lamination of the µPAD used in the current study contributes to controlling the sample volume that can be absorbed by the device. In order to evaluate the tolerance against small variations in the applied sample liquid volume (±0.1 µL), HSA response curves were recorded between 0.7–0.9 µL of applied sample. The resulting data is summarized in Table 4 and Figure 7.

The results demonstrate an acceptable tolerance against minor variations in the applied sample volume, since slopes and intercepts of the calibration curve are not significantly different. However, despite full lamination of the µPADs, the colorimetric signal does remain dependent on the amount of applied sample, and total volume independence is not achieved. It has to be kept in mind that paper is a rather heterogeneous material and that, therefore, variations in liquid volumes adsorbed by the cellulose fiber network are inevitable.

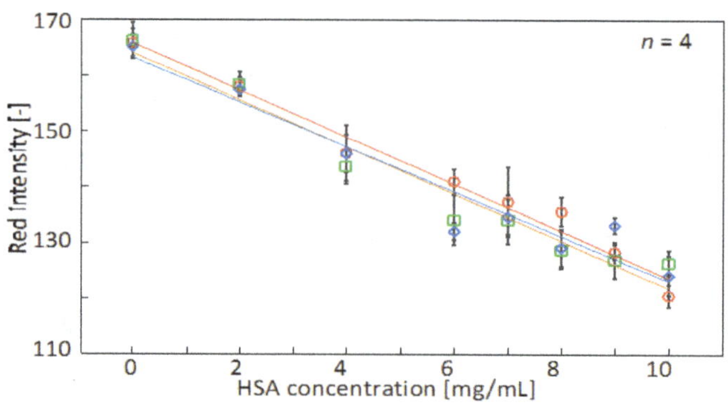

Figure 7. Calibration curves for HSA obtained with variable sample volumes: 0.7 µL (red circles), 0.8 µL (green squares), and 0.9 µL (blue diamonds). Data plots and error bars represent mean red intensities and corresponding standard deviations extracted from four detection zones (parameters of regression lines shown in Table 4).

4. Conclusions

The current study has demonstrated the possibility of using the popular wax-printing technology combined with the most widely used filter paper substrate to obtain miniaturized paper-based analytical devices with microfluidic structures, enabling the analysis of samples only available in small volumes. To reach this goal, the consideration of several factors influencing the resolution of the microfluidic patterning process was necessary. It has been experimentally demonstrated that replacing the hot plate by simple hot lamination as a heating method for melting printed wax into the paper substrate is a key factor in achieving increased patterning resolution. This approach does not significantly alter the costs for device fabrication and is expected to be fully compatible with roll-to-roll mass-production methods. The device presented is just one possible model. It is clear that other factors not evaluated here, such as the length of microfluidic channels or the integration of additional reaction zones required to perform a specific assay, are essential in determining the minimal volume of sample required in a µPAD. However, we believe that the current work demonstrates that the applicability of µPADs is not limited to samples available in amounts of several microliters, but can be extended into the sub-microliter volume range.

Acknowledgments: This research has been partially supported by the Medical Research and Development Programs Focused on Technology Transfer: Development of Advanced Measurement and Analysis Systems (SENTAN) (Japan Agency for Medical Research and Development; AMED). K.Y. kindly acknowledges a research grant provided by the Graduate School of Science and Technology, Keio University.

Author Contributions: R.O., K.Y., T.H., and D.C. conceived and designed the experiments; K.T. and K.Y. performed the experiments; K.T. analyzed the data; K.S. attended the discussions and edited the final version of the paper; K.T. and D.C. wrote the paper; D.C. is the principal investigator of the supporting grant.

References

1. Xia, Y.; Si, J.; Li, Z. Fabrication techniques for microfluidic paper-based analytical devices and their applications for biological testing: A review. *Biosens. Bioelectron.* **2016**, *77*, 774–789. [CrossRef] [PubMed]
2. Cate, D.M.; Adkins, J.A.; Mettakoonpitak, J.; Henry, C.S. Recent Developments in Paper-Based Microfluidic Devices. *Anal. Chem.* **2015**, *87*, 19–41. [CrossRef] [PubMed]
3. Yamada, K.; Henares, T.G.; Suzuki, K.; Citterio, D. Paper-Based Inkjet-Printed Microfluidic Analytical Devices. *Angew. Chem. Int. Ed.* **2015**, *54*, 5294–5310. [CrossRef] [PubMed]
4. He, Y.; Wu, Y.; Fu, J.-Z.; Wu, W.-B. Fabrication of paper-based microfluidic analysis devices: A review. *RSC Adv.* **2015**, *5*, 78109–78127. [CrossRef]
5. Ahmed, S.; Bui, M.-P.N.; Abbas, A. Paper-based chemical and biological sensors: Engineering aspects. *Biosens. Bioelectron.* **2016**, *77*, 249–263. [CrossRef] [PubMed]
6. Yetisen, A.K.; Akram, M.S.; Lowe, C.R. Paper-based microfluidic point-of-care diagnostic devices. *Lab Chip* **2013**, *13*, 2210–2251. [CrossRef] [PubMed]
7. Tanaka, H.; Fiorini, P.; Peeters, S.; Majeed, B.; Sterken, T.; Op de Beeck, M.; Yamashita, H. Sub-micro-liter electrochemical single-nucleotide-polymorphism setector for lab-on-chip system. *Jpn. J. Appl. Phys.* **2012**, *51*, 04DL02. [CrossRef]
8. Huang, S.-B.; Lee, G.-B. Pneumatically driven micro-dispenser for sub-micro-liter pipetting. *J. Micromech. Microeng.* **2009**, *19*, 035027. [CrossRef]
9. Li, L.; Boedicker, J.Q.; Ismagilov, R.F. Using a multijunction microfluidic device to inject substrate into an array of preformed plugs without cross-contamination: comparing theory and experiments. *Anal. Chem.* **2007**, *79*, 2756–2761. [CrossRef] [PubMed]
10. Yamada, K.; Takaki, S.; Komuro, N.; Suzuki, K.; Citterio, D. An antibody-free microfluidic paper-based analytical device for the determination of tear fluid lactoferrin by fluorescence sensitization of Tb^{3+}. *Analyst* **2014**, *139*, 1637–1643. [CrossRef] [PubMed]
11. Jain, S.; Rajasingham, R.; Noubary, F.; Coonahan, E.; Schoeplein, R.; Baden, R.; Curry, M.; Afdhal, N.; Kumar, S.; Pollock, N.R. Performance of an Optimized Paper-Based Test for Rapid Visual Measurement of Alanine Aminotransferase (ALT) in Fingerstick and Venipuncture Samples. *PLOS ONE* **2015**, *10*, e0128118. [CrossRef] [PubMed]
12. Martinez, A.W.; Phillips, S.T.; Butte, M.J.; Whitesides, G.M. Patterned paper as a platform for inexpensive, low-volume, portable bioassays. *Angew. Chem. Int. Ed.* **2007**, *46*, 1318–1320. [CrossRef] [PubMed]
13. Carrilho, E.; Martinez, A.W.; Whitesides, G.M. Understanding wax printing: a simple micropatterning process for paper-based microfluidics. *Anal. Chem.* **2009**, *81*, 7091–7095. [CrossRef] [PubMed]
14. Lu, Y.; Shi, W.; Jiang, L.; Qin, J.; Lin, B. Rapid prototyping of paper-based microfluidics with wax for low-cost, portable bioassay. *Electrophoresis* **2009**, *30*, 1497–1500. [CrossRef] [PubMed]
15. Jeong, S.-G.; Lee, S.-H.; Choi, C.-H.; Kim, J.; Lee, C.-S. Toward instrument-free digital measurements: A three-dimensional microfluidic device fabricated in a single sheet of paper by double-sided printing and lamination. *Lab Chip* **2015**, *15*, 1188–1194. [CrossRef] [PubMed]
16. Sun, M.; Johnson, M.A. Measurement of total antioxidant capacity in sub-μL blood samples using craft paper-based analytical devices. *RSC Adv.* **2015**, *5*, 55633–55639. [CrossRef]
17. Yeh, S.-H.; Chou, K.-H.; Yang, R.-J. Sample pre-concentration with high enrichment factors at a fixed location in paper-based microfluidic devices. *Lab Chip* **2016**, *16*, 925–931. [CrossRef] [PubMed]
18. Sariri, R.; Ghafoori, H. Tear proteins in health, disease, and contact lens wear. *Biochem. Mosc.* **2008**, *73*, 381–392. [CrossRef]

19. Rantamäki, A.H.; Seppänen-Laakso, T.; Oresic, M.; Jauhiainen, M.; Holopainen, J.M. Human tear fluid lipidome: from composition to function. *PLOS ONE* **2011**, *6*, e19553. [CrossRef] [PubMed]
20. Xu, Y.; Enomae, T. Paper substrate modification for rapid capillary flow in microfluidic paper-based analytical devices. *RSC Adv.* **2014**, *4*, 12867–12872. [CrossRef]
21. Suzuki, Y. Theoretical analysis concerning the characteristics of a dye-binding method for determining serum protein based on protein error of pH indicator: effect of buffer concentration of the color reagent on the color development. *Anal. Sci.* **2005**, *21*, 83–88. [CrossRef] [PubMed]
22. Zhou, L.; Beuerman, R.W.; Foo, Y.; Liu, S.; Ang, L.P.; Tan, D.T. Characterisation of human tear proteins using high-resolution mass spectrometry. *Ann. Acad. Med. Singap.* **2006**, *35*, 400–407. [PubMed]

14

A Method of Three-Dimensional Micro-Rotational Flow Generation for Biological Applications

Yaxiaer Yalikun [1,2,*], Yasunari Kanda [3] and Keisuke Morishima [1,4,*]

1 Department of Mechanical Engineering, Osaka University, 2-1 Yamadaoka, Suita, Osaka 565-0871, Japan
2 Laboratory for Integrated Biodevice, Quantitative Biology Center, RIKEN, 1-3 Yamadaoka, Suita, Osaka 565-0871, Japan
3 Division of Pharmacology, National Institute of Health Sciences, 1-18-1 Kamiyoga, Setagaya, Tokyo 158-8501, Japan; kanda@nihs.go.jp
4 Global Center for Advanced Medical Engineering and Informatics, Osaka University, 2-2 Yamadaoka, Suita, Osaka 565-0871, Japan
* Correspondence: yaxiaer.yalikun@riken.jp (Y.Y.); morishima@mech.eng.osaka-u.ac.jp (K.M.)

Academic Editors: Manabu Tokeshi and Kiichi Sato

Abstract: We report a convenient method to create a three-dimensional micro-rotational fluidic platform for biological applications in the direction of a vertical plane (out-of-plane) without contact in an open space. Unlike our previous complex fluidic manipulation system, this method uses a micro-rotational flow generated near a single orifice when the solution is pushed from the orifice by using a single pump. The three-dimensional fluidic platform shows good potential for fluidic biological applications such as culturing, stimulating, sorting, and manipulating cells. The pattern and velocity of the micro-rotational flow can be controlled by tuning the parameters such as the flow rate and the liquid-air interface height. We found that bio-objects captured by the micro-rotational flow showed self-rotational motion and orbital motion. Furthermore, the path length and position, velocity, and pattern of the orbital motion of the bio-object could be controlled. To demonstrate our method, we used embryoid body cells. As a result, the orbital motion had a maximum length of 2.4 mm, a maximum acceleration of 0.63 m/s^2, a frequency of approximately 0.45 Hz, a maximum velocity of 15.4 mm/s, and a maximum rotation speed of 600 rpm. The capability to have bio-objects rotate or move orbitally in three dimensions without contact opens up new research opportunities in three-dimensional microfluidic technology.

Keywords: three-dimensional microfluidic platform; micro-rotational flow; non-contact; open space

1. Introduction

With the development of integrated microfluidic technology, the methodology of biological and chemical experiments has become more space saving, more efficient, and requires the use of smaller amounts of reagents and cells. Most of the conventional applications using integrated microfluidic technology are based on a two-dimensional microfluidic system or platform. However, a three-dimensional microfluidic system has been shown to have certain advantages over two-dimensional flow in such fields as tissue engineering and cell differentiation mainly because it requires to create a growth environment that mimics the native tissue as closely as possible [1,2]. Cells cultured in three-dimensional systems enable a larger cell structure and a longer-term incubation owing to the delivery of nutrients to the entire cell structure and tissues [3,4]. However, no use of a three-dimensional culturing system has been reported yet without a scaffold [5], hydrogel [6], specifically designed channel [7,8], additional particle [9] or chamber structure [10]. In the field of cell differentiation, the shear stress generated by the three-dimensional flow can aid in the formation of three-dimensional vascular tubes by increasing the organization of endothelial cells, resulting in a cell response to flow changes according to specific flow parameters [11–13] However, in most cases, a stimulation system applies the fluid stimulation on only one side of the cells which means their functionality, phenotype and responses to environmental cues might be altered [14]. In some cases, such as when red blood cells in a vessel are exposed to an alternative flow environment of either a high flow rate (artery, m/s) and a low flow rate (capillary, μm/s) environment, such a large difference in flow rate may induce different phenomena of cell differentiation.

Size-based cell sorting by micro-rotational flow in a closed channel has been previously reported and it is considered to be a high-throughput sorting method [15]. Cells of a specific size can be captured in the micro-rotational flow and the others will flow out. We considered that a three-dimensional micro-rotation flow sorting system in which flow style, acceleration rate, and velocity features are controllable is more efficient and capable of capturing cells in a wider range, and the open environment of sorting makes it easy to retrieve the sorted sample.

For cell manipulation, conventional manipulation systems [16–20] require a complex channel or supportive structure to guide the flow. In addition, fabrication and construction of conventional systems require specific skills in micro-photolithography and use of a complex control system. For example, in our previous work, we reported a three-dimensional microfluidic manipulation system consisting of nine high-specification pumps, a multiple-layer structure microfluidic chip with nine orifices, a control system with a D/A board, a computer and a self-made complex controller [21]. The system functioned by using a self-made control algorithm [22]. However, for practical use in the biological field, a more convenient system requiring no extra chip design, multiple components or complex control system to realize the purposes of capturing, moving and rotating cells is desired.

In this study, we significantly simplified our previous system and we developed a method for creating a three-dimensional micro-rotational fluidic environment by using only a single orifice in a microfluidic chip and a conventional syringe pump. The micro-rotational environment is stable and has a good potential for applications such as culturing, stimulating, sorting, and manipulating cells.

2. Methods and Materials

2.1. Principles

The concept of our method is based on micro-rotational flow that is generated in an open space. The micro-rotational flow caused by sudden changes in geometries when the fluid flows through has been well documented. Figure 1 shows a schematic representation of cells in a three-dimensional micro-rotational flow. When pushing out solution from a micro-orifice of a microchip to an expansion area, a micro-rotational flow composed of a three-dimensional stable vortex occurs between the air–liquid interface and the surface of the microchip.

In this paper, we consider that formation of micro-rotational flow depends on two main factors: flow velocity and the height of the air–liquid interface. A cell captured by this flow preforms self-rotational motion or orbital motion. By tuning the flow rate, it is potentially possible to rotate the cell that is in a high rate flow, or to achieve a controllable orbital motion over a wide range of sizes and species. Target cells can either be more or less dense than the surrounding medium.

To demonstrate this method, estimation of the force necessary for capturing and rotating is required. In our previous research [21], we proved calculation model to estimate driving force for manipulating object and ignored the lift force because object not floating in solution. In this paper, we considered and modified a model of a solid object in vortex flow. When moving uniformly upstream, a rotating sphere in the range of Reynolds number (R_e) < 100, the motion of an individual object obeys Newton's second law and the force situation is estimated as follows:

$$F_I = F_H + F_L + F_B + F_G + F_{add} \tag{1}$$

where F_I is the total external force exerted on the object in the vortex flow stream, and F_H, F_L,F_B, F_G, and F_{add} are the hydrodynamic force, lift force, buoyancy force, gravity force, and added mass force, respectively, as shown in Figure 2.

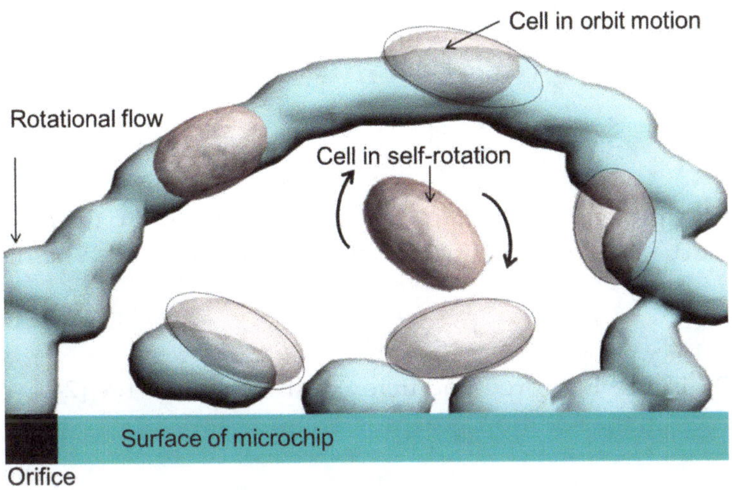

Figure 1. Conceptual illustration of the motion of a cell captured by micro-rotational flow. A cell in the center of the micro-rotational flow performs self-rotational motion. A cell along the rotation flow performs orbital motion.

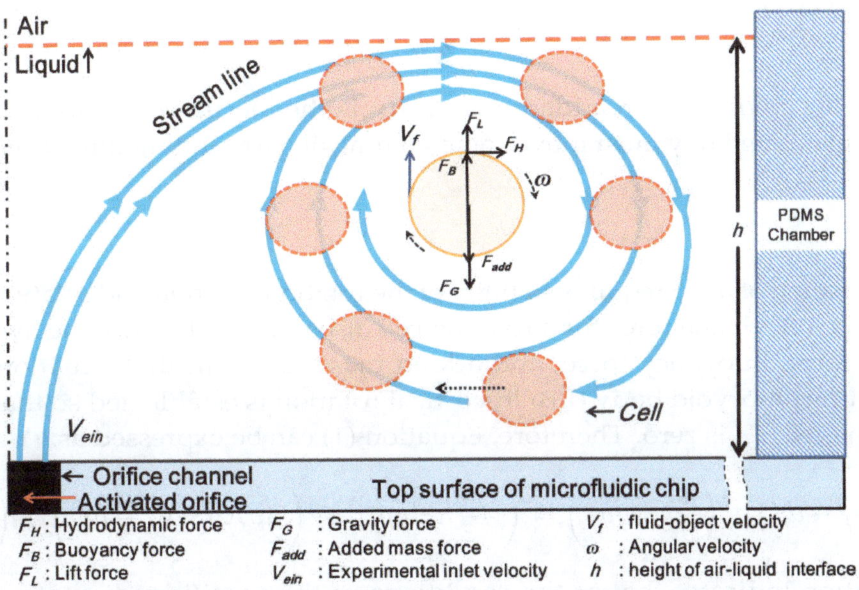

Figure 2. Schematic diagram showing the forces acting on a cell in an assumed equilibrium position of rotational motion and in orbital motion. The cell is captured and moved to the center of the micro-rotational flow in the polydimethylsiloxane (PDMS) chamber. The cell continuously rotates in the zone, where it is exposed to several forces.

The micro-rotational flow exploits a property of fluid itself on its vertical direction. The hydrodynamic force and lift force applied on the fluid–cell interface generate a centripetal force to capture and rotate the cell at the middle of the micro-rotational flow or in length-controlled orbital motion. The hydrodynamic force in the operating area can be expressed by:

$$F_H = \frac{1}{2}\rho_f C_D S(V_f - v)^2 \quad (2)$$

In Equation (2), ρ_f is the density of liquid, C_D is the hydrodynamic force coefficient, S is the projected section of the cell, V_f is the relative velocity between the flow and the rotating object, and v is the object velocity, which was assumed to be 0 m/s in this study. Because the Reynolds number is generally small in a micro-scale device, the hydrodynamic force coefficient can be approximately calculated as follows, when the Reynolds number (R_e) is 2×10^5 or less. Moreover, in a previous study [23], it was found that the drag coefficient C_D (Equation (3)) was not significantly influenced by the rotation of the object.

$$C_D \approx \frac{24}{R_e} + \frac{6}{1+\sqrt{R_e}} + 0.4 \tag{3}$$

A cell in a shear field experiences a lift force that is perpendicular to the direction of flow. The shear lift originates from the inertia effects in the viscous flow around the cell. The lift force present on a cell can be described as:

$$F_L = \frac{1}{2}\rho_f C_L S {V_f}^2 \tag{4}$$

In Equation (4), C_L is the lift force coefficient estimated as 1.4 [24] when R_e is 100 or less. The gravity force of the cell in the solution is expressed as:

$$F_G = \frac{4}{3}\pi r^3 g \rho_o \tag{5}$$

In Equation (5), r is the radius of the cells, g is gravity force, and ρ_o is the density of cells. The buoyancy force of the cell in the solution is expressed as:

$$F_B = \frac{4}{3}\pi r^3 g \rho_f \tag{6}$$

In a microfluidic environment, added mass force is the inertia added to an object because an accelerating or decelerating body must move some volume of surrounding fluid as it moves through it:

$$F_{add} = \frac{2}{3}\pi r^3 g \rho_f \tag{7}$$

Continuous rotation of a cell requires stability of the rotation position and rotation speed. To satisfy this requirement, the centripetal force for a rotating cell also needs to balance the hydrodynamic force, lift force, gravity force, buoyancy force, and added mass force. In that situation, an equilibrium situation of a rotating embryoid body (EB) in vertical rotation is established so that the sum of each force present on the cell, F_I, is zero. Therefore, Equation (1) can be expressed as:

$$(\frac{1}{2}\rho_f C_d S (V_f - v)^2) + \left(\frac{4}{3}\pi r^3 g \rho_f\right) = \left(-\frac{4}{3}\pi r^3 g \rho_o\right) + \left(\frac{1}{2}\rho_f C_l S (V_f - v)^2\right) + \left(\frac{2}{3}\pi r^3 \rho_f\right) \tag{8}$$

This formulation indicates that at the conditions of the specific radius of a cell (diameter of 250 μm), the density of medium solution, velocity of the flow, and lift force coefficient, the bio-object can be rotated. The steady relative velocity between the stream and rotating object V_f is calculated to be 0.002217 m/s with a cell rotation speed of 50 rpm. The aim of the calculation is to determine the critical equilibrium velocity condition to estimate the flow rate from the computational fluid dynamics (CFD) software (Ansys, 14.0, Ansys, Inc., Canonsburg, PA, USA) and use it in an experiment. In order to make a bio-object rotate, we calculated the flow rate used in the experiment from Equation (8).

2.2. Material

For demonstration of the self-rotational and orbital motions, we focused on iPS (induced pluripotent stem) cells and embryonic stem (ES) cells. After comparing the different uses and culture conditions [25], we chose ES cells. ES cells are obtained from the inner cell mass of the blastocyst of a developing embryo [26]. Because ES cells are pluripotent cells, they can become any type of cell [27]. They are also self-renewable and can be used for tissue regeneration and cellular replacement therapies. Control and manipulation of the numerous differentiation pathways in ES cells have been a topic investigated by numerous researchers [28–31]. Because an embryoid body (EB) cell is usually large, heavy, and distinct from other bio-objects when exposed to mechanical strain, possibly altering gene expression, most present three-dimensional microfluidic platforms for EBs have been a closed environment; however, retrieving these EBs from the closed environment or cancelling the limitation from around the device structure (such as chamber size limitation) is difficult. Therefore, we demonstrated our method as an option of the three-dimensional microfluidic platform for EB research.

The EBs were generated using a standard protocol, as previously described [32]. First, mouse R1 ES cells (American Type Culture Collection, Manassas, VA, USA) were thawed on mitomycin C-inactivated mouse embryonic fibroblasts (MEFs; Millipore, Darmstadt, Germany) and cultured on gelatin-coated dishes in ES medium Dulbecco's modified Eagle's medium (DMEM; Sigma–Aldrich, St. Louis, MO, USA) supplemented with 10% fetal bovine serum (FBS; Biological Industries, Cromwell, CT, USA), 1000 units/mL leukemia inhibitory factor (LIF; ESGRO, Millipore, Darmstadt, Germany), 1 mM sodium pyruvate (Invitrogen, Thermo Fisher Scientific K.K., Yokohama, Japan), 0.1 mM 2-mercaptoethanol (Sigma–Aldrich), 0.1 mM non-essential amino acids (Invitrogen), 100 units/mL penicillin, and 100 µg/mL streptomycin (Invitrogen)) at 37 °C and 5% CO_2. After ES cells were dissociated to the single cell suspension using 0.25% Trypsin-EDTA (Gibco, Thermo Fisher Scientific K.K., Yokohama, Japan), one hundred cells were seeded into a non-adherent 96-well U-bottom plate (PrimeSurface; Sumitomo Bakelite, Tokyo, Japan) in DMEM with 10% FBS for 2 days.

2.3. Experimental Device and Rotational and Orbital Motion

The microfluidic chip (Figure 3A) described in our previous work was used in this study [21]. It was a two-layer fusion-bonded glass structure chip; the center of the chip is shown in Figure 3B. Layer one (orifice layer) contained eight orifices (orifice diameter, 100 µm) fabricated in a circle (circle diameter, 500 µm) and one orifice fabricated in the middle of the circle by a mechanical drill. In layer two (channel layer), eight channels (each contained three types of channel) were fabricated with a mechanical end mill; one had a width of 500 µm and length of 30 mm, one had a width of 200 µm and length of 1 mm and the last one, a width of 100 µm and length of 250 µm. The two layers were aligned and fusion-bonded to complete the chip assembly. The design details are shown in Figure 3D. In our previous study [21], to achieve the linear motion of an object for a wide range of sizes, and to control the direction of the object, multiple orifices in different positions were used. In addition, at least two orifices and pumps were required to manipulate the object. This kind of control required a complex system design and was not suitable for practical use in biological applications. In this paper, we only activated one orifice and one pump for generation of the micro-rotational flow (Figure 3B). A polydimethylsiloxane (PDMS) chamber was placed on the microfluidic chip, and its topside was open. The actual diameter of the PDMS chamber was 15.7 mm, and its total volume was approximately 600 µL (Figure 3E). The initial height of air-liquid (L) was 1.5 mm (Figure 3F). The experiment was completed before the chamber overflowed, and therefore no outlet port was used.

Experimental images were taken by charge-coupled devices (CCDs, Lu075C, Lumenera, Ottawa, ON, Canada) with zoom lenses (KCM-Z, Tokina, Tokyo, Japan) placed at the top and side of the PDMS chamber (Figure S1), the reflection by the air-liquid interface and the surface wave clearly influence the quality of these images. In the near future, we plan to consider use of a semi-closed or a closed chamber to solve the problems of the reflection and surface wave to achieve better image quality.

The demonstrations of self-rotational motion and orbital motion of the EB were conducted using different processes. In the demonstration of rotational motion, use of an improper flow rate may pull the EBs away from the orifices. Therefore, we initially used the flow rate estimated from the equilibrium position in the CFD simulation. In the demonstration of orbital motion, we used a flow rate that was lower or higher than the one used in the equilibrium position of rotational motion.

We observed orbital motion of the EB within a Φ2.5-mm area from the center of the microfluidic chip when the height of the air-liquid interface was L = 1.5 mm, as shown in Figure 3F.

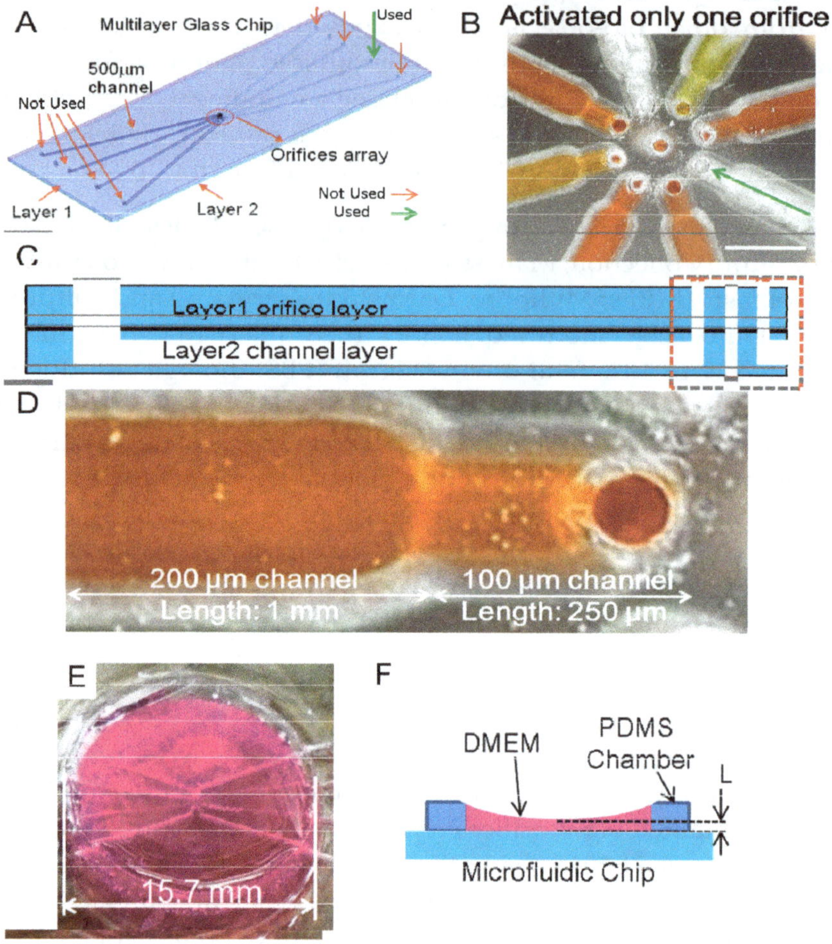

Figure 3. Microfluidic chip used in this study. (**A**) Schematic diagram of the microfluidic chip design. Although there are eight inlets (and channels), we did not use seven of them (indicted by **red** arrows), the inlet and channel we used are indicted by the **green** arrow; (**B**) the orifices in the center area of the microfluidic chip. Each channel was filled with colored solution. The white scale bar is 500 μm; (**C**) schematic diagram of the layer structure of the chip. Layer one had nine orifices, layer two had eight channels; (**D**) there were two channel widths: 100 and 500 μm. The channel depth was 100 μm for all. The diameter of orifice was 100 μm; (**E**) the chamber around the orifices was filled with Dulbecco's modified Eagle's medium (DMEM). The designed chamber diameter was 20 mm and the actual size was 15.7 mm); and (**F**) cross-sectional view of the chamber. The air–liquid interface height (L) was 1.5 mm.

2.4. CFD Simulation

In order to provide theoretical guidance and understand the field of micro-rotational flow, we conducted the CFD simulation using ANSYS Fluent software (Ansys 14.0, Ansys, Inc.). The goal of the simulation was to reveal the velocity distribution, the micro-rotational flow pattern, and velocity of the particles. A three-dimensional two-phase model was applied to calculate the micro-rotational flow pattern in the PDMS chamber. To simplify the simulation, only flow from the inlet in the liquid domain was considered, and velocity of flow at the orifice V_{ein} was used. The simulation domain we used was 15.7 mm (diameter) × 1.5 mm (depth). The bottom surfaces were set as free convective boundaries at room temperature of 25 °C.

3. Results

3.1. Experimental Confirmation of the Generation of Micro-Rotational Flow

Because our system was an open space environment, the velocity of flow from orifice V_{ein} dispersed significantly, and it was not equal to the relative velocity between the stream and rotating object V_f. Therefore, to experimentally confirm the velocity property of the three-dimensional rotation flow, an experiment was conducted using Fluoro Spheres (20 μm in diameter; Molecular Probes, Invitrogen, Carlsbad, CA, USA). The results showed that the velocity of the flow clearly decreased when being pushed from the orifice. In the maximum velocity area (Figure 4A), the velocity was the same as that calculated from the flow rate; however, in the area of the center of the micro-rotational flow, the velocity was much lower. In addition, flow visualization (Figure 4B) showed the area of the micro-rotational flow was several millimeters from the activated orifice. The velocity measured from the maximum and average velocity areas is shown in Figure 4C. To reach the V_f which was calculated to be 0.002217 m/s, a flow rate approximately around 45–85 μL/min was required according to Figure 4C. The flow velocity from the orifice V_{ein} was calculated to be 0.095–0.18 m/s.

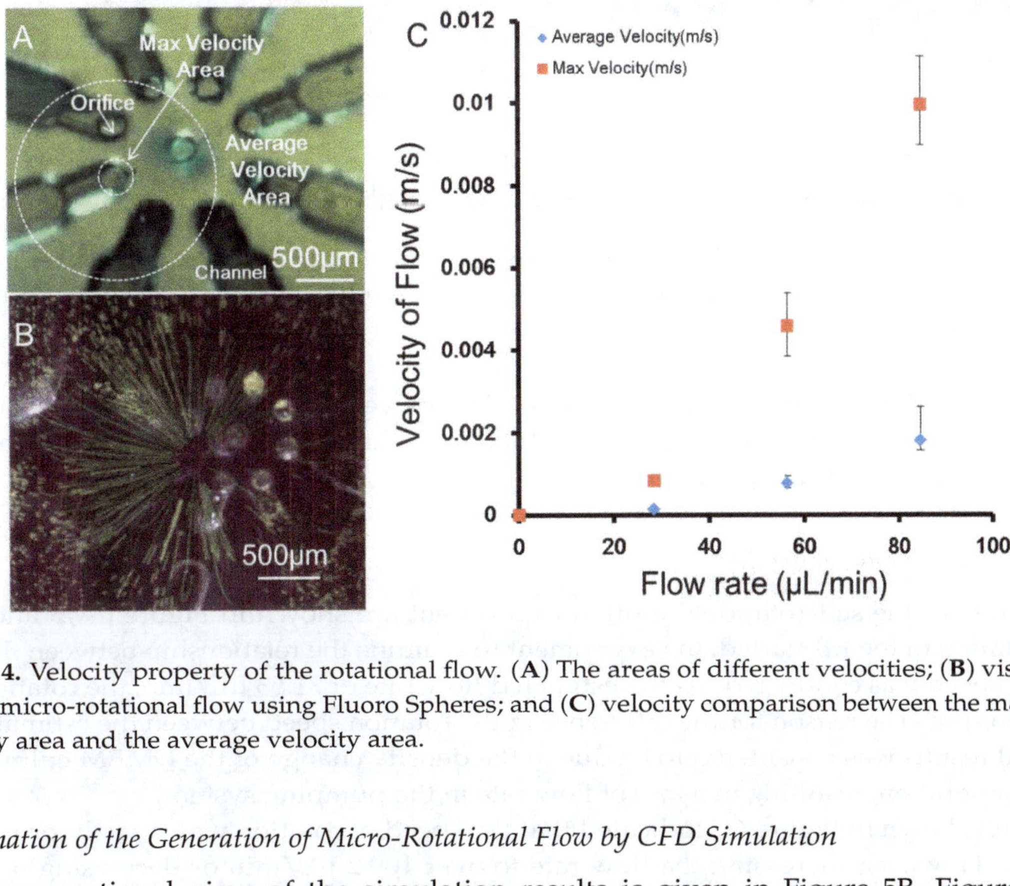

Figure 4. Velocity property of the rotational flow. (**A**) The areas of different velocities; (**B**) visualized area of micro-rotational flow using Fluoro Spheres; and (**C**) velocity comparison between the maximum velocity area and the average velocity area.

3.2. Confirmation of the Generation of Micro-Rotational Flow by CFD Simulation

The cross-sectional view of the simulation results is given in Figure 5B. Figure 5C shows details of the simulated micro-rotational flow in the white dashed line in Figure 5B. The generation of micro-rotational flow with velocity larger than V_f was confirmed at a flow rate condition of 71.63 μL/min.

The streamlines indicated that the object in the micro-rotational flow could be exposed to shear velocity. In this situation, a lift force was generated on the object due to the significant difference in velocity and pressure exerted by the stream on opposite sides of the object. When the lift force was powerful enough to balance the hydrodynamic force in the tangential direction and in the direction of the other forces such as the gravity force, added mass force, and buoyancy force, the equilibrium position of the object was obtained. In addition, the velocity of the area near the orifice was clearly faster than the area far from the orifice, and the different flow velocity led to the different rotation speed of the object that was captured.

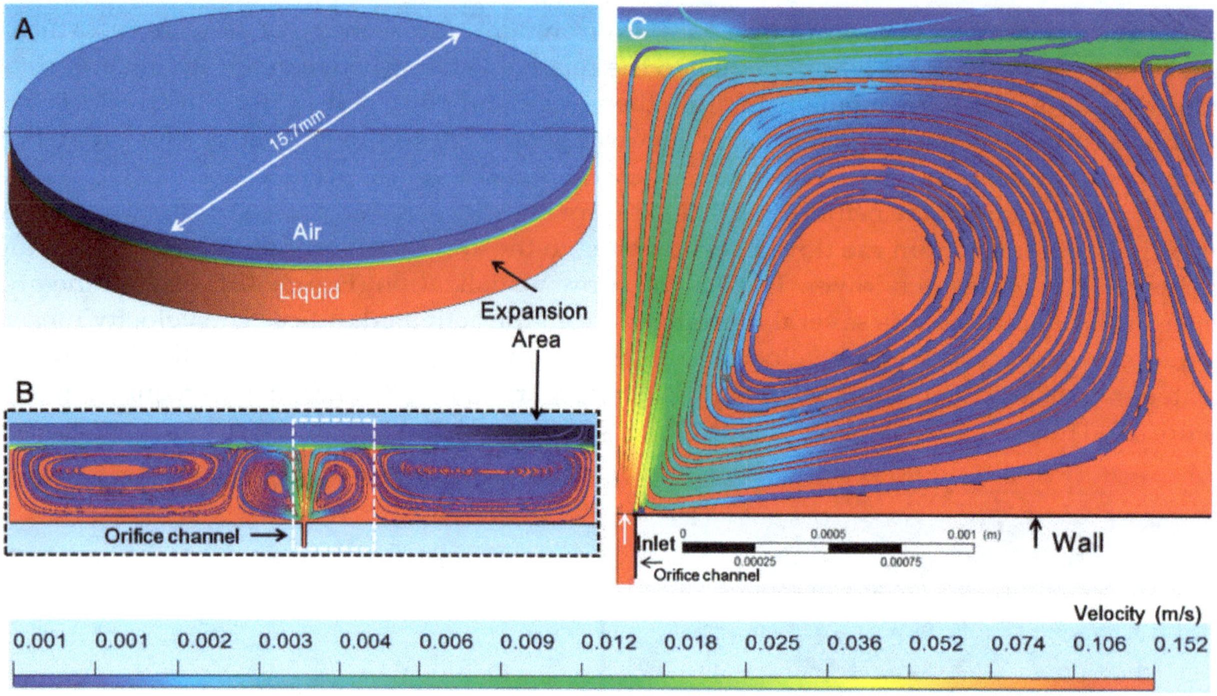

Figure 5. Simulation domain and results. (**A**) The simulation domain contained two phases (air and liquid); (**B**) cross-sectional view of the simulation results; and (**C**) enlarged view of the generated micro-rotational flow.

For the calculated conditions, the diameter of the low velocity core of the micro-rotational flow was from 200 to several hundred micrometers; therefore, the object diameter was set to be several hundred micrometers for the experiment.

3.3. Self-Rotational Motion of the EB

The results of the self-rotational motion experiment are shown in Figure 6A,B and Video S1. When the rotation of the EB started, the experiment to examine the relationship between the flow rate and rotation speed was conducted. At the estimated flow rate of 71.63 μL/min, the rotation speed of the EB was 66 rpm. The reason for the difference in the rotation speed between the estimated and the experimental results was considered to be due to the density change of the DMEM culture medium caused by evaporation, resulting in a loss of flow rate in the pumping system.

The result shown in Figure 6C indicated that the speed of rotation was directly proportional to the flow rate. However, increasing the flow rate to over 109.2 μL/min or decreasing it below 57.4 μL/min induced the self-rotational motion of the EB to become orbital motion. The reason was considered to be that the excessive flow rate caused a significant increase in the lift force, so that the EB was pulled into the stream from the orifice and flowed to the edge of the rotational flow, turning into a slow long-distance orbital motion. In the case of an insufficient flow rate, the lift force was decreased by the change in flow rate, and the force balance was no longer maintained in the equilibrium position. Figure 7A, from Video S2, shows an image at the moment that the rotation shifted to the fast short-distance orbital motion. Moreover, we used two cotton fibers (1.54 g/cm^3) (diameter of 10 μm, length 200 μm) to demonstrate the system was capable of rotating different objects and also to indicate the possibility that it was capable of manipulating multiple objects. (Figure 7B and Video S3).

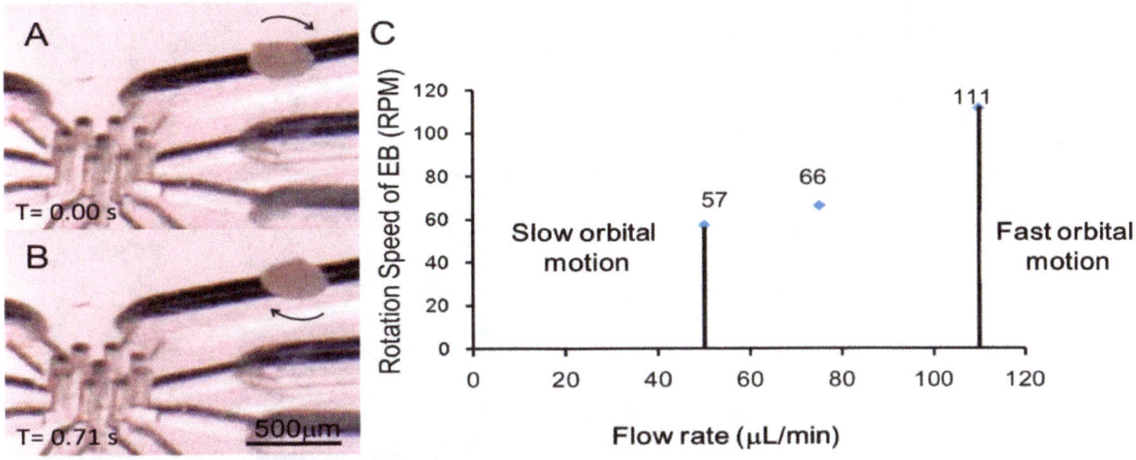

Figure 6. Self-rotational motion of the embryoid body (EB) in DMEM solution. (**A**,**B**) Images from the video footage of EB self-rotation seen from a side and (**C**) the dependence between flow rate and rotation speed.

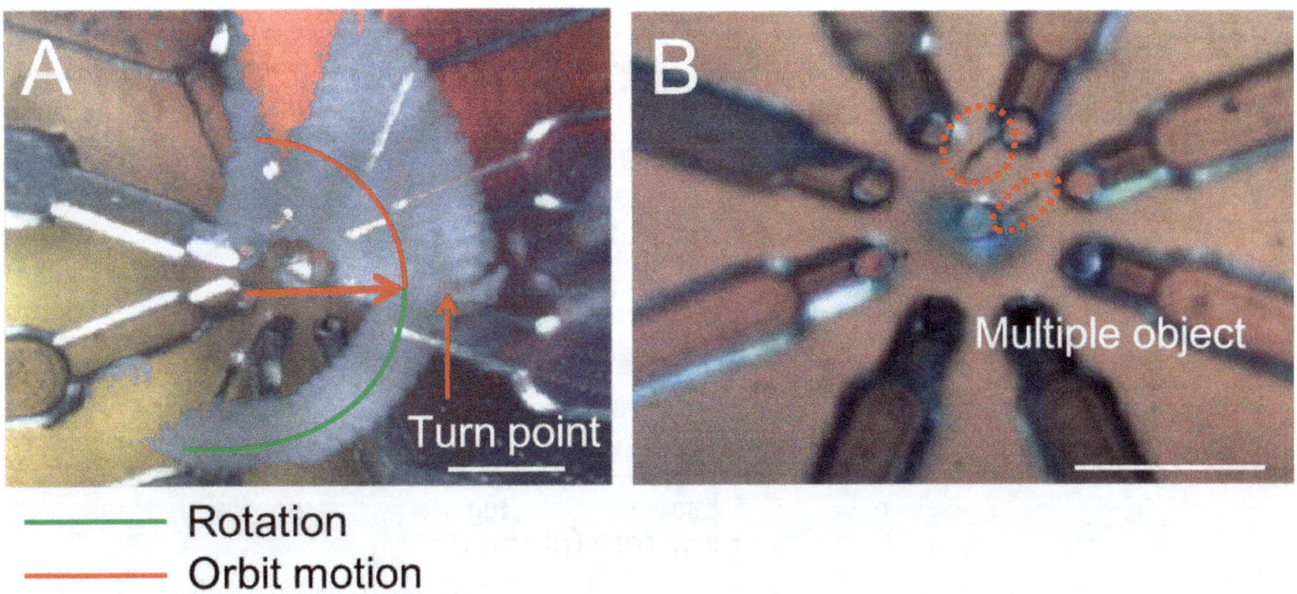

Figure 7. Turning point of rotation. (**A**) The moment that the rotation of the EB became orbital motion; and (**B**) objects of different sizes and species can also be rotated. The scale bar is 500 μm.

3.4. Orbital Motion

The results of the experimental demonstration of the orbital motion of the EB (diameter, 200 μm) are shown in Figure 8 (taken from Video S4). At the estimated flow rates of 36.2 μL/min, 72.5 μL/min, and 108.7 μL/min, the expected distances on the *x*-axis were calculated by the simulation as 419, 837, and 997 μm, respectively. The experimental results are shown in Figure 8B–D, and the distance (in length) is shown in Figure 9; distances of 422.8, 495.8, 734.5 μm were obtained. The reason for the difference between theoretical and experimental values was considered to be the loss of flow rate in the pumping system.

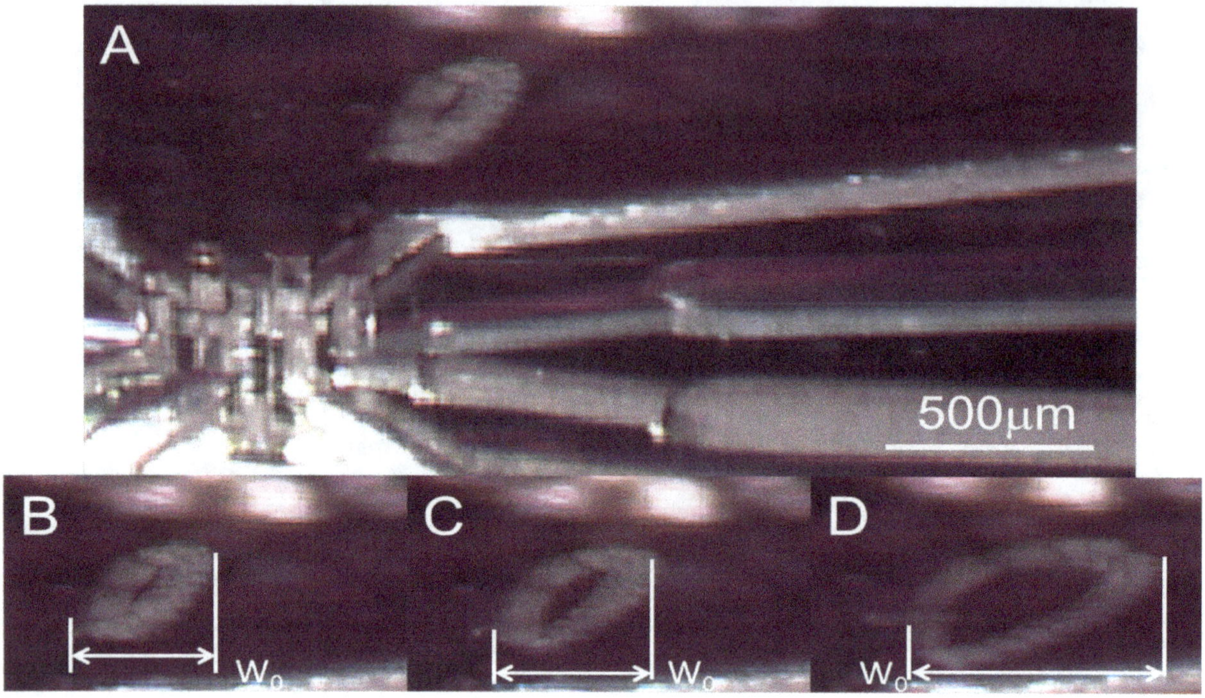

Figure 8. Trajectories of the EB in orbital motion. (**A**) The orbital motion was close to the center area; and (**B**–**D**) images of trajectories obtained under different flow rates of 36.2, 72.5, and 108.7 μL/min.

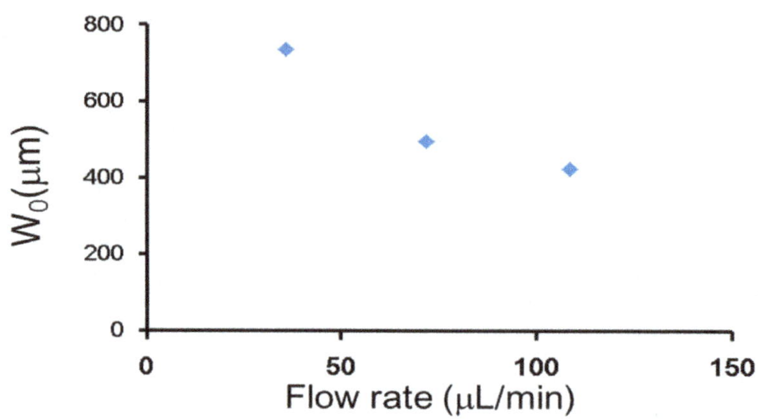

Figure 9. The length on the *x*-axis of trajectories obtained at different flow rates.

In addition, orbital motion in our system was a variable velocity motion. The acceleration, which is relative to the shear force, is important for the function of the force stimulator for a bio-object. Therefore, the orbital motion experiment of the EB was conducted at the flow rate of 154.9 μL/min, which is the maximum output of our system.

The trajectory obtained by an imaging process from the recorded video (Video S5) of the orbital motion is shown in Figure 10A, and the relationship between the acceleration in the vertical direction and flow rate is shown in Figure 10B. The maximum acceleration was approximately 0.63 m/s^2, and the frequency was approximately 0.45 Hz. The maximum speed of the EB was approximately 15.4 mm/s. As shown in Figure 11, cyclic variations in acceleration also induced cyclic variations in the shear force due to velocity. The frequency and acceleration were controllable, which means that the proposed system could function as the force stimulator for the bio-object.

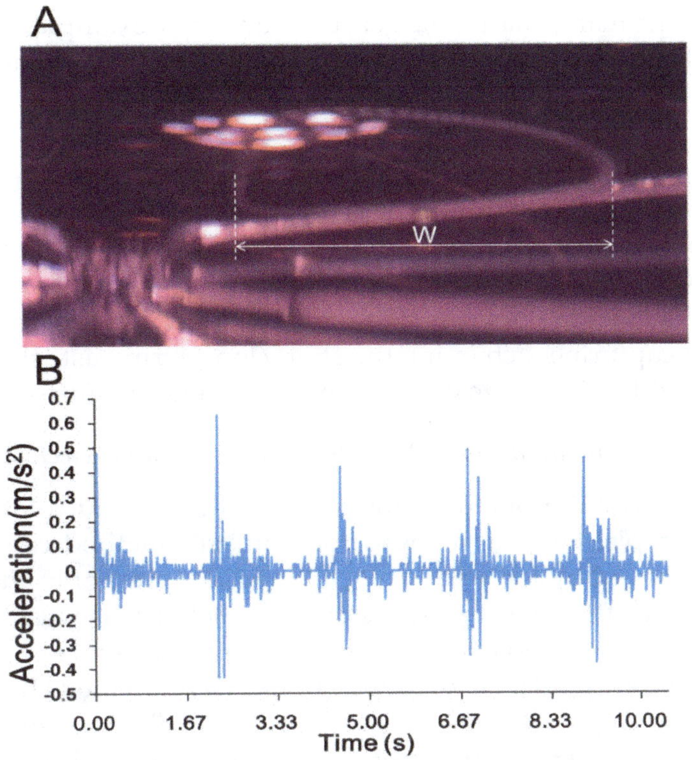

Figure 10. Acceleration of the EB in orbital motion. (**A**) The length on the *x*-axis of trajectories obtained by the imaging process; and (**B**) the acceleration was obtained while the EB was in orbital motion.

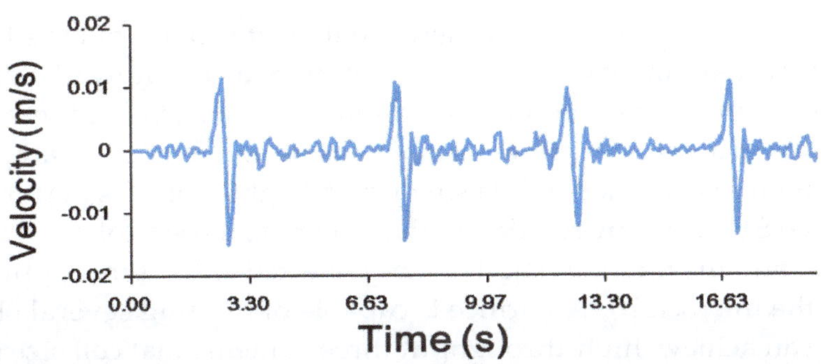

Figure 11. Velocity of the EB in orbital motion. The velocity was obtained from the video recording of the EB in orbital motion.

4. Discussion

The use of three-dimensional rotational flow enables various possible applications in a microfluidic system. Most of these applications are difficult to obtain with a two-dimensional microfluidic system.

4.1. *Self-Rotational Motion*

The method proposed in this paper is different from other methods based on the use of electric fields, optical forces, surface acoustic wave force, magnetic force, and mechanical tools. We achieved the self-rotational motion of an EB by controlling the micro-rotational flow, and we could determine the relationship between the rotation speed and flow rate. In addition, by regulating the height of the air-liquid interface, the position of rotation was also controllable. Compared with the conventional methods that rotate cells using a complex system, a closed space, and have size and species limitations for the manipulation target, our present method is convenient, conducted in an open space, and allows for a wider size range of the manipulation target (μm–mm). For example, we achieved the rotation speed of 600 rpm of the EB (Video S6) or the orbital motion of the 20 μm diameter EB (Video S7).

4.2. Three-Dimensional Culturing Mode

Three-dimensional culturing of a cell or a small amount of cells was possible if the cell had a slow orbital motion or self-rotational motion at a lower rotation speed. Rotating the object cells ensured the supply of nutrition to every part of the cell or the whole cell aggregate. Three-dimensional aggregates of different kinds of cells also can be made and cultured in our system without mechanical agitation. However, unlike other methods using a complex scaffold, microchannel, or micro-chamber, our method is capable of capturing, achieving the short/long-term, fast/slow rotational and orbital motions of an object in a wider size range of the manipulation target (μm–mm) (Video S8).

4.3. Three-Dimensional Flow Stimulation Mode: High-Speed Rotation and Acceleration of Flow

This method also showed the potential for practical high-speed rotational motion or orbital motion of a cell. For example, with high-speed rotation, this method can provide the controlled shear stress, which is produced by the steep velocity gradient present in the micro-rotational flow, to the whole bio-object. The effect of shear stress on endothelial cells was reported previously [33]. This method offers a powerful tool capable of presenting shear stress on not only the suspended cells but also on non-suspended bio-objects. Furthermore, unlike other methods that make time-consuming observations of physical changes of cells [34] induced from the free high shear stresses at the bio-object–fluid interface, our system can observe the changes in real time, which may offer new opportunities to study stress-induced signaling events for cells.

4.4. Three-Dimensional Cell Sorting Mode

Chen et al. [35] have reported the two-dimensional orbit motion of cells for different purposes In the present study, the three-dimensional orbital motion of a cell showed a greater rotational flow area than the two-dimensional orbital motion. We found that objects of different sizes and densities performed different motions at the same flow rate (Videos S1 and S9). For example, larger and heavier objects performed self-rotational motion, while smaller and lighter objects performed orbital motion at the same flow rate (Video S1). Thus, by regulating the flow rate, the size of a cell to be rotated in a fixed position can be selected, and the rotating object can be collected easily (on stopping the flow, the object sinks to the surface of the microchip). An orifice is capable of rotating several objects, and increasing the number of orifices can achieve high-throughput three-dimensional cell size sorting.

4.5. Three-Dimensional Manipulation

Compared with other conventional methods using hydrodynamic force [16–19], our method is capable of capturing bio-objects up to several millimeters in size. The capturing process is fast (less than 1 s), and the rotation and moving of the bio-object is simple and effective. Furthermore, the process requires no complex equipment or tools, relying simply on an orifice and a pump.

4.6. Comparison with Other Methods used in above Applications

In our evaluated applications, all the operations were conducted using rotational flow. The flow velocity is the most important factor that generates a possible negative influence on the cells. We do not have direct evidence to prove our system has less negative influence because we cannot clarify the reason for the difference between EB cells in a normal situation and after self-rotational motion or orbital motion. However, if we compare the velocity applied on cells in other methods, we can assume our system is bio-friendly and suited for these applications.

As shown in Table 1, in the culturing mode, velocities from 0.127–3 mm/s were previously used; in this paper, we applied velocities up to 2.217 mm/s (slow orbital and rotational motion) on the EBs.

The presently applied velocities are clearly lower than those used in previous research, indicating our method does not have a bigger negative influence on cells compared with previous methods. In addition, we observed the EB cells after we had rotated them for 10 min, orbitally moved them for 10 min, and then cultured them for 12 h. For comparison, we prepared EB cells by the same process of culturing them on a glass bottom dish coated with gelatin under the same conditions (Figure S2A,B). There were no significant differences in shape, and no dark zone appeared in the EBs we rotated (Figure S2C,D).

Table 1. Comparison of environment situation and applied velocity (For embryoid bodies having diameters of 200–300 μm).

Applications	Environment Situation		Applied Velocity (mm/s)	
	Other Methods	This Paper	Other Methods *	This Paper
EB Culturing	Closed	Open	0.127 [28]–3 [29]	Up to 2.217
EB Stimulation	Closed	Open	0.833–3 [30]	Up to 15.4
EB Sorting	Closed	Open	0.383 [31]	Up to 15.4
EB Manipulation	Closed	Open	0.383 [31]	Up to 2.217

* Velocity was calculated from the flow rate and microfluidic chip dimensions that were used in other previous methods

In the stimulating mode, larger flow velocities increase the effects of fluidic shear stress; however, previous methods were unable to use a flow velocity of more than 3 mm/s due to the risk of EBs being flushed away [30]. In our system, a maximum speed of 15 mm/s can be applied without losing the target EB.

In the sorting and manipulating modes, the flow velocity generating the drag force is normally used; however, in a closed space, velocity-induced pressure effects are not ignorable because our open space system has a lower pressure than a closed system. During the processes, only the force required to cancel the calculated gravity will be applied on the EBs; therefore our method has a smaller pressure effect.

Overall, our system is capable of providing most of the functions that other methods are capable of providing, as well as having a wider range of applied velocities and a smaller pressure effect in an open space.

5. Conclusions

We here reported a convenient method of three-dimensional micro-rotational flow generation. This method has potential for such biological applications as culturing, stimulating, sorting, and manipulating different kinds of cells. A controllable micro-rotational flow was generated near an orifice when the solution was pushing from the orifice. The size, velocity, and position of the micro-rotational flow core could be controlled by tuning the parameters such as the flow rate and the liquid-air interface height. A bio-object captured by the micro-rotational flow performed self-rotational motion or orbital motion. In addition, speed and position of rotation, velocity, frequency, and length of orbital motion of cells in the micro-rotational flow could also be controlled. As a typical biological target, EB cells were used to demonstrate our method. We obtained the maximum distance of orbital motion of 2.4 mm, maximum acceleration of 0.63 m/s^2, frequency of approximately 0.45 Hz, velocity of 15.4 mm/s, and maximum rotation speed of 600 rpm. The capability to have objects rotate or be in orbital motion in three dimensions without contact opens up new research opportunities in three-dimensional microfluidic technology.

Supplementary Materials
Figure S1: Schematic drawing to show how the image was taken from two sides. There was a CCD camera on the top and another on one side. The side camera was at an angle of 19.5° to the center part of the chip; Figure S2: Images of EBs taken after preparation and culturing—(A) the EBs were prepared from ES cells; (B) after culturing for 48 h, the ES cells were aggregated as EBs on a 96-well U-bottom plate; (C) EBs were removed from the U-bottom plate, and cultured on a gelatin glass bottom dish for 12 h. The image was taken by phase-contrast microscopy; and (D) EBs after rotating for 10 min, orbitally moving them for 10 min, and then culturing them for 12 h. The image was taken by bright field microscopy. Video S1: Self-rotation of EB; Video S2: Rotation speed of an EB at 60 rpm as it changes to orbital motion; Video S3: A single object and two objects were rotated in the micro-rotational flow; Video S4: Short orbital motion of an EB performed in low flow rate; Video S5: Long orbital motion of an EB performed in high flow rate; Video S6: Rotation speed control of an EB; Video S7: Self-rotational motion and orbital motion of an EB (diameter, 20 μm); Video S8: Processes of capturing and performing short and long orbital motions of an EB; Video S9: Different size objects performing orbital motions of different lengths.

Acknowledgments: This work was partly supported by J MEXT/JSPS KAKENHI Grant Numbers JP15K11918, JP21676002, JP26249027, JP26560210, JP15H00819, the Fluid Power Technology Promotion Foundation, the Sasagawa Scientific Research Grant, Japan Science Society and the Marubun Research Promotion Foundation.

Author Contributions: Y.Y. and K.M. conceived and designed the experiments; Y.Y. performed the experiments; Y.Y. analyzed the data; Y.K. contributed materials; Y.Y., Y.K. and K.M. wrote the paper.

References

1. Ramalingam, A.V.; Shi, S. *Stem Cell Biology and Tissue Engineering in Dental Sciences*; Academic Press: Cambridge, MA, USA, 2014; Volume 5.
2. Haycock, J.W. 3D Cell Culture: A Review of Current Approaches and Techniques. In *3D Cell Culture Methods and Protocol*; Humana Press: New York City, NY, USA, 2011; Volume 695.
3. Xie, Y.; Hardouin, P.; Zhu, Z.; Tang, T.; Dai, K.; Lu, J. Three-dimensional flow perfusion culture system for stem cell proliferation inside the critical-size beta-tricalcium phosphate scaffold. *Tissue Eng.* **2006**, *12*, 3535–3543. [CrossRef] [PubMed]
4. Edmondson, R.; Broglie, J.J.; Adcock, A.F.; Yang, L. Three-dimensional cell culture systems and their applications in drug discovery and cell-based biosensors. *Assay Drug Dev. Technol.* **2014**, *12*, 207–218. [CrossRef] [PubMed]
5. Lee, J.; Cuddihy, M.J.; Kotov, N.A. Three-dimensional cell culture matrices: State of the art. *Tissue Eng. Part B Rev.* **2008**, *14*, 61–86. [CrossRef] [PubMed]
6. Lee, K.Y.; Mooney, D.J. Hydrogels for tissue engineering. *Chem. Rev.* **2001**, *101*, 1869–1879. [CrossRef] [PubMed]
7. Huh, D.; Hamilton, G.A.; Ingber, D.E.; Program, B. From Three-Dimensional Cell Culture to Organs-on-Chips. *Trends Cell Biol.* **2011**, *21*, 745–754. [CrossRef] [PubMed]
8. Choi, N.W.; Cabodi, M.; Held, B.; Gleghorn, J.P.; Bonassar, L.J.; Stroock, A.D. Microfluidic scaffolds for tissue engineering. *Nat. Mater.* **2007**, *6*, 908–915. [CrossRef] [PubMed]
9. Haisler, W.L.; Timm, D.M.; Gage, J.A.; Tseng, H.; Killian, T.C.; Souza, G.R. Three-dimensional cell culturing by magnetic levitation. *Nat. Protoc.* **2013**, *8*, 1940–1949. [CrossRef] [PubMed]
10. Tung, Y.-C.; Hsiao, A.Y.; Allen, S.G.; Torisawa, Y.; Ho, M.; Takayama, S. High-throughput 3D spheroid culture and drug testing using a 384 hanging drop array. *Analyst* **2011**, *136*, 473–478. [CrossRef] [PubMed]
11. Garvin, K.A.; Dalecki, D.; Yousefhussien, M.; Helguera, M.; Hocking, D.C. Spatial patterning of endothelial cells and vascular network formation using ultrasound standing wave fields. *J. Acoust. Soc. Am.* **2013**, *134*, 1483–1490. [CrossRef] [PubMed]
12. Tzima, E.; Del Pozo, M.A.; Shattil, S.J.; Chien, S.; Schwartz, M.A. Activation of integrins in endothelial cells by fluid shear stress mediates Rho-dependent cytoskeletal alignment. *EMBO J.* **2001**, *20*, 4639–4647. [CrossRef] [PubMed]
13. Stolberg, S.; McCloskey, K.E. Can Shear Stress Direct Stem Cell Fate? *Biotechnol. Prog.* **2009**, *25*, 10–19. [CrossRef] [PubMed]
14. Zhang, H.; Dai, S.; Bi, J.; Liu, K.-K. Biomimetic three-dimensional microenvironment for controlling stem cell fate. *Interface Focus* **2011**, *1*, 792–803. [CrossRef] [PubMed]
15. Mach, A.J.; Kim, J.H.; Arshi, A.; Hur, S.C.; Di Carlo, D. Automated cellular sample preparation using a Centrifuge-on-a-Chip. *Lab Chip* **2011**, *11*, 2827–2834. [CrossRef] [PubMed]

16. Hagiwara, M.; Kawahara, T.; Arai, F. Local streamline generation by mechanical oscillation in a microfluidic chip for noncontact cell manipulations. *Appl. Phys. Lett.* **2012**, *101*, 074102. [CrossRef]
17. Lutz, B.R.; Chen, J.; Schwartz, D.T. Hydrodynamic tweezers: 1. Noncontact trapping of single cells using steady streaming microeddies. *Anal. Chem.* **2006**, *78*, 5429–5435. [CrossRef] [PubMed]
18. Tanyeri, M. Hydrodynamic trap for single particles and cells. *Appl. Phys. Lett.* **2010**, *96*, 224101. [CrossRef] [PubMed]
19. Hayakawa, T.; Sakuma, S.; Arai, F. On-chip 3D rotation of oocyte based on a vibration-induced local whirling flow. *Microsyst. Nanoeng.* **2015**, *1*, 15001. [CrossRef]
20. Guo, F.; Mao, Z.; Chen, Y.; Xie, Z.; Lata, J.P.; Li, P.; Ren, L.; Liu, J.; Yang, J.; Dao, M.; et al. Three-dimensional manipulation of single cells using surface acoustic waves. *Proc. Natl. Acad. Sci. USA* **2016**, *113*, 1522–1527. [CrossRef] [PubMed]
21. Yalikun, Y.; Akiyama, Y.; Hoshino, T.; Morishima, K. A Bio-Manipulation Method Based on the Hydrodynamic Force of Multiple Microfluidic Streams. *J. Robot. Mechatron.* **2013**, *1*, 611–618.
22. Yalikun, Y.; Akiyama, Y.; Asano, T.; Morishima, K. System Integration, Modelling, and Simulation for Automation of Multiple Microfluidic Stream Based Bio-manipulation. In Proceedings of the 2014 5th International Conference on Intelligent Systems, Modelling and Simulation, Langkawi, Malaysia, 27–29 January 2014; pp. 209–214.
23. Ben Salem, M.; Oesterle, B. A Shear Flow Around a Spinning Sphere: Numerical Study at Moderate Reynolds Numbers. *Int. J. Multiph. Flow* **1998**, *24*, 563–585. [CrossRef]
24. Changfu, Y.; Haiying, Q.; Xuchang, X. Lift force on rotating sphere at low Reynolds numbers and high rotational speeds. *Acta Mech. Sin.* **2003**, *19*, 300. [CrossRef]
25. Ferro, F.; Baheney, C.S.; Spelat, R. Three-Dimensional (3D) Cell Culture Conditions, Present and Future Improvements. *Razavi Int. J. Med.* **2014**, *2*, e17803. [CrossRef]
26. Thomson, J.; Itskovitz-Eldor, J.; Shapiro, S.; Waknitz, M.; Swiergiel, J.; Marshall, V.; Jones, J. Embryonic stem cell lines derived from human blastocysts. *Science* **1998**, *282*, 1145–1147. [CrossRef] [PubMed]
27. Odorico, J.S.; Kaufman, D.S.; Thomson, J.A. Multilineage differentiation from human embryonic stem cell lines. *Stem Cells* **2001**, *19*, 193–204. [CrossRef] [PubMed]
28. Kang, E.; Choi, Y.Y.; Jun, Y.; Chung, B.G.; Lee, S.-H. Development of a multi-layer microfluidic array chip to culture and replate uniform-sized embryoid bodies without manual cell retrieval. *Lab Chip* **2010**, *10*, 2651–2654. [CrossRef] [PubMed]
29. Kim, C.; Lee, K.S.; Bang, J.H.; Kim, Y.E.; Kim, M.-C.; Oh, K.W.; Lee, S.H.; Kang, J.Y. 3-Dimensional cell culture for on-chip differentiation of stem cells in embryoid body. *Lab Chip* **2011**, *11*, 874–882. [CrossRef] [PubMed]
30. Fung, W.-T.; Beyzavi, A.; Abgrall, P.; Nguyen, N.-T.; Li, H.-Y. Microfluidic platform for controlling the differentiation of embryoid bodies. *Lab Chip* **2009**, *9*, 2591–2595. [CrossRef] [PubMed]
31. Lillehoj, P.B.; Tsutsui, H.; Valamehr, B.; Wu, H.; Ho, C.-M. Continuous sorting of heterogeneous-sized embryoid bodies. *Lab Chip* **2010**, *10*, 1678–1682. [CrossRef] [PubMed]
32. Yasuda, S.; Hasegawa, T.; Hosono, T.; Satoh, M.; Watanabe, K.; Ono, K.; Shimizu, S.; Hayakawa, T.; Yamaguchi, T.; Suzuki, K.; et al. AW551984: A novel regulator of cardiomyogenesis in pluripotent embryonic cells. *Biochem. J.* **2011**, *437*, 345–355. [CrossRef] [PubMed]
33. Conway, D.E.; Breckenridge, M.T.; Hinde, E.; Gratton, E.; Chen, C.S.; Schwartz, M.A. Fluid shear stress on endothelial cells modulates mechanical tension across VE-cadherin and PECAM-1. *Curr. Biol.* **2013**, *23*, 1024–1030. [CrossRef] [PubMed]
34. Shelby, J.P.; Chiu, D.T. Controlled rotation of biological micro- and nano-particles in microvortices. *Lab Chip* **2004**, *4*, 168–170. [CrossRef] [PubMed]
35. Chen, Y.; Chung, A.J.; Wu, T.-H.; Teitell, M.A.; Di Carlo, D.; Chiou, P.-Y. Pulsed laser activated cell sorting with three dimensional sheathless inertial focusing. *Small* **2014**, *10*, 1746–1751. [CrossRef] [PubMed]

15

Microfluidic Autologous Serum Eye-Drops Preparation as a Potential Dry Eye Treatment

Takao Yasui [1,2,3,*], Jumpei Morikawa [1], Noritada Kaji [1,2], Manabu Tokeshi [2,4], Kazuo Tsubota [5] and Yoshinobu Baba [1,2,6,*]

1 Department of Applied Chemistry, Graduate School of Engineering, Nagoya University, Furo-cho, Chikusa-ku, Nagoya 464-8603, Japan; morikawa.jumpei@gmail.com (J.M.); kaji@apchem.nagoya-u.ac.jp (N.K.)
2 ImPACT Research Center for Advanced Nanobiodevices, Nagoya University, Furo-cho, Chikusa-ku, Nagoya 464-8603, Japan; tokeshi@eng.hokudai.ac.jp
3 Japan Science and Technology Agency (JST), PRESTO, 4-1-8 Honcho, Kawaguchi, Saitama 332-0012, Japan
4 Division of Applied Chemistry, Faculty of Engineering, Hokkaido University, Sapporo 060-8628, Japan
5 Department of Ophthalmology, School of Medicine, Keio University, Tokyo 160-8582, Japan; tsubota@z3.keio.jp
6 Health Research Institute, National Institute of Advanced Industrial Science and Technology (AIST), Takamatsu 761-0395, Japan
* Correspondence: yasui@apchem.nagoya-u.ac.jp (T.Y.); babaymtt@apchem.nagoya-u.ac.jp (Y.B.)

Academic Editors: Andrew J. deMello and Nam-Trung Nguyen

Abstract: Dry eye is a problem in tearing quality and/or quantity and it afflicts millions of persons worldwide. An autologous serum eye-drop is a good candidate for dry eye treatment; however, the eye-drop preparation procedures take a long time and are relatively troublesome. Here we use spiral microchannels to demonstrate a strategy for the preparation of autologous serum eye-drops, which provide benefits for all dry eye patients; 100% and 90% removal efficiencies are achieved for 10 μm microbeads and whole human blood cells, respectively. Since our strategy allows researchers to integrate other functional microchannels into one device, such a microfluidic device will be able to offer a new one-step preparation system for autologous serum eye-drops.

Keywords: dry eye; autologous serum eye-drops; spiral microchannel

---

Dry eye is a problem in tearing quality and/or quantity, mainly due to overusing personal computers, tablets, and smartphones, air-drying, and wearing contact lenses. Nowadays, the number of persons suffering from dry eyes may well be over a hundred million worldwide and it increases daily. Dry eyes may be seen as a lack of tears on the corneal epithelial layer induced by corneal damage, and it is also a symptom of problems such as meibomian gland dysfunction and Sjögren's syndrome. Since the reasons for dry eyes are not straightforward, commercially available eye-drops are generally insufficient to treat dry eyes completely; they only can lubricate the front surface of the eye.

Autologous serum eye-drops are a good candidate for dry eye treatment since they contain epidermal growth factor (EGF), vitamin A, and so on, which is essential for cell differentiation and division [1–3]. Treatment using the autologous serum eye-drops is based on the concept that dry eye worsening is not due to drying out the front surface of the eye, but rather to poorly supplying essential components to the cornea; therefore, the autologous serum eye-drops can treat dry eyes comprehensively, by not only lubricating the front surface of the eye but also promoting cornea regrowth by the EGF [4]. Autologous serum eye-drops have two features. One is that users can reduce the chance of infection because the person's own blood is utilized, and the other is that the eye-drops can be stored for up to three months at −80 °C. The autologous serum eye-drops are prepared as follows: first, a patient's blood is collected in a heparin-unmodified blood collection tube; secondly, the collected blood is centrifuged at 3000 rpm for 10 min; thirdly, the supernatant is filtered through a 0.45-μm-pore-size filter; and finally, the filtered serum is diluted to reach a target concentration using saline. However, the preparation is relatively troublesome and takes a long time due to the centrifugation, filtration, and dilution steps.

Here we demonstrated a strategy for the preparation of autologous serum eye-drops using a microfluidic technique. Microfluidics has shown great promise for significantly improving diagnostics, as well as biological and medical research studies [5]. Microfluidics has been variously used for passive blood cells separation approaches [6], such as hydrodynamic separation [7–17], sedimentation-based separation [18–21], and filtration-based separation [22–33]. Considering the desire for high throughput and the need for a further dilution process, we fabricated a spiral microchannel (Figure 1a) to realize inertial migration, one of the hydrodynamic separation techniques [34]. In curving microchannels, particles experience a combination of inertial lift force and Dean drag force; inertial lift force acts to focus microbeads at an equilibrium position between the channel wall and centerline [35,36], and Dean drag force acts to entrain microbeads as two counter-rotating vortices with flow directed toward the outer bend at the midline of the channel and inwards at the channel edges [37,38]. A ratio of these forces (inertial lift, F_L/Dean drag, F_D) would be a key parameter to determining the equilibrium positions of the microbeads [39,40]. An inertial force ratio, $R = F_L/F_D \approx a^3 \approx 1/H^3$, where a is the particle diameter and H is the channel height, is obtained by dividing the dimensional scaling of the inertial lift force with the scaling of the Dean drag force [13,40,41]. This force ratio shows that particles with a larger diameter migrate to inertial equilibrium positions, and particles in a channel of larger height do not migrate to inertial equilibrium positions but remain entrained in the channel vortices. We demonstrated the focusing of 10-μm-diameter microbeads (2.65%, Polyscience, Inc., Warrington, UK) at the equilibrium position close to the inner wall of the spiral microchannel (Figure 1b). Using the spiral microchannel, we performed blood cell removal for the microfluidic autologous serum eye-drops preparation as a potential dry eye treatment.

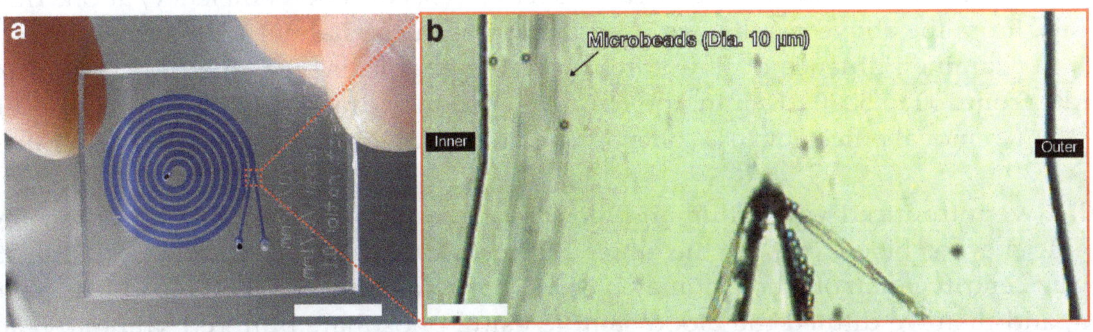

Figure 1. A spiral microfluidic device. (**a**) Photograph of a spiral microfluidic device; scale bar, 10 mm. Microchannels are highlighted by Trypan blue dye solution. Channel width and height are 707 and 70.7 μm, respectively. Distance between two adjacent microchannels is 303 μm; (**b**) A magnified micrograph of part of a spiral microchannel, enclosed by the red dotted box in Figure 1a; scale bar, 100 μm. Ten-fold diluted microbeads (10 μm diameter) in phosphate buffered saline were focused at an equilibrium position close to the inner wall of the microchannel.

For the fabrication of microfluidic devices with a spiral microchannel, we used poly(dimethylsiloxane) (PDMS; silpot 184, Dow Corning Toray Co., Ltd., Tokyo, Japan) replication techniques from an SU-8 mold (SU-8 3050, Nippon Kayaku Co., Ltd., Tokyo, Japan). First, photo-curable SU-8 resin was spin-coated on Si substrates (Silicon Technology Co., Ltd., Tokyo, Japan) and pre-baked at 95 °C for 20 min. The thickness of the SU-8 resin was controlled by spinner rotation speed and time. The SU-8 microchannel was patterned by a mask aligner (MJB4, SÜSS MicroTec AG., Munich, Germany) through emulsion photomasks (Topic Co., Ltd., Kawaguchi, Japan). In addition, the patterned

SU-8 resin was post-baked at 95 °C for more than 4 min and developed using a SU-8 developer (Nippon Kayaku Co., Ltd.). The developed SU-8 mold was finished by putting it into a vacuum chamber under a trichloro(1H, 1H, 2H, 2H-perfluorooctyl)silane atmosphere for 3 h. PDMS was poured into the silanized SU-8 mold and cured at 80 °C for 2 h. After peeling off the cured PDMS, via holes were made for one inlet and two outlets. The PDMS with the via holes and glass slides were bonded to each other after plasma treatment (SDP-1012, Meiwafosis Co., Ltd., Tokyo, Japan). Removal efficiency (collection efficiency) was calculated by dividing the number of introduced microbeads or blood cells by collected ones. In addition, the number of microbeads or blood cells was calculated using collected sample volume and concentrations, which are estimated from a calibration curve (optical density vs. concentrations).

The spiral microchannels showed 100% removal efficiency for 10-μm-diameter microbeads, which is a model material for blood cells (Figure 2). The features of the spiral microchannels, such as the aspect ratio, the number of microchannel spirals, and flow rates, should be candidate parameters governing removal efficiency. Since maximum channel velocity, which is determined by the cross-sectional area of the microchannel, is known to affect removal efficiency [34,40–42], we supposed that the cross-sectional area should be 50,000 μm^2.

By changing the aspect ratio from 0.1 to 1.0 under other fixed conditions, we concluded that the aspect ratio from 0.1 to 0.2 was suitable for 10 μm particle removal; in particular, the 0.1 ratio gave a 99% removal efficiency (1% collection efficiency) at the outer outlet (Figure 2a). This meant that a smaller aspect ratio had higher removal efficiency, which was in good agreement with the behavior predicted by the inertial force ratio: particles in a smaller height channel migrated to inertial equilibrium positions. Next, we considered the effect of the number of microchannel spirals, ranging from 0.5 to 7.5 spirals, on removal efficiency (Figure 2b).

Figure 2c showed that the removal efficiency increased with an increase of the number of microchannel spirals, leading to 99% removal efficiency (1% collection efficiency) at one outer outlet in 7.5 spirals. From the above results, we used the spiral microchannel with a 0.1 aspect ratio and 7.5 spirals to examine influence of flow rates on removal efficiency (Figure 2d). As we increased the flow rate from 100 to 5000 μL/min, the removal efficiency drastically improved, and finally we achieved 100% removal efficiency (0% collection efficiency) at the flow rate of 5000 μL/min.

Finally, we introduced whole human blood into the spiral microchannels and achieved 90% removal efficiency of blood cells at the outer outlet (10% collection efficiency) (Figure 3). After sampling and centrifugation of whole human blood, we mixed blood cells and blood plasma to be 50% hematocrit, and then we diluted the blood sample using phosphate buffered saline to reach target hematocrit values. As for the 10 μm microbeads, the removal efficiency of blood cells increased as the flow rate increased; however, we could not attain 100% efficiency due to the disc shape of the red blood cells which had an 8 μm diameter and 2.5 μm thickness (Figure 3a).

Considering the inertial force ratio, it made sense that removal efficiency was degraded for the smaller particle diameter. It is well known that the inertial lift force drops with a decrease in the Reynolds number [34–36], and as we expected, the viscosity of the blood samples affected removal efficiency, and the removal efficiency at the outer outlet increased to 90% (10% collection efficiency) as the concentration decreased (Figure 3b). Figure 3c shows photographs of collected blood samples at the inner and outer outlets; hemolyzed blood was not observed.

We confirmed that hemolyzed blood was not observed at any of the concentrations used (Figure 3b). From these results, we concluded that inertial force in the spiral microchannels at the concentrations used had no hemolyzing property.

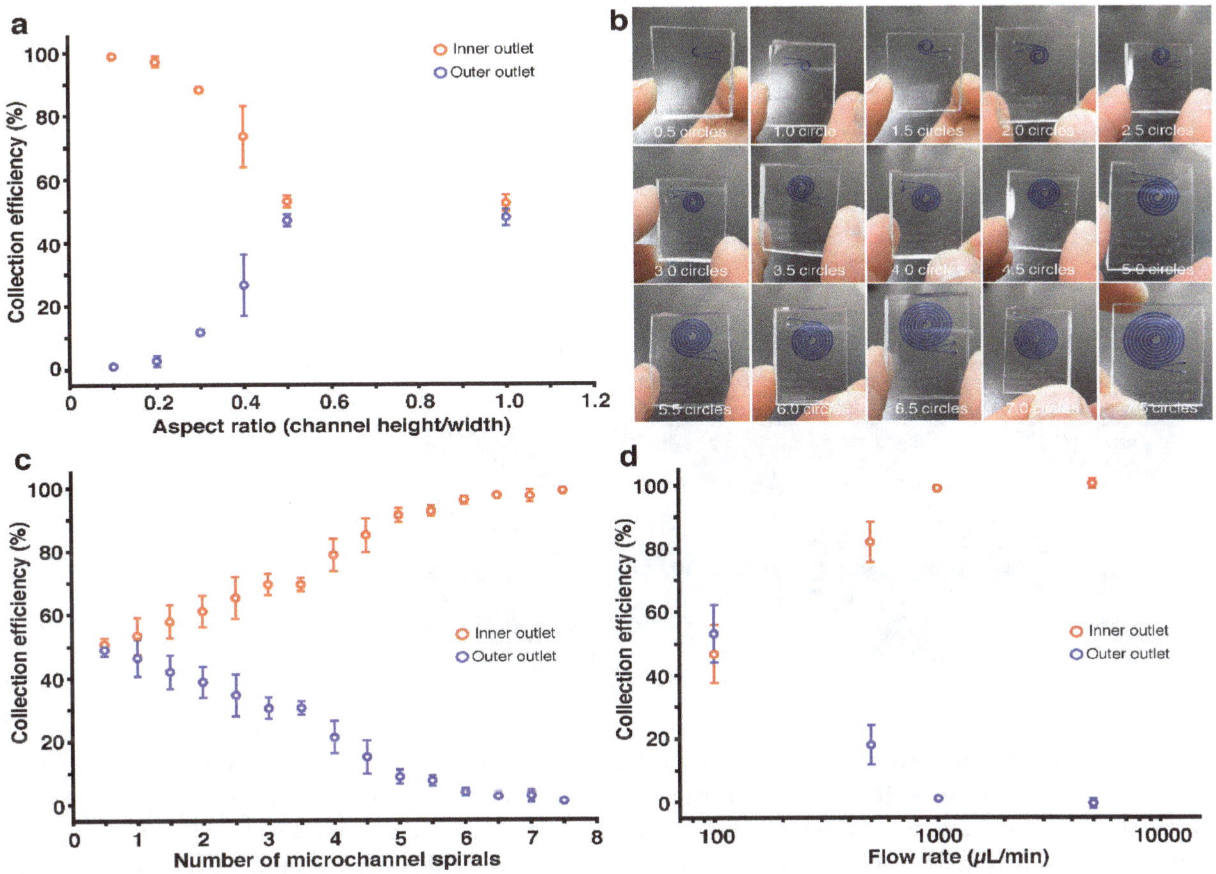

Figure 2. Collection efficiency of 10 μm particles. Cross-sectional area was 50,000 μm^2. Ten-fold diluted microbeads (10 μm diameter) in phosphate buffered saline were used. Error bars are the standard deviation for a series of measurements ($N = 3$). (**a**) Collection efficiency vs. aspect ratio of spiral microchannels. The aspect ratio is the ratio of channel height to width. The number of microchannel spirals was 7.5, and flow rate was 1000 μL/min; (**b**) Photographs of fabricated spiral microchannels with 0.5 to 7.5 circles. One circle is one spiral. The microchannels are highlighted by Trypan blue dye solution; (**c**) Collection efficiency vs. number of microchannel spirals. The aspect ratio of the microchannels was 0.1, and flow rate was 1000 μL/min; (**d**) Collection efficiency vs. flow rate. The aspect ratio of the microchannels was 0.1, and the number of microchannel spirals was 7.5.

To achieve the 100% removal efficiency of blood cells, we can propose two methods: increasing the inertial lift force and decreasing the Dean drag force. Both ways lead to increasing the inertial force ratio. For increasing the inertial lift force, we should increase the Reynolds number by increasing the flow rates [36,43,44]. In this approach, we could apply 10,000 μL/min for a maximum flow rate due to a deformability issue of PDMS. Since Si, glass or polymethyl methacrylate (PMMA) are much harder materials than PDMS, these microchannels can be good candidates for applying more than 10,000 μL/min. Note that we should confirm the hemolysis issue of blood cells when we apply more than 10,000 μL/min. For decreasing the Dean drag force, we should decrease the Dean number by reducing the channel height or increasing the curvature ratio [38–40]. In this approach, we used the microchannels with a 0.1 aspect ratio and 7.5 spirals due to a roof collapse issue of PDMS and a size issue of glass slides. Si, glass or polymethyl methacrylate (PMMA) microchannels would also help researchers to avoid the roof collapse issue and reduce the aspect ratio, and a larger size of the glass slides would allow researchers to avoid the size issue and increase the number of microchannel spirals. Note that we should confirm a clogging issue of blood cells when we use lower aspect ratio microchannels.

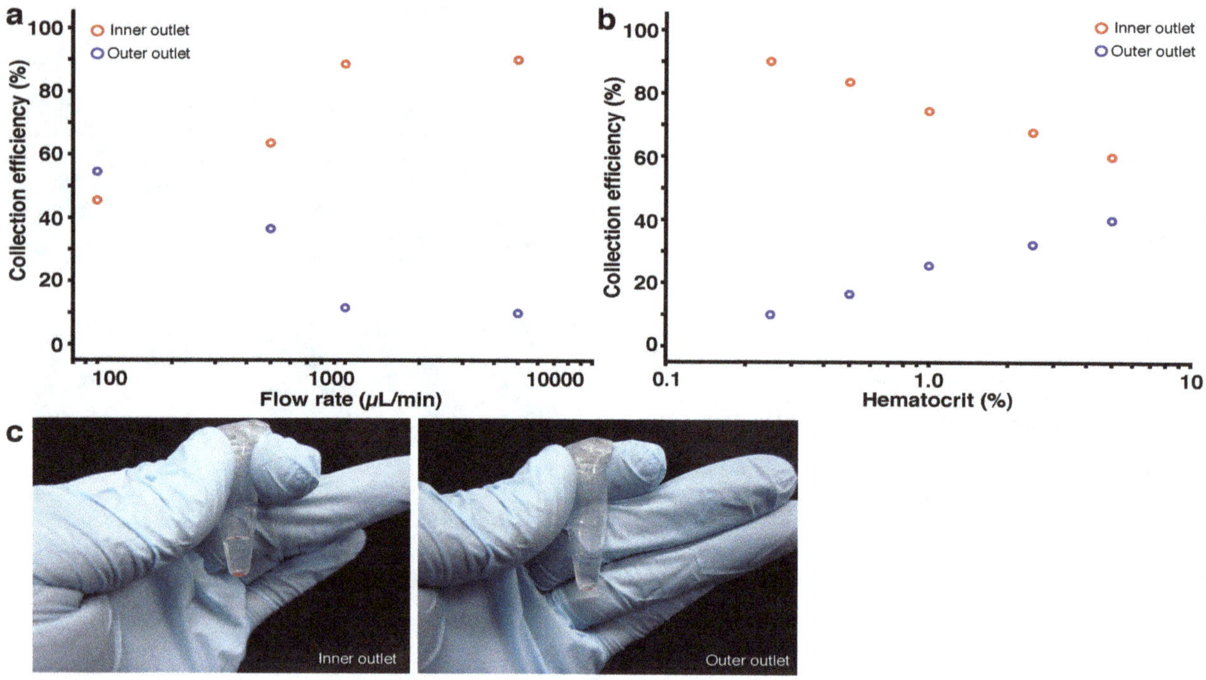

Figure 3. Collection efficiency of whole human blood cells. Cross-sectional area was 50,000 μm^2, the aspect ratio was 0.1, and the number of microchannel spirals was 7.5. (**a**) Collection efficiency vs. flow rate. Initial hematocrit of blood samples was 0.25%; (**b**) Collection efficiency vs. whole blood concentration. Flow rate was 5000 μm/min; (**c**) Photographs of collected samples from inner and outer outlets after centrifugation. Flow rate was 5000 μm/min, and initial hematocrit of blood samples was 0.25%. Hemolyzed blood was not observed.

In summary, we have demonstrated a strategy for the preparation of autologous serum eye-drops based on spiral microchannels, which enables passive blood cell removal. The spiral microchannels achieved complete removal of 10 μm microbeads as a model sample, and 90% removal of whole human blood cells. While the current removal efficiency is not yet enough to make autologous serum eye drops, flow rates with more than 10,000 μL/min (up to a flow rate without hemolysis), which can increase the inertial lift force, and lower aspect ratio microchannels (down to an aspect ratio without clogging) over eight spirals, which can decrease the Dean drag force, have the potential for application in preparation devices for blood cell removal, with the eventual goal of realizing the dry eye treatment. Since the present strategy allows researchers to make a further integration with a separation microchannel for platelets and clotting factors and a dilution microchannel, such microfluidic devices can offer a new path for the development of a one-step preparation system for autologous serum eye-drops.

Acknowledgments: This research was supported by the JSPS Grant-in-Aid for Scientific Research (A) 16H02091, Nanotechnology Platform Program (Molecule and Material Synthesis) of the Ministry of Education, Culture, Sports, Science and Technology (MEXT), and PRESTO, JST.

References

1. Noda-Tsuruya, T.; Asano-Kato, N.; Toda, I.; Tsubota, K. Autologous serum eye drops for dry eye after LASIK. *J. Refract. Surg.* **2006**, *22*, 61–66. [PubMed]
2. Ubels, J.L.; Foley, K.M.; Rismondo, V. Retinol secretion by the lacrimal gland. *Investig. Ophthalmol. Vis. Sci.* **1986**, *27*, 1261–1268.
3. Ohashi, Y.; Motokura, M.; Kinoshita, Y.; Mano, T.; Watanabe, H.; Kinoshita, S.; Manabe, R.; Oshiden, K.; Yanaihara, C. Presence of epidermal growth factor in human tears. *Investig. Ophthalmol. Vis. Sci.* **1989**, *30*, 1879–1882.

4. Tsubota, K. New approaches in dry eye management: Supplying missing tear components to the ocular surface epithelium. In *Current Opinions in the Kyoto Cornea Club*; Kugler Publications: Amsterdam, The Netherlands, 1997; Volume 1, pp. 27–32.
5. Sackmann, E.K.; Fulton, A.L.; Beebe, D.J. The present and future role of microfluidics in biomedical research. *Nature* **2014**, *507*, 181–189. [CrossRef] [PubMed]
6. Tripathi, S.; Kumar, Y.V.B.V.; Prabhakar, A.; Joshi, S.S.; Agrawal, A. Passive blood plasma separation at the microscale: A review of design principles and microdevices. *J. Micromech. Microeng.* **2015**, *25*, 083001. [CrossRef]
7. Yang, S.; Undar, A.; Zahn, J.D. A microfluidic device for continuous, real time blood plasma separation. *Lab Chip* **2006**, *6*, 871–880. [CrossRef] [PubMed]
8. Jaggi, R.D.; Sandoz, R.; Effenhauser, C.S. Microfluidic depletion of red blood cells from whole blood in high-aspect-ratio microchannels. *Microfluid. Nanofluid.* **2007**, *3*, 47–53. [CrossRef]
9. Faivre, M.; Abkarian, M.; Bickraj, K.; Stone, H.A. Geometrical focusing of cells in a microfluidic device: An approach to separate blood plasma. *Biorheology* **2006**, *43*, 147–159. [PubMed]
10. Sollier, E.; Cubizolles, M.; Fouillet, Y.; Achard, J.L. Fast and continuous plasma extraction from whole human blood based on expanding cell-free layer devices. *Biomed. Microdevices* **2010**, *12*, 485–497. [CrossRef] [PubMed]
11. Marchalot, J.; Fouillet, Y.; Achard, J.L. Multi-step microfluidic system for blood plasma separation: Architecture and separation efficiency. *Microfluid. Nanofluid.* **2014**, *17*, 167–180. [CrossRef]
12. Rodriguez-Villarreal, A.I.; Arundell, M.; Carmona, M.; Samitier, J. High flow rate microfluidic device for blood plasma separation using a range of temperatures. *Lab Chip* **2010**, *10*, 211–219. [CrossRef] [PubMed]
13. Kersaudy-Kerhoas, M.; Kavanagh, D.M.; Dhariwal, R.S.; Campbell, C.J.; Desmulliez, M.P.Y. Validation of a blood plasma separation system by biomarker detection. *Lab Chip* **2010**, *10*, 1587–1595. [CrossRef] [PubMed]
14. Tripathi, S.; Prabhakar, A.; Kumar, N.; Singh, S.G.; Agrawal, A. Blood plasma separation in elevated dimension T-shaped microchannel. *Biomed. Microdevices* **2013**, *15*, 415–425. [CrossRef] [PubMed]
15. Prabhakar, A.; Kumar, Y.V.B.V.; Tripathi, S.; Agrawal, A. A novel, compact and efficient microchannel arrangement with multiple hydrodynamic effects for blood plasma separation. *Microfluid. Nanofluid.* **2015**, *18*, 995–1006. [CrossRef]
16. Lee, M.G.; Choi, S.; Kim, H.J.; Lim, H.K.; Kim, J.H.; Huh, N.; Park, J.K. Inertial blood plasma separation in a contraction-expansion array microchannel. *Appl. Phys. Lett.* **2011**, *98*, 253702. [CrossRef]
17. Xiang, N.; Ni, Z.H. High-throughput blood cell focusing and plasma isolation using spiral inertial microfluidic devices. *Biomed. Microdevices* **2015**, *17*, 110. [CrossRef] [PubMed]
18. Sun, M.; Khan, Z.S.; Vanapalli, S.A. Blood plasma separation in a long two-phase plug flowing through disposable tubing. *Lab Chip* **2012**, *12*, 5225–5230. [CrossRef] [PubMed]
19. Zhang, X.B.; Wu, Z.Q.; Wang, K.; Zhu, J.; Xu, J.J.; Xia, X.H.; Chen, H.Y. Gravitational sedimentation induced blood de lamination for continuous plasma separation on a microfluidics chip. *Anal. Chem.* **2012**, *84*, 3780–3786. [CrossRef] [PubMed]
20. Tachi, T.; Kaji, N.; Tokeshi, M.; Baba, Y. Simultaneous separation, metering, and dilution of plasma from human whole blood in a microfluidic system. *Anal. Chem.* **2009**, *81*, 3194–3198. [CrossRef] [PubMed]
21. Dimov, I.K.; Basabe-Desmonts, L.; Garcia-Cordero, J.L.; Ross, B.M.; Ricco, A.J.; Lee, L.P. Stand-alone self-powered integrated microfluidic blood analysis system (SIMBAS). *Lab Chip* **2011**, *11*, 845–850. [CrossRef] [PubMed]
22. Li, C.Y.; Liu, C.; Xu, Z.; Li, J.M. Extraction of plasma from whole blood using a deposited microbead plug (DMBP) in a capillary-driven microfluidic device. *Biomed. Microdevices* **2012**, *14*, 565–572. [CrossRef] [PubMed]
23. Moorthy, J.; Beebe, D.J. In situ fabricated porous filters for microsystems. *Lab Chip* **2003**, *3*, 62–66. [CrossRef] [PubMed]
24. Thorslund, S.; Klett, O.; Nikolajeff, F.; Markides, K.; Bergquist, J. A hybrid poly(dimethylsiloxane) microsystem for on-chip whole blood filtration optimized for steroid screening. *Biomed. Microdevices* **2006**, *8*, 73–79. [CrossRef] [PubMed]
25. Wang, S.Q.; Sarenac, D.; Chen, M.H.; Huang, S.H.; Giguel, F.F.; Kuritzkes, D.R.; Demirci, U. Simple filter microchip for rapid separation of plasma and viruses from whole blood. *Int. J. Nanomed.* **2012**, *7*, 5019–5028.

26. Chung, K.H.; Choi, Y.H.; Yang, J.H.; Park, C.W.; Kim, W.J.; Ah, C.S.; Sung, G.Y. Magnetically-actuated blood filter unit attachable to pre-made biochips. *Lab Chip* **2012**, *12*, 3272–3276. [CrossRef] [PubMed]
27. Aran, K.; Fok, A.; Sasso, L.A.; Kamdar, N.; Guan, Y.L.; Sun, Q.; Undar, A.; Zahn, J.D. Microfiltration platform for continuous blood plasma protein extraction from whole blood during cardiac surgery. *Lab Chip* **2011**, *11*, 2858–2868. [CrossRef] [PubMed]
28. Crowley, T.A.; Pizziconi, V. Isolation of plasma from whole blood using planar microfilters for lab-on-a-chip applications. *Lab Chip* **2005**, *5*, 922–929. [CrossRef] [PubMed]
29. VanDelinder, V.; Groisman, A. Separation of plasma from whole human blood in a continuous cross-flow in a molded microfluidic device. *Anal. Chem.* **2006**, *78*, 3765–3771. [CrossRef] [PubMed]
30. Chen, X.; Cui, D.F.; Liu, C.C.; Li, H. Microfluidic chip for blood cell separation and collection based on crossflow filtration. *Sens. Actuators B Chem.* **2008**, *130*, 216–221. [CrossRef]
31. Kim, Y.C.; Kim, S.H.; Kim, D.; Park, S.J.; Park, J.K. Plasma extraction in a capillary-driven microfluidic device using surfactant-added poly(dimethylsiloxane). *Sens. Actuators B Chem.* **2010**, *145*, 861–868. [CrossRef]
32. Geng, Z.X.; Ju, Y.R.; Wang, Q.F.; Wang, W.; Li, Z.H. Multi-component continuous separation chip composed of micropillar arrays in a split-level spiral channel. *RSC Adv.* **2013**, *3*, 14798–14806. [CrossRef]
33. Kang, T.G.; Yoon, Y.J.; Ji, H.M.; Lim, P.Y.; Chen, Y. A continuous flow micro filtration device for plasma/blood separation using submicron vertical pillar gap structures. *J. Micromech. Microeng.* **2014**, *24*, 087001. [CrossRef]
34. Di Carlo, D. Inertial microfluidics. *Lab Chip* **2009**, *9*, 3038–3046. [CrossRef] [PubMed]
35. Ho, B.P.; Leal, L.G. Inertial migration of rigid spheres in 2-Dimensional unidirectional flows. *J. Fluid Mech.* **1974**, *65*, 365–400. [CrossRef]
36. Matas, J.P.; Morris, J.F.; Guazzelli, E. Inertial migration of rigid spherical particles in Poiseuille flow. *J. Fluid Mech.* **2004**, *515*, 171–195. [CrossRef]
37. Berger, S.A.; Talbot, L.; Yao, L.S. Flow in curved pipes. *Annu. Rev. Fluid Mech.* **1983**, *15*, 461–512. [CrossRef]
38. Di Carlo, D.; Irimia, D.; Tompkins, R.G.; Toner, M. Continuous inertial focusing, ordering, and separation of particles in microchannels. *Proc. Natl. Acad. Sci. USA* **2007**, *104*, 18892–18897. [CrossRef] [PubMed]
39. Di Carlo, D.; Edd, J.F.; Irimia, D.; Tompkins, R.G.; Toner, M. Equilibrium separation and filtration of particles using differential inertial focusing. *Anal. Chem.* **2008**, *80*, 2204–2211. [CrossRef] [PubMed]
40. Gossett, D.R.; Di Carlo, D. Particle focusing mechanisms in curving confined flows. *Anal. Chem.* **2009**, *81*, 8459–8465. [CrossRef] [PubMed]
41. Di Carlo, D.; Edd, J.F.; Humphry, K.J.; Stone, H.A.; Toner, M. Particle segregation and dynamics in confined flows. *Phys. Rev. Lett.* **2009**, *102*, 094503. [CrossRef] [PubMed]
42. Kuntaegowdanahalli, S.S.; Bhagat, A.A.; Kumar, G.; Papautsky, I. Inertial microfluidics for continuous particle separation in spiral microchannels. *Lab Chip* **2009**, *9*, 2973–2980. [CrossRef] [PubMed]
43. Asmolov, E.S. The inertial lift on a spherical particle in a plane Poiseuille flow at large channel Reynolds number. *J. Fluid Mech.* **1999**, *381*, 63–87. [CrossRef]
44. Schonberg, J.A.; Hinch, E.J. Inertial migration of a sphere in poiseuille flow. *J. Fluid Mech.* **1989**, *203*, 517–524. [CrossRef]

16

Microfluidic Approaches for Manipulating, Imaging and Screening *C. elegans*

Bhagwati P. Gupta [1,*] and Pouya Rezai [2]

1 Department of Biology, McMaster University, Hamilton, ON L8S 4K1, Canada
2 Department of Mechanical Engineering, York University, Toronto, ON M3J 1P3, Canada; prezai@yorku.ca
* Correspondence: guptab@mcmaster.ca

Academic Editors: Manabu Tokeshi and Kiichi Sato

Abstract: The nematode *C. elegans* (worm) is a small invertebrate animal widely used in studies related to fundamental biological processes, disease modelling, and drug discovery. Due to their small size and transparent body, these worms are highly suitable for experimental manipulations. In recent years several microfluidic devices and platforms have been developed to accelerate worm handling, phenotypic studies and screens. Here we review major tools and briefly discuss their usage in *C. elegans* research.

Keywords: *C. elegans*; microfluidics; live imaging; electrotaxis; neurobiology; high-throughput screening; drug discovery

---

1. Introduction

C. elegans, commonly referred to as the worm, is the leading animal model for biomedical research and screening of drugs and drug targets [1,2]. As a multicellular system, *C. elegans* offers many experimental advantages including small size, transparency, ease of culturing, rapid growth, large brood size, cell lineage, and a relatively compact genome (~100 Mb) that is fully sequenced. The animal consists of two sexes: hermaphrodites (that are essentially females but produce a limited number of sperm initially before switching to make oocytes) and males. Hermaphrodites produce progeny using their own sperm as well as from males following mating. Fertilized eggs hatch to become L1 larvae, which then transition through L2, L3, and L4 larval stages, each separated by molting, to become adults in less than three days at 20 °C. The nervous system of the adult worm consists of 302 neurons whose interconnections are fully mapped [3,4]. Research has shown that many of the cellular and molecular processes in worms are conserved across almost all eukaryotes [5,6], making it a highly relevant system to understand human biology, investigate the mechanisms of diseases, and carry out drug discovery experiments. However, the small size of *C. elegans* (approximately 1 mm length and 60 μm width) and its continuous undulatory locomotion imposes a significant challenge in manipulations that involves automated and high throughput methods. The research community has therefore been in continuous search for technologies/techniques to address this issue.

Miniature devices such as microfluidic chips and Micro-Electro-Mechanical Systems are increasingly being used in biological research and drug discovery. These devices offer many advantages such as the small size, low consumption of reagents, ability to manipulate small objects and automate operations, increased throughput, small footprints, and excellent safety and reliability. Microfluidics in particular deals with the study and control of fluids and nano- to micro-scale objects inside miniaturized environments via incorporation of operation automation tools such as micro-pumps and micro-valves. Microfluidic tools have been used successfully to manipulate biological fluids, cells, tissues, and even whole organisms [7,8]. In this review, we specifically focus on the applications of microfluidics in research involving *C. elegans*.

The usage of microfluidic chips for *C. elegans* experiments has grown rapidly in the last decade. The material commonly used in the fabrication of these devices is polydimethylsiloxane (PDMS), a polymer that is transparent, flexible, gas permeable, and is not toxic to worms. PDMS parts could be joined or bonded with other components made of glass, silicon, and steel to create complex devices. These devices can handle almost every step of worm manipulation and analysis including culturing, different treatments, and continuous monitoring of cellular and behavioral phenotypes. The existing *C. elegans* microfluidic systems could be classified into roughly three categories based upon their use: (1) devices to assist with routine laboratory procedures (e.g., culturing worms); (2) devices to perform particular types of experiments (e.g., monitoring neuronal activities); and (3) devices for high-throughput screens. This review summarizes developments in these categories, highlighting unique features of different devices and their advantages in accelerating research.

2. Microfluidic Devices to Assist with Routine Procedures

Many routine experiments in *C. elegans* labs such as phenotypic observations require collecting worms of a certain age or phenotype, and culturing them for a set duration before imaging. Routine protocols to process animals for these purposes are tedious due to their manual nature, which is labor-intensive, time-consuming, and does not readily scale up. For example, collecting animals of a specific developmental stage from a culture of mixed stages requires picking them individually which could take hours. Depending on the operator, progress and quality of output could vary significantly. Microfluidic approaches can overcome many of these limitations by automating common steps, leading to increased throughput, higher accuracy, reproducibility, and lower cost of operations. Some of the most common procedures are discussed below.

2.1. Worm Sorting

One of the most common needs in *C. elegans* labs is sufficient quantities of animals of a certain type, e.g., developmental stage, size, sex, or phenotype. In late 1990s, Union Biometrica (http://www.unionbio.com) developed an automated worm sorting platform, termed COPAS (Complex Object Parametric Analyzer and Sorter), for automated sorting of worms based on their size and certain other features. Subsequently an advanced sorting platform, termed "BioSorter", was released that offers a rich set of features and modularity. Although both these non-microfluidic platforms are useful in *C. elegans* studies, they are quite sophisticated in terms of operation and maintenance. Furthermore, their cost is prohibitive for the majority of laboratories. As an alternative, several microfluidic sorting devices have been fabricated in the last decade that are affordable and easy to operate. Table 1 provides an overview of these techniques and their main features.

Table 1. Overview of microfluidic devices to enable sorting of *C. elegans*.

Device	Method	Throughput	Sorting Capabilities	References
PDMS single channel device	Electrotaxis	78 worms per minute	Specific developmental stages and adults; mutants from wild type	Rezai et al., 2010 and 2012 [9,10]
Agarose gel box containing a single long channel	Electrotaxis	Unknown	Adults of different age; mutants from wild type	Maniere et al., 2011 [11]
PDMS device containing interconnected mazes	Flow filtration	200–300 worms per minute	Larvae from adults	Solvas et al., 2011 [12]
PDMS multichannel device	Electrotaxis	4 worms per minute	All stages including adults	Han et al., 2012 [13]
PDMS micro-pillar device	Flow filtration	130–180 worms per minute	All stages including adults	Ai et al., 2014 [14]
PDMS device containing optical fiber and laminar flow switch	Fluorescence filtration	12 worms per minute	Fluorescent animals	Yan et al., 2014 [15]
PDMS-Agarose fan-shaped device	Electrotaxis	56 worms per minute	L2–L4 larvae and adult; size-based separation	Wang et al., 2015 [16]
PDMS device containing adjustable filter	Pressure-based filtration	200 worms per minute	Certain developmental stages	Dong et al., 2016 [17]

Our group was the first to demonstrate worm sorting based on the electrotaxis speed of the animals [9,10]. Electrotaxis is the tendency of worms to respond inherently to desirable electric signals and move towards the favorable electrode (negative electrode in case of *C. elegans*). Using a device (Figure 1A) that consisted of a worm loading and a collection chamber interconnected with a series of parallel thin microchannels that acted as local electric field traps, we succeeded in obtaining synchronized cultures of L3, L4, and adult animals from populations of certain mixed stages (e.g., L3 mixed with L4). The mixed population of worms was passed slowly across the loading chamber while being exposed to a perpendicular electric field that induced them to move laterally towards the narrow traps. However, since the electric field in the traps was unfavorable to older (larger) animals, they ended up restraining from entering the trap while the younger (smaller) worms conveniently passed through the channel and became separated from the mix. The device could reach a throughput of 78 worms per minute and also separate neuron- or muscle-defective worms from normal worms.

Subsequently, other labs also reported electrotaxis-based microfluidic sorters. Maniere et al. [11] constructed a "worm electrophoresis" unit that is essentially a gel box containing a 10 cm long agar track filled with buffer. Worms were placed at one end and allowed to swim under the influence of the DC electric field. The animals were spatially sorted along the runway based on their speed. Although the device was slow to operate (up to an hour for each run to complete), it could distinguish movement-defective worms from wild type as well as separate older animals from younger ones. Han et al. [13] fabricated a device containing multiple parallel micro sinusoidal channels. The channel was designed such that they presented physical barriers and required worms to make efforts as they moved in the presence of an electric field. Using their setup, authors succeeded in sorting worms from a mixed culture of all stages and reported a throughput of roughly 250 adults an hour with 95% accuracy. An earlier investigation [9] showed that electric field does not severely affect the viability of worms, thus supporting the use of electrotaxis approach in long-term post-sorting assays.

While the above microdevices are useful, they lack the ability to sort all developmental stages simultaneously. To this end, a PDMS-agarose hybrid device was developed by Wang et al. [16] that relied on the sensitivity of *C. elegans* to electric field strength. It was earlier shown that worms, when exposed to the electric field on an open gel surface, prefer to move towards the cathode at an angle rather than in a straight line [18]. Wang et al. [16] found that this angular movement (termed "deflecting electrotaxis") varies for different stages of worms even at a constant electric field. The older worms tend to deflect more than younger stages. Based on the observation of deflecting electrotaxis, they designed a fan-shaped structure, from $-50°$ to $+50°$, consisting of channels at $5°$ intervals that originate from a central worm loading spot. Mixed stages of L2, L3, L4, and adult worms were successfully sorted using this setup at a throughput of ~56 worms per minute. The device was also able to sort worms of different sizes as well as separate males from hermaphrodites.

Besides electrotaxis, researchers have developed microfluidic devices that utilize mechanical methods to sort animals. Four reports [12,14,17,19] have described the use of channels, pillars, variable size filters, and thermosensitive hydrogels to achieve a similar goal. Solvas et al. [12] developed a device (Figure 1B), termed "smart mazes," that consisted of a main channel containing a variety of micro-features such as pillar arrays, pools, and "smart" filter and mazes of different dimensions to passively sort worms based on their size. Up to 94% accuracy was observed in sorting adults from larvae. Typically 200–300 worms were processed per minute, which could go up to 1200 worms per minute under some conditions. Although the device is limited in applications, e.g., due to fewer stages being sorted, and the lack of mutant sorting, it could still be useful in many routine assays. Another study by Ai et al. [14] incorporated micro-pillar structures in their devices to sort worms using a flow filtration approach. Each device contained geometrically optimized arrays of pillars such that the spacing was suitable for each developmental stage. While each of these devices could be operated alone to separate two populations that were mixed together, when connected in a serial fashion, they allowed efficient sorting of worms containing a mixture of L1–L4 larvae and adults.

The throughput ranged roughly between 130 and 180 worms per minute, which is suitable for most routine biological assays. Investigations of physiological processes (pharyngeal pumping rate and body bend frequency) and three-day survival in L4-stage worms sorted in the device showed that the sorting experiments had no detrimental effect on worms' physiology and fecundity. Finally, earlier this year, Dong et al. [17] reported a sorting device that was based on the principle of adjustable filter to allow passage of worms of certain diameter. The device (Figure 1C) consisted of a straight microchannel with one inlet and one outlet, two worm collection chambers, and four deflectable membrane valves that act as adjustable filter. The filter parameters were altered using pressure applied onto the membrane through a syringe pump. Mixtures of worms of two adjacent developmental stages, e.g., L3 and L4, could be efficiently and rapidly (roughly 200 worms per minute) separated. Although authors suggest that multiple stages of mixed cultures can also be sorted, it was not demonstrated in their study. Another feature of this setup was the separation of embryos from adults, which can be useful in many studies.

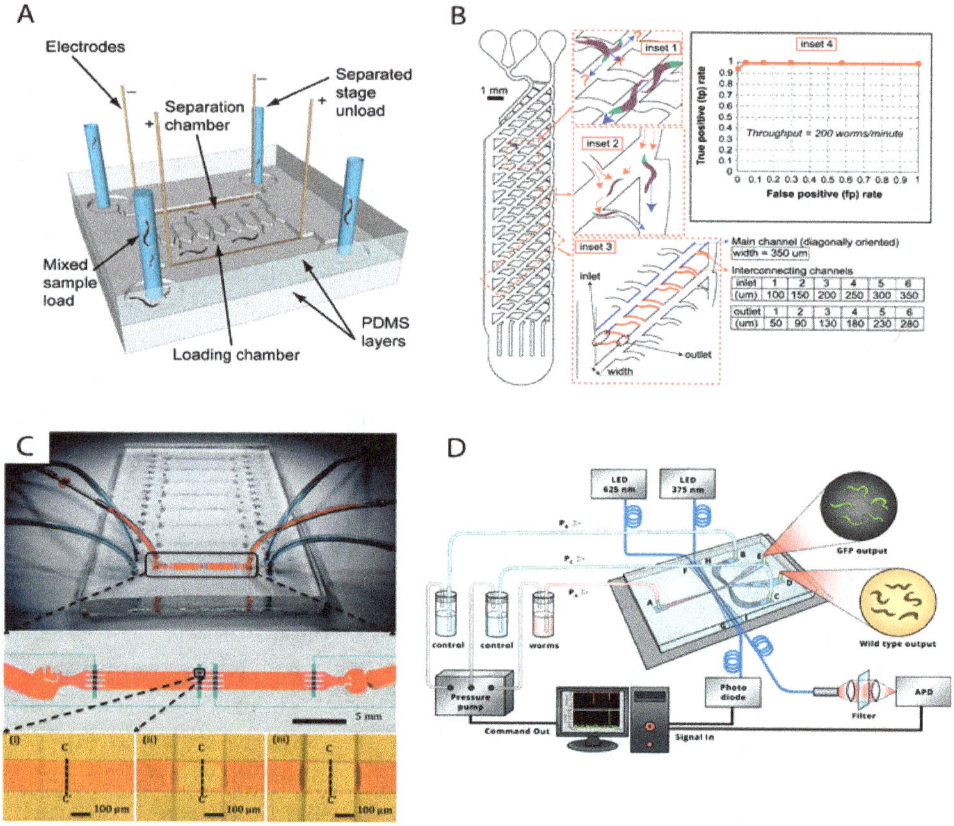

Figure 1. Microfluidic devices for *C. elegans* sorting using (**A**) electrotaxis (Rezai et al. [10], (**B**) mechanical microstructures (Solvas et al. [12], (**C**) deflectable membranes (Dong et al. [17], and (**D**) fiber-based fluorescent detection (Yan et al. [15]). Reproduced with permission from The Royal Society of Chemistry. Panel (**A**) shows the electric trap-based sorting device. Loading chamber contains mixed stage worms. Sorted worms accumulate in the separation chamber and are recovered via unload channels. The smart maze concept is shown in panel (**B**). The four insets show worm orientation (inset **1**); flushing of small larvae (inset **2**); dimensions of the successful design (inset **3**); and successful recovery of adults in an experiment (inset **4**). The deflectable membrane device in panel (**C**) shows eight individual worm selection units (one of these connected with tubes). The fluidic and valve control channels are enlarged to show details. The fluidic path is squeezed upon activation of the control valve. The device in panel (**D**) contains inlets and outlets for worms and buffer. The optical fiber channels (LED 625 and 375 nm) are used to differentiate between wild-type and fluorescing worms. Refer to respective references for more details.

The approaches discussed so far deal with isolating synchronized animals by relying on differences in their sizes and movement responses. Since many *C. elegans* experiments require transgenic animals that express Green Fluorescent Protein (GFP) reporters, it would be highly desirable to perform sorting based on their fluorescence. A number of automated systems for fluorescent-based detection and sorting in microfluidic devices have been reported [15,20,21]. The device by Yan et al. [15] (Figure 1D) can sort fluorescing animals using optical fiber detection and laminar flow switching. The system used two pairs of closely placed fibers, one for sensing the presence of nematodes as they pass through the light path and the other for detecting fluorescence. Sorting then occurred into the arms of a downstream Y-shaped channel that was controlled by flow pressure that guided worms to a desired arm based on their fluorescence status. The entire operation was controlled by customized software and did not require a microscope. This platform could sort worms at an optimum speed of 700 worms per hour with almost 100% accuracy.

In summary, the passive sorting methods that involve mechanical obstacles in microfluidic channels allow separation of animals based on their size at reasonably high throughputs. However, they are unable to perform sorting using other characteristics such as marker gene expression or behavioral responses. In these cases active methods such as electrotactic or fluorescence-based sorting are more desirable.

2.2. Worm Culturing

Many of the in vivo studies of biological processes in *C. elegans* require maintaining cultures for an extended period of time. Worms need to be monitored and periodically examined for changes in cellular processes. Microfluidic devices have been fabricated to streamline culturing and observations of worms [22–31]. In one of the earliest studies by Kim et al. [32], authors reported a compact disc (CD)-shaped device to culture worms for several days. This simple tool relied on the centrifugal force to drive the food diagonally in a rotating CD from inner nutrient reservoirs into cultivation chambers to feed the worms and to eject the waste from these chambers to outer waste reservoirs. For observation purposes, a small aliquot of the culture was removed periodically and examined under the microscope. The device was able to grow animals for up to three generations (in two weeks) without affecting their growth and behavior in any obvious way. However, due to its simple design, it could not distinguish between progeny and parents.

While useful, the compact disc system has some limitations. For example, it lacks the ability to track individual animals and to perform automated on-chip imaging. To address this, Hulme et al. [23] developed a new device. The device (Figure 2A) consists of chambers for long-term culturing and locomotion studies, connected to side narrow microchannels used for immobilization, imaging, and body size measurements at different time slots. While the device could not monitor early larval stages, it was capable of lifelong culturing of single worms from L4 stage onwards by trapping them individually in chambers. Up to 16 adults could be cultured at a time in parallel chambers, which accelerated studies of age-related behavioral and physiological changes. The authors successfully reported that there is a close correlation between the end of growth day and decline in swimming frequency with the lifespan of worm.

In addition to cultivating the worms in chambers and moving them to immobilization channels for imaging, researchers have also used the concept of droplets and responsive reversible gels to perform long-term investigations [19,29–31]. For instance, Krajniak and Lu [29] developed an integrated microfluidic device (Figure 2B), that consisted of an array of eight microchambers controlled by surrounding channels and valves, and successfully showed culturing, immobilization and imaging of animals on a single platform. Animals were loaded from a single inlet into eight culturing chambers. For imaging, Pluronic F127 (PF127), an amphiphilic block copolymer (PEO99-PPO67-PEO99, PEO: Poly(ethylene oxide) and PPO: poly(propylene oxide)), was injected into the chambers from the same inlet and gelled by heating. This polymer is highly viscous at low temperatures (e.g., 15 °C) but acquires gel-like properties upon raising the temperature (e.g., 21 °C), making it possible to reversibly immobilize animals. Since PF127 does not exhibit autofluorescence and has no effect on the viability and development of worms, animals could be successfully monitored from the L1 stage to adulthood. The gel-based immobilization technique was also utilized in a recent publication by Cornaglia et al. [17] to monitor in vivo protein aggregation in *C. elegans* models of two neurodegenerative diseases, namely amyotrophic lateral sclerosis (ALS) and Huntington's disease (HD). The authors demonstrated the progression of mutated human superoxide dismutase 1 (SOD-1) tagged with Yellow Fluorescent Protein (SOD1-YFP) into the body wall muscles of individual worms over several days. They also precisely localized fluorescent proteins in tissues, and monitored the time-lapse and sub-cellular evolution of single aggregates over time.

Based on the above discussions, it is clear that the selection of an appropriate microfluidics method for culturing *C. elegans* will depend on the unit operations to be performed for monitoring processes and the length of assay. For instance, if there are multiple chemical exchanges desired in the assay or if the experiments involve offspring, then the chamber-based methods that incorporate nutrient and waste exchange are better choices. However, if studies require observing animals for a short duration at a lower throughput then hydrogel immobilization approach may be more desirable.

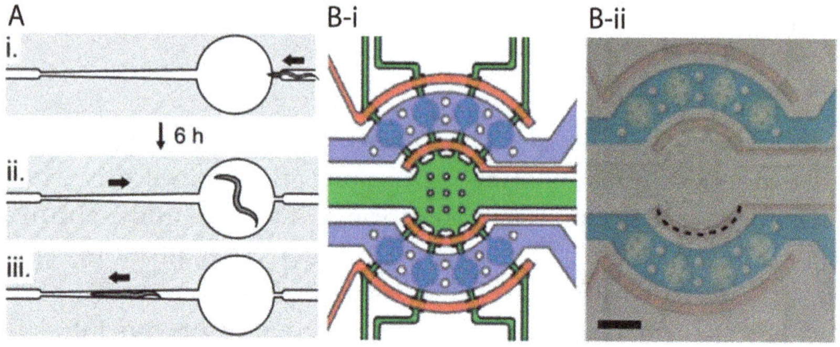

Figure 2. Microfluidic devices for culturing and long-term studies of worms inside cultivation chambers while immobilization and imaging is performed by (**A**) tapered microchannels (Hulme et al. [23] or (**B**) responsive hydrogels (Krajniak et al. [29]. Reproduced with permission from The Royal Society of Chemistry. The tapered microchannels connected to growth chambers (panel (**A**)) allow single worms (early L4 stage) to enter into each chamber. Arrows indicate the direction of liquid flow. Once the worm has grown it is unable to escape the chamber. For imaging purposes, the worm is temporarily immobilized in the tapered region. Panel (**B**) The two sub-panels **B-i** and **B-ii** show the device that contains valves (**red**) to control fluid flow, channel for flowing heating liquid (**light blue**), and eight worm culturing chambers (two sets of four) and a central waste outlet tube connected to a loading channel (**green**). Refer to respective references for more details.

2.3. Worm Immobilization Methods for Developmental, Morphological, and Physiological Studies

Because of their microscopic size and continuous movement, worms need to be immobilized for routine observations of live processes. For developmental studies, imaging is often performed at multiple time points throughout the life of animal, so it is important that immobilization is reversible and does not cause harm. This is typically achieved by the use of agents such as anesthetics and glue. However, both these approaches are less than satisfactory due to their potential toxicity and slow pace. Also, the use of glue precludes monitoring the same animal at multiple time points. Alternative approaches such as instant cooling, compression, thermosensitive hydrogels, and exposure to gases have been successfully attempted.

Krajniak et al. [29] used the temperature-sensitive PF127 gel (mentioned in Section 2.2) to immobilize animals for developmental studies. Up to eight worms could be studied simultaneously in their device (Figure 2B), each maintained in individual chambers and supplied with food. For imaging purposes, a solution of PF127 was flooded in culture chambers. Temperature increase was achieved by circulating hot water via separate channels passing above the worm cultivation channel, shown in

Figure 2B. Although the platform makes it possible to perform time-lapse studies at a high speed, the addition of temperature control modules makes the system somewhat complex.

Another attractive approach to image worms in a microfluidic setup is to physically trap them in a narrow space. Gilleland et al. [33] have described a detailed protocol for setting up a pressure-based immobilization imaging platform that uses a flexible PDMS membrane to restrain the motion of worms in the channel while they are being imaged (Figure 3A). The device consisted of two layers of PDMS microchannels bonded together with the flexible membrane sandwiched in between. The bottom layer allows for loading and unloading the worms into the immobilization section of the device. The top layer is used to apply a pneumatic pressure on the membrane in order to deflect it down onto the loaded worm. Animals can be recovered from the device upon removing pneumatic pressure from the flexible membrane. Subsequent studies showed no visible sign of stress or damage in such worms following recovery.

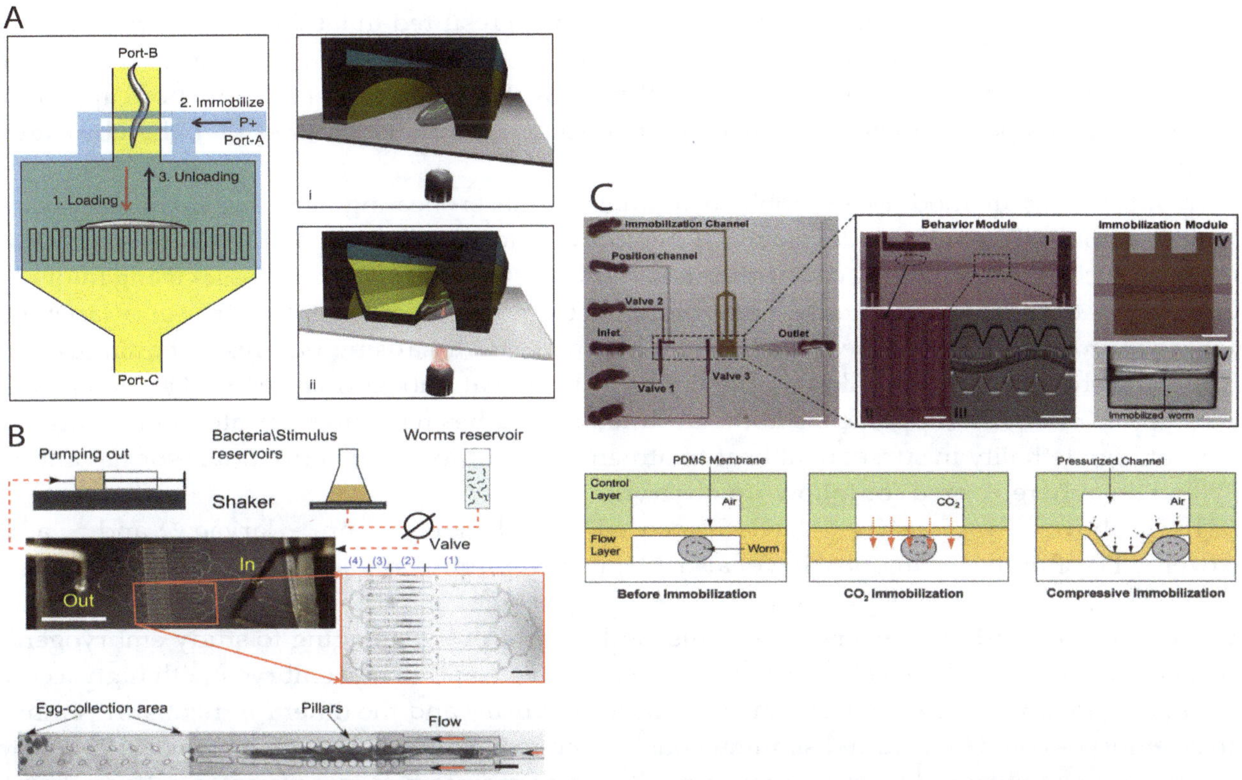

Figure 3. Microfluidic devices to immobilize *C. elegans* using (**A**) deflectable membrane (Gilleland et al. [33]; (**B**) tapered microchannels (Kopito and Levine [34]); or (**C**) CO_2 exposure (Chokshi et al. [35]. Reproduced with permissions from The Royal Society of Chemistry and Macmillan Publishers Ltd. Nature Protocols. Panel (**A**) shows the chip containing an array of narrow channels to apply suction pressure. Worm is loaded/removed through port-B and restrained by the narrow channel array. Pressure through port-A causes the compression layer to move downwards and immobilize the worm (explained on the right). Releasing the pressure allows the worm to be recovered. The WormSpa device, in panel (**B**), contains four regions for worm loading and distribution (**1**), egg chambers (**2**), egg collection (**3**), and outflow (**4**). The device for CO_2 based immobilization is shown in panel (**C**). It contains modules for behaviour assay (first row of pictures) and immobilization (second row of pictures). Refer to respective references for more details.

An alternative mechanical method to keep worms steady at one location is to use tapered microchannels, as shown in Figure 2A. This method does not require a moving part in the device; instead the worm is gently pushed longitudinally to a narrow region of the channel such that it has no room to move any further. Such a confinement approach is desired in scenarios where animals are exposed to chemicals in a controlled manner or there is a need to collect embryos. For instance, Kopito and Levine [34] used the principle of tapered channels in their device that held animals in parallel units, each consisting of two rows of micro-pillar sidewalls instead of solid walls (Figure 3B). The pillars allowed worms to wiggle slightly without compressing while making it easier for embryos to escape by frequent washes. Their chips, termed "WormSpa," could be used to simultaneously culture up to 64 adults in the long term while monitoring physiological changes and gene expression under a microscope. Our group has also developed a simple micro-structured device to instantaneously confine several animals on a regular glass slide in order to image them for up to several hours [36]. These confinement approaches do not appear to affect the growth and reproduction of the animals, making them suitable for many *C. elegans* applications.

Two papers [35,37] have compared the pressure-based approach with CO_2 exposure through a thin membrane as an anesthetic gas to immobilize animals in microfluidic devices (Figure 3C). Chokshi et al. [35] showed that 1–2 h long exposure of CO_2 caused no apparent harm to animals and they recovered successfully. However, longer exposures (3 h and beyond) resulted in lethality. The pressure-based immobilization, using a deflectable PDMS membrane, implemented in both these studies allowed short term (minutes) immobilization and imaging without affecting the recovery. While both approaches are useful, CO_2 is reported to have a detrimental effect on neuronal processes [37], thereby limiting its applicability.

An interesting method to immobilize animals by laser heating was recently described by Chuang et al. [38]. The authors developed a new device to rapidly immobilize worms for routine applications such as morphological characterizations, in vivo study of cellular processes using fluorescent reporter, and targeted cell ablations. The unique feature of the device is that it uses a laser beam combined with electric field to heat the liquid medium to 31 °C causing paralysis of animals in less than 10 s, a technique termed "addressable light-induced heat knockdown (ALINK)". While useful in some applications, this approach is not well suited for studies requiring repeated immobilizations because of high lethality in subsequent heat treatment cycles. Moreover, individual worms cannot be easily and reliably recovered for follow-up observations.

The above approaches have focused exclusively on larvae and adults. However, microfluidic techniques have also been successfully applied to manipulate embryos, allowing for automated and high throughput arraying of many embryos at predetermined locations on a chip, physiological maintenance over time, and long-term live imaging to study embryogenesis. Cornaglia et al. [39] recently developed a new device to isolate and image embryos with high accuracy. Their device contained two chambers, one for culturing worms and the other for incubating embryos. Up to 20 embryos could be cultured simultaneously, each resting in stable positions in little incubators. The embryos were obtained from adult worms that were maintained in a separate culture chamber. Embryos could be imaged over time, which was useful for morphological and gene expression studies. Gene expression and developmental events were captured using a fully automated multi-dimensional imaging system.

All in all, the immobilization methods summarized in this section are suitable for a wide range of applications. The choice of a specific approach in an experiment will depend on the immobilization duration and the readout signal desired. For instance, CO_2 exposure can be used to immobilize animals for up to 2 h, but if neuronal processes or pharyngeal movements are to be investigated then it is better to avoid anaesthetics. For experiments requiring manipulations such as neuronal ablation, one could utilize mechanical, chemical, and even temperature-based approaches to quickly restrain worms for short durations.

2.4. Microinjection

Among routine procedures in *C. elegans* laboratories is the injection of DNA, proteins, and chemicals for the purpose of examining biological processes. When DNA pieces are injected into the germline, they may be incorporated in the genome resulting in the generation of transgenic animals. Traditional microinjection protocol involves placing animals in a particular orientation on a glass coverslip containing a thin layer of dried agarose and immersing them in oil to slow down desiccation. This step is rather difficult because it requires that animals have sufficiently adhered to the slide and do not wiggle too much. Next, they are placed on a microscope and an injection needle, filled with the desired solution, is carefully brought into the field of view and manually inserted in the worm to inject the contents. Finally, worms are recovered from the slide. Due to the slow and labor-intensive nature of this procedure, it can take hours even for an experienced person to inject a sufficient number of worms for a single experiment.

In recent years, attempts have been made to develop microfluidic devices to automate this protocol. In 2013, we reported such a device (Figure 4A) [40,41]. This "T" shape unit was fabricated by pre-loading the injection needle into the T-channel before bonding the device layers together, hence ensuring that the needle tip is aligned properly with the narrow worm trapping channel. Injections were performed by confining worms in the narrow trap using suction pressure. The movement of the needle and injection were achieved by applying 200 kPa pressure pulses with 1 s duration using an attached micromanipulator. Success with the generation of transgenic animals was also demonstrated [41]. The process of introducing worms into the device and recovery is still performed manually, but could be automated in the future. Another device by Zhao et al. [42] used a suction-based approach to immobilize animals while they were being injected. The injection step was controlled by hand. Earlier this year, Song et al. [43] reported an automated system capable of injecting 6–7 worms per minute with a success rate of 77.5%. The device (Figure 4B) consisted of microvalve-controlled loading and unloading channels connected to a narrow microchannel to house the worm, a set of perpendicular side suction channels for immobilizing animals in the narrow channel, and a 5 µm-tip micropipette controlled automatically by a three degree of freedom micromanipulator for precise injection of chemicals into worms. The throughput of this device was increased by incorporating a robotic system to automate the operation, real-time image processing to operate the needle, and the process of injection. Overall, future application of the microfluidic approach to microinjection is promising.

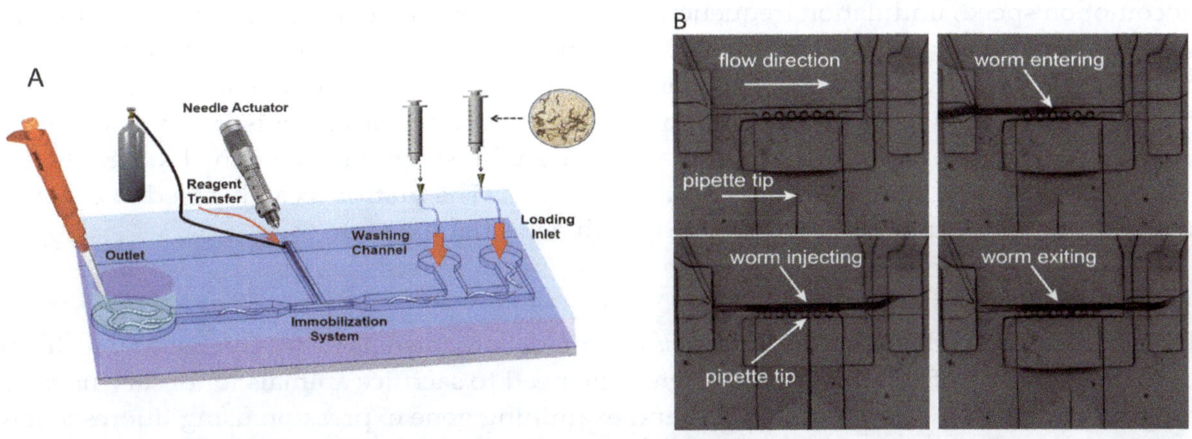

Figure 4. Microfluidic devices for microinjection in (**A**) closed microchannels (Ghaemi [41] and (**B**) open chambers (Song et al. [43]. Panel (**B**) reproduced with permission from American Institute of Physics Publishing. The device in panel (**A**) contains worm loading and washing channels (on the right) and an outlet for collecting injected worms. Worm is immobilized in the middle region for injection. The image frames in panel (**B**) show a sequence of worm loading, injection, and flushing. Refer to respective references for more details.

3. Microfluidic Devices to Assist with Specialized Assays

In addition to the routine procedures of handling *C. elegans*, as described in the previous section, a large number of microfluidic tools have been developed to facilitate research in this animal model. These include investigations of specific processes such as movement, cell morphology, gene expression, learning and memory, toxicology, and aging. Key applications are summarized below.

3.1. Study of Movement Responses in Microstructured Environments

The natural habitat of *C. elegans* is complex and requires animals to probe the environment in order to live, feed and reproduce. Mechanosensation, i.e., the response to touch, plays a vital role in these processes. Microfluidic devices have been designed to investigate the mechanosensory response of animals in the laboratory. One of the earlier devices, reported by Park et al. [44] was constructed of agar and not PDMS. It contained 300 μm diameter circular posts in a 200 mm × 20 mm × 0.11 mm chamber arranged in a square grid configuration. The device was placed in a petri dish with the topside open and accessible. The chambers were filled with buffer prior to placing worms inside them. Examination of movement responses of different worms revealed that animals rely on touch to move efficiently in such an environment since mechanosensory mutants *mec*-4 and *mec*-10 were slower than the wild type. Increasing the post spacing from 400 to 475 μm resulted in higher swimming speed but any further increase in the spacing had an opposite effect. The device could be used to study the movement of worms, and possibly screen for new mechanosensory and uncoordinated mutants.

Another device, reported by Parashar et al. [45] contained micro-sinusoidal shaped structures to examine movement responses of worms. Channels of different shapes (ascending and descending in amplitudes from 135 to 399 μm and from 135 to 10 μm, respectively) were tested, and it was observed that animals moved faster in certain configurations but slower in others. The speed was the highest (average of approximately 300 μm/s) in a channel with sinusoidal shape amplitude of around 200 μm. Analysis of movement defective strains (*lev*-8 and *unc*-38) revealed an almost 50% reduction in velocities of worms in these channels, suggesting that such a setup may be useful in screening animals for locomotion defects.

Subsequently, two devices were reported that contained micro-pillar structures for measuring the forces exerted by worms on their microenvironment [46,47]. Johari et al. [46] used PDMS pillar elastic deflection concept to measure the force, and found that despite changes in the placements of pillars in a "honeycomb" or "lattice" configuration, worms continued to move in a sinusoidal pattern but their locomotion speed, undulation frequency, and body force changed significantly. Measurement of forces applied on pillars revealed that the mid-body region generates the maximum force. The device by Qiu et al. [47] was designed to expose worms to a particular frequency of light that they tend to avoid and measure force generated during forward or backward movements, or omega turns in the structured environment. A LabVIEW program (National Instruments, Austin, TX, USA) was also developed to assist with image capture and data analysis. The platform can be used to examine the locomotive behavior of animals and correlate it with neuronal function.

3.2. Study of Cellular Processes

One of the biggest advantages of *C. elegans* as a model organism is its transparency, which enables real-time in vivo study of cellular events. There is no need to sacrifice animals for observing cells and tissues, performing surgical manipulations, and examining gene expression using fluorescent (GFP) reporters. In fact, the worm model allows these experiments to be performed in live condition and longitudinal manner. Over the years, several groups have reported microfluidic devices to automate, streamline, and accelerate such assays.

Studies of neuronal activities require keeping animals stationary for several hours during imaging processes. It is critical that animals are not paralyzed, which precludes the use of anesthetics, and that they are not subjected to stress during the period of observation. To achieve this goal, Mondal et al. [37] fabricated a device consisting of a pressure-deflecting PDMS membrane to immobilize animals. The device permitted long-term imaging of neuronal transport processes (up to an hour) with high resolution without causing harm to neurons, something that is not possible with the traditional anesthetic-based immobilization methods. The authors demonstrated the capability of their device to investigate subcellular events such as synaptic vesicle transport, Q neuroblast divisions, and mitochondrial transport in early larval stages, which was not observable in worms of the same age that were anesthetized with chemicals like levamisole. In addition to worms, authors also used the device to successfully study processes in *Drosophila* larvae. However, this setup has very low throughput and is only suited for detailed cell biological studies in a small set of animals.

Another way to restrain animals is to trap just part of the animal body. This approach was recently demonstrated by Hwang et al. [48] in a series of optogenetic experiments. In such a system, worms were illuminated with a specific wavelength of light that causes an escape response and observed while trapping them at both ends using two pneumatically-controlled microvalves. To accelerate data collection the device allows simultaneous image captures from 16 parallel channels. The authors used this setup to examine the activities of body wall muscles. They successfully studied the roles of 15 genes that encode sarcomere proteins in striated body wall muscles and showed that these genes are required in various steps of muscle function. Several key parameters of muscle kinetics and rate constants of contractions and relaxations were determined. The findings led to an improved understanding of how muscles control locomotion in worms.

In addition to imaging, the pressure-based immobilization technique can also be useful in other procedures such as laser-assisted surgeries. In one of the earliest reports [49], the authors described a device to perform long-term observations of neuronal processes with or without laser-assisted ablation of synapses. The device contained several parallel channels that were large enough at one end to allow easy uploading of worms using a syringe pump-driven fluid flow. The channels were tapered to immobilize animals as they move down the channels. Time-lapse imaging was performed using a CCD camera. To ablate synapses, short pulses of a UV laser were used. The setup was successfully used to monitor neuronal processes for up to 4 h. Another surgical approach, developed in recent years, involved the use of a genetic system. Lee et al. [50] used a KillerRed (KR) system to perform rapid neuronal ablation in a device and image behavior of animals for up to 24 h. The KillerRed protein is a genetically-encoded photosensitizer that is activated upon exposure to green light of the wavelength range 540–590 nm. Activation of KR in a cell causes production of reactive oxygen species, resulting in its death [51].

While the above approaches of imaging immobilized individual animals offer a detailed view of cells and cellular processes, they lack the ability to capture dynamic changes in real time as animals are interacting with the environment. Two papers [31,52] have reported new systems to address this limitation. Larsch et al. [52] developed a microfluidic arena of 3.28×3.28 mm^2 and 50 μm depth that could be monitored using a high numerical aperture objective (2.5×/0.12 N.A. or 5×/0.25 N.A.) and sensitive low-noise CCD camera. This device could track up to 20 worms in real time and allow measurements of neuronal activities using GFP-based genetically encoded calcium indicators [53]. The setup enabled the authors to expose animals to different odors and monitor neuronal activities following changes in their movement responses. Overall, the approach is powerful because it allows for mapping of neuronal signals while the organism is responding to external signals and making decisions. For example, the results revealed that odor-induced Ca^{2+} signals in the AWA chemosensory neuron were variable among animals, possibly due to differences in their development, epigenetic modifications, or other processes. Measuring such signals can help explain the mechanistic basis of behavior in animals as they encounter stimuli in the environment. In the other system [31], animals of specific developmental stages were introduced in a culture chamber and maintained by supplying with food and buffer exchange. For imaging purposes, a PF127 sol-gel polymer was utilized. As mentioned earlier in this review (Section 2.2), the viscosity of this polymer changes with temperature. When the chip temperature is raised to 25 °C using a thermoelectric module, it triggers the gelation of PF127, causing worms to immobilize and neuronal processes to be examined in detail. Two different human disease models (ALS and HD) were successfully studied using this platform, which revealed changes in protein aggregation in cells over time.

The three-dimensional shape of worms does not allow all cellular structures to be visible in any one given focal plane. Therefore, visualization of entire cellular structures requires either optical sectioning technique using expensive confocal systems or orienting worms in different directions while imaging. Ardeshiri et al. [54] were the first to demonstrate full rotation and multi-directional imaging of *C. elegans* in a microfluidic device (Figure 5A), using a rotary glass capillary that was used to pneumatically grab the worm from the anterior region and rotate it in a narrow microchannel. The authors successfully showed the 360° rotation of adult worms in the channel with on-demand immobilization and imaging of organs (vulva) and fluorescing neurons. Another study, reported earlier this year, came up with an elegant acoustic-based microfluidics approach to precisely and rapidly rotate worms and image them [55]. The device (Figure 5B) enabled acoustofluidic rotational manipulation (ARM) of *C. elegans* as well as single cells. During the process of worm loading, microbubbles were trapped into small microcavities created within sidewalls of the channel. This channel was placed in the vicinity of a piezoelectric transducer fabricated on a glass slide. The transducer generated acoustic waves that caused the microbubbles to initiate oscillatory motion resulting in steady microvortices, which rotated the animals that were in contact. By controlling the duration of waves (a few milliseconds), rotation angles could be controlled with high precision. For example, a worm could be fully rotated (360°) within 60 milliseconds with 5-millisecond pulses each causing 4° rotations.

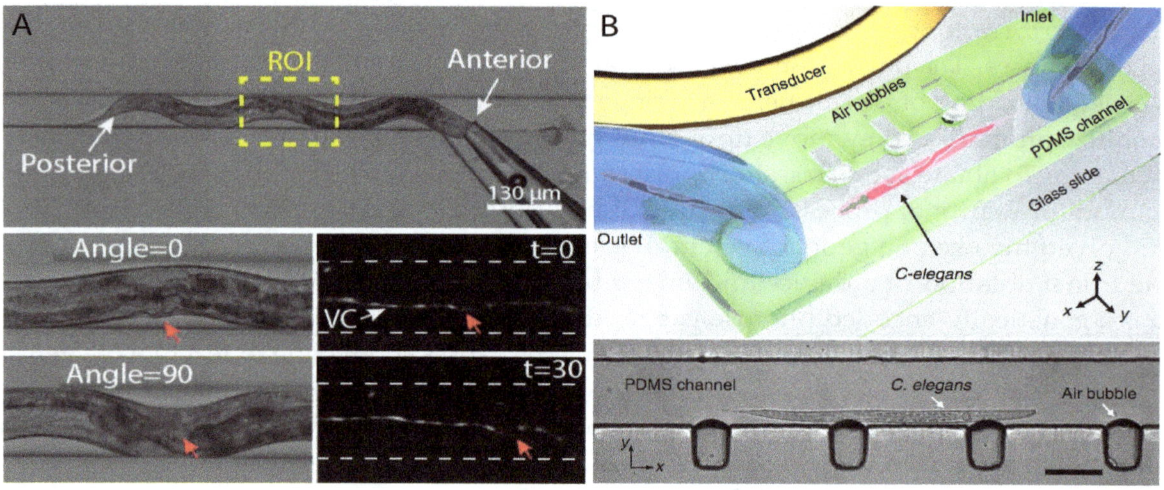

Figure 5. Microfluidic devices for multidirectional orientation and imaging of *C. elegans* using (**A**) rotatable glass capillaries (Ardeshiri et al. [54] and (**B**) acoustofluidic rotational manipulation (ARM) (Ahmed et al. [55]. Panel (**B**) reproduced with permission from Adapted by permission from Macmillan Publishers Ltd. Nature Communications. Panel (**A**) shows an adult worm inside the channel with the region of interest (ROI) in the middle. The worm is held by the negative pressure in the glass capillary. The two sets of brightfield and fluorescent images below show pre- and post-rotated views of specific neuronal processes (VC). Schematic view of the ARM device (**B**). It contains a piezoelectric transducer to generate acoustic waves. Air bubbles within sidewall cavities cause worms to rotate. The image below shows a mid-L4 worm trapped by oscillating bubbles. Refer to respective references for device details.

Two drawbacks of the acoustofluidic technique are that the worms need to be anesthetized for imaging and that they cannot be accessed by external objects for further manipulation such as microinjection. The other approach involving micro-capillary to orient the worm [54] (Figure 5A) allows additional treatments, e.g., exposure to external stimuli, to be performed on animals.

3.3. *Study of Chemotaxis, Electrotaxis, and Other Behaviors*

As a metazoan, *C. elegans* responds to environmental stimuli or internal physiological changes by modifying its behavior. Because such responses depend on the functioning of the nervous system, assaying behavior provides valuable information about the state of neuronal signaling. Some of the commonly studied examples of *C. elegans* behavioral responses include chemotaxis (response to chemicals), thermotaxis (response to heat), phototaxis (response to light), and electrotaxis (response to electric field).

Traditional behavioral assays in *C. elegans* are carried out on agar-containing Petri plates. The procedures are slow to perform because they require a worker to carefully execute multiple steps including preparation of assay plates, handling worms, and data analysis. Furthermore, it is not possible to observe neurons in individual animals in such assays. In 2007 the Bargmann lab reported fabrication of two microfluidic chips to characterize neuronal and behavioral responses in worms at a high resolution [56,57]. One of these, the behavioral chip [56], had a single channel whose width was gradually decreased to 40 μm at one end (similar to the tapered channels discussed in Section 2.3). Worms were introduced into the channel one at a time and gently pushed towards the narrow end until they were trapped. Such animals could still generate a sinusoidal wave along the body length in an attempt to move, although they could not escape the small opening of the channel. Neuronal activities were measured in real time using GCaMP protein (genetically encoded Calcium indicator containing GFP, Calmodulin and M13 peptide sequences) that acts as a Ca^{2+} sensor. The chip allowed researchers to demonstrate that AVA neurons play important role in locomotion. The other chip, the olfactory chip [56,57], was used to trap worms in a narrow channel in such a way that their noses protruded into another connected channel where odors were delivered. This setup was used successfully to measure activities of several neurons including ASH and AWA to chemical stimuli.

In recent years, several new microfluidic devices have been developed to further expand *C. elegans* studies. Albrecth and Bargmann [58] used microstructured areans consisting of hexagonally arranged cylindrical posts to examine odor-induced movement of worms. A custom automated tracking software was used to classify behavior into five distinct states: forward, pause, reverse, piroutette reverse (the reversal before an omega turn), and pirouette forward (forward motion after an omega turn). The responses were characterized, which revealed new quantitative and statistical insights into the odor-induced behavior of worms. This setup will enable future studies to understand neural circuits as animals encounter environmental stimuli and make decisions. McCormick et al. [59] reported a pair of microfluidic devices (Figure 6A), termed chemosensory and thermosensory devices, to measure the behavior of individual worms in response to chemical, thermal and osmotic stimuli. The unique feature of these devices was that responses were determined based on the head movement of semi-restrained animals. The lower half of such animals was fixed in the microfluidic channel while the upper half was free to perform side-to-side swings. Using these devices, one can reliably deliver chemical and thermal stimuli to animals and determine attractive and repulsive responses in the form of head swings. A new finding from this work was that both a rise and a fall in osmolarity lead to reversals in worms. The authors suggested that, with some modifications, it might be possible to perform neuronal imaging while recording the behavior of live and intact animals.

While the above setup enables single animal-based studies, chemotaxis assays are performed typically in batches consisting of hundreds of animals. A microfluidic device, developed by Hu et al. [60], simplifies such population-based experiments (Figure 6B). Repetitive flow splitting and mixing microchannels arranged in a tree-like design were used to generate linear gradients of sodium chloride (NaCl) across the device. Worms were loaded from the central outlet and their movement to chambers of different chemical concentrations was quantitatively studied. Although the preparation and loading of worms were manual in their configuration, the method offered advantages in setting up the assay and testing 0–300 mM of NaCl for attractive or repulsive responses. The authors showed that L3-stage worms exhibited a stronger response to low concentration NaCl as opposed to adult animals. Overall, the device can accelerate chemotaxis measurements in worms.

For detailed neurobiological studies of behavior, there is a need to observe changes in neuronal activities as worms interact with their environment. To this end, Hu et al. [61] fabricated a comb-shaped device that allowed in vivo monitoring of neurons using the GCaMP sensor in single immobilized worms as they were exposed to gases and odors. As a proof of principle, the authors examined two different types of neurons—URX that respond to oxygen (O_2) and BAG that are sensitive to CO_2 levels. In both cases, changes in Ca^{2+} signals were observed as expected. Additional sensory neurons were also successfully tested following exposure to odors such as 1-Octanol. Overall, the device is useful in studying neuronal responses in individual animals. In a different study Ca^{2+} sensor was used to identify magnetosensory neurons. For this a two-layer microdevice was fabricated containing a flow layer and valve layer. Worms were introduced via one end of the flow layer. For observation purposes they were fully immobilized by applying pressure in the valve layer. The animals were then exposed to a 65-Gauss (100× earth) rotating (2 Hz) magnetic field stimulus for 8 s and imaging was performed [62]. The result revealed that AFD neurons are specifically responsible for sensing the magnetic field. This mechanism may play a role in mediating the burrowing behavior of worms as they search for food in their natural environment.

Another stimulus that *C. elegans* is strongly sensitive to is the electric field. This was reported in 1978 by Sukul and Croll [63], who observed preferential movement of animals with an angle towards cathode. Prior to this discovery, electrotaxis phenomenon was described in other nematodes including *Panagrellus redivivus* and *Trichostrongylus retortaeformis* [64,65]. In spite of these early studies, no significant progress was made in this area until 2007 when Gabel et al. [18] reported amphid neurons and genes that mediate electrotaxis behavior using an open agar gel surface setup. Our group was the first to incorporate the electrotaxis assay in a microfluidic setup shown in Figure 6C. In the paper by Rezai et al. [9] we presented a straight 300 μm-wide microfluidic channel with end electrodes and used it to show that *C. elegans* responds to DC electric field and moves towards the cathode with a characteristic speed (roughly 350 μm/s in case of young adults). We also showed that a response to electric field is present starting at the L3 larval stage and that exposure to the field causes no harm to animals.

Microfluidic electrotaxis approach offers many advantages in investigating neuronal signaling and movement-related neuronal disorders (e.g., Parkinson's disease-like phenotype [66]). It is the only non-invasive on-demand tool to induce instant directional movement with a defined speed. In subsequent studies we further characterized the electrotaxis phenomenon and showed that a shorter pulse of DC can also be effective in inducing robust movement in worms [67]. Interestingly, we found that worms are sensitive to the AC field as well but that the AC field produces a different response. Specifically, a symmetric square wave AC field of 1 Hz frequency was able to restrain animals effectively at one location [68], making it an attractive approach to implementing in a high throughput system where there is a need to localize and concentrate many worms. With the goal of accelerating microfluidic electrotaxis assays, we recently developed a semi-automated system [69]. At a throughput of 20 worms an hour, this system allows fast screening of mutants and drug-exposed worms with altered neuronal function. A custom LabVIEW program has been written to manipulate the steps of loading, capturing, flushing, releasing, electrotaxis screening, and channel cleaning. As a proof of principle, a mutant affecting dopamine signaling was successfully characterized.

3.4. Antimicrobial and Toxicological Studies

Other applications of microfluidic technology in *C. elegans* research include assessing the impact of toxic chemicals and harmful bacteria on living systems. As a whole organism model, the worm is frequently used as a bio-indicator in toxicological and environmental studies [70,71]. Typical exposure assays involve measuring growth rate, reproduction, movement, shape and size, neurodegeneration, and survival. The paper by Jung et al. [72] describes a device to measure the body volume of animals following exposure to toxic chemicals. Animals were immobilized in a tapered channel and size was measured by capacitance change using a pair of electrodes at the two sides of the tapered channel. As expected, the authors observed a reduction in size following exposure to cadmium. While useful, a limitation of this setup was that worms needed to be treated on standard Petri dish culture plates prior to doing measurements in the device. To address this, the authors later developed a new design that included on-chip exposure and collection chambers [73]. Animals were kept in one chamber and as they passed, one by one, through a narrow region containing electrodes, changes in capacitance were recorded and attributed to animals' body sizes. This system accelerated chemical treatments and measurements of resulting changes in the body size.

The electrotaxis assay has also been extended to toxicological research. In one study, we examined the movement responses of worms in a microchannel device following exposure to toxins affecting dopaminergic (DAergic) neurons [74]. Animals with excessive neurodegeneration showed abnormal electrotaxis, demonstrating the utility of our assay in neurotoxicology studies. Ongoing cell biological experiments in our lab have identified components of the DA signaling pathway in mediating the electrotaxis behaviour [75].

Other researchers have also investigated the health of DAergic neurons of *C. elegans* in biosensor devices. Zhang et al. [76] used a microfluidic device consisting of eight culture chambers arrayed radially and connected in the center to a microfluidic chemical gradient generator similar in operation to the one shown in Figure 6B. The device could be used to measure survival and body stroke frequency of animals, as well as fluorescence intensities of neurons. Worms were exposed to varying concentrations of a heavy metal Manganese (Mg) (0 to 100 mM concentration range) and effects on movement and neurodegeneration were monitored. Exposure to natural antioxidants such as vitamin E, resveratrol, and quercetin were shown to rescue defects effectively.

Besides heavy metals and toxins, microfluidic has also been successfully implemented in experiments involving exposure to bacterial pathogens and screening of antimicrobial compounds. Yang et al. [77] developed a device consisting of 32 culture chambers. Each chamber accommodated approximately 15 worms. The animals were fed with cultures of *Staphylococcus aureus* bacteria for about 6 h and monitored subsequently for survival. As expected, the exposure caused the death of the entire population within three days, demonstrating the suitability of the device in such assays. Natural compounds were also screened for antimicrobial activities.

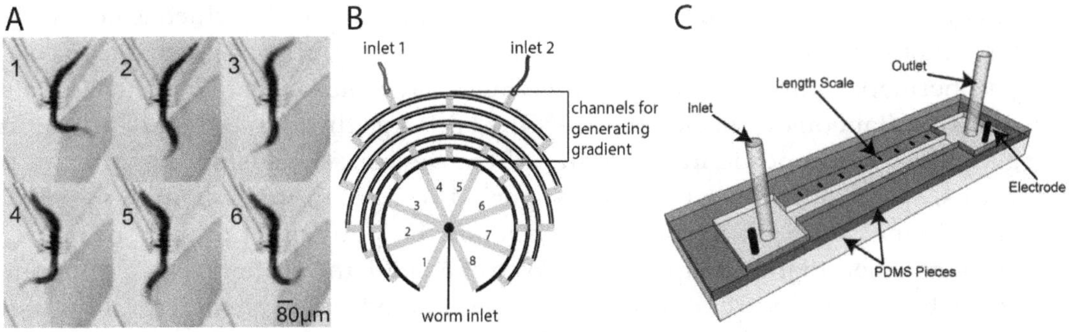

Figure 6. Microfluidic devices to investigate *C. elegans* behavior in response to (**A**) chemicals and heat (McCormick et al. [59]); (**B**) chemical gradients (schematic drawing of the device used by Hu et al. [60]); and (**C**) electric field (Rezai et al. [9]). Panel (**C**) reproduced with permissions from the Royal Society of Chemistry. Panel (**A**) shows head swinging of the worm in response to chemical exposure. In panel (**B**), the circular channel pattern used to generate the chemical gradient is shown. Worms enter into channels 1–8, which are 300 μm wide, 80 μm high, and 10 μm long depending upon their attractive responses to the NaCl gradient. The electrotaxis device in panel (**C**) contains electrodes to apply a DC electric field and a long channel for worm swimming. Refer to respective references for more details.

3.5. Learning and Memory Studies

Research has shown that *C. elegans* have the ability to learn from past experiences and can alter subsequent responses accordingly. For example, when animals were conditioned with NaCl in the absence of food they showed reduced chemotaxis response [78,79]. In 2005, Zhang et al. [80] reported for the first time a microdevice consisting of an eight-arm maze to test the learning preference of worms to bacterial strains. Results revealed that worms exposed to pathogenic bacteria learn to avoid them within a few hours by recognizing associated odors while increasing their attraction to odors from familiar nonpathogenic bacteria. This learning process is mediated by serotonin in a specific chemosensory neuron, the ADF. In another study Qin and Wheeler [81] fabricated new devices to test olfactory learning of worms. These devices, consisting of T-shaped and U-shaped patterns, were made to examine the behavior of worms as they explore the environment and choose to go in a particular direction. It was found that worms use olfactory stimuli to make biased movement. This behavior was also dependent on the dopamine neurotransmitter.

3.6. Reproductive Aging and Lifespan Studies

As discussed above, microfluidic technology is well suited for long-term culturing and observations of *C. elegans*. Given that these animals typically produce offspring during the first 3–4 days of their life and live for roughly three weeks, microfluidics holds significant potential to accelerate reproductive aging and lifespan-related studies.

The device by Hulme et al. [23], discussed above (Figure 2A), is not only useful in culturing worms but can also allow monitoring of age-associated changes. Two parameters, body size and locomotion, were monitored and found to be in agreement with the results derived from standard plate-based studies. Another study by Xian et al. [82] described an integrated multilayer PDMS device platform, termed WormFarm, for aging studies. This setup consisted of eight chambers (each $3 \times 10\ mm^2$) that were separated to avoid cross-contamination. Each chamber could hold approximately 40 worms. Chambers were flushed periodically to remove progeny and supply fresh bacteria to feed adults. Worms were counted using a custom algorithm. Additionally, GFP fluorescence in aging animals could also be quantified as average signal intensity per pixel. The device allowed rapid quantification of changes in lifespan caused by environmental and genetic manipulations.

Two other devices have also been reported recently. One of these, described in Li et al. [83], was used to automate the counting of progeny from several worms simultaneously, which is normally a manual procedure that is tedious to perform. The device (Figure 7A) consists of 16 microchambers, each housing a single *C. elegans* hermaphrodite (wild type or *daf*-2 insulin receptor mutant). The chambers were fed with bacteria and simultaneously washed to direct the laid eggs through a filter to an automated and real-time progeny counting module in the center of the device. The system re-confirmed previous published findings, specifically the reproductive period of *daf*-2 mutants being longer than that of wild-type animals. In the future this promises to accelerate the study of reproduction and reproductive aging in a real-time manner. In the other paper, by Wen et al. [30], the authors took a different approach that involved formation of fluorocarbon oil (FC-40)-based microdroplets to monitor individual animals (Figure 7B). Worms were encapsulated into droplets (one per droplet) at the L1 stage using a clever design and growth was monitored over time. Fast exchange of media and substance were possible several times a day to allow worms to live normally till adulthood. The device can be useful in studying changes such as body shape, length, and developmental delays in large synchronous populations following certain manipulations. For instance, the authors showed that the worms overexpressing *hif*-1 (hypoxia-inducible factor) had a short body length and a slow growth rate, which indicates the possibility of HIF-1 being involved in developmental processes.

Chuang et al. [84] recently reported an interesting application of an electrotaxis-based microfluidic flow chamber to study the effect of exercise on age-related degenerative changes. Their device, termed a "worm treadmill," involved inducing directed movement in worms using a DC electric field stimulus. Worms were allowed to swim for 10 min without interruption once every day for a total of eight days, and the effects on various age-related processes were subsequently monitored. The results showed that exercise improved the overall health of animals, as judged by a number of factors such as increased size and density of mitochondria, reduced oxidative stress, and prolonged life span.

In summary, microchamber environments hold a significant promise for a range of lifespan-related and developmental assays. Advantages offered include phenotypic studies at the single animal level, micro-compartments to manage the number of animals for a given study, effective control of microenvironments, being able to separate offspring from mothers, and the ability to monitor behavioral and neuronal activities over an extended period of time.

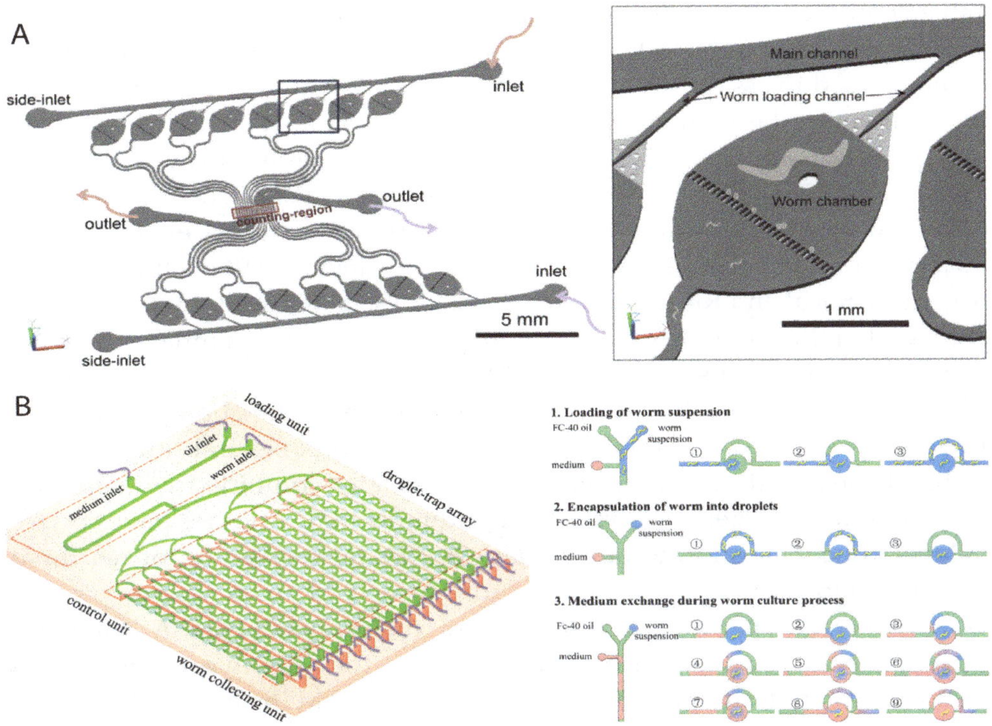

Figure 7. Microfluidic devices to investigate *C. elegans* (**A**) egg-laying (Li et al. [83]) and (**B**) development (Wen et al. [30]), by isolating worms inside microchambers with renewable chemical environment. Reproduced with permission from the Royal Society of Chemistry. The left-hand diagram in panel (**A**) shows eight chambers on each side (one of which is enlarged on the right). Inlets are used to load worms and outlets for bacterial flow. The middle counting region (2 mm × 2 mm), indicated by the red rectangle, is monitored by camera. Panel (**B**) shows the droplet chip. Schematics of worm encapsulation and substrate exchange in each droplet are shown in three steps on the right side. The amount of substrate exchange is indicated by the color change. Refer to respective references for more details.

4. Microfluidic Devices to Perform Fast, Automated, and High-Throughput Screening

The small size and short life cycle of *C. elegans* make it highly suitable for high-throughput screening to investigate mechanisms of diseases and identify candidate drugs for further validations in other model organisms. As the *C. elegans* genome contains more than half of human disease-related gene orthologs including genes implicated in cancers, diabetes, and neurodegenerative diseases [85], microfluidic tools hold great potential to automate *C. elegans* screening, thereby accelerating the development of effective treatments for human diseases. Via integration of automation micro-components (e.g., micropumps and microvalves) and a variety of nontoxic worm-compatible materials (e.g., PDMS elastomer, glass, and hydro- and responsive gels), several high-throughput devices have been developed for phenotypic screening, microsurgery, and drug exposure of *C. elegans*.

4.1. Rapid and High-Throughput Screening

Microfluidic devices are appropriate tools for high-throughput neuro-behavioral screening of worms and subsequent sorting based on size, mutation, or response to stimuli. Neuronal screening has been achieved by immobilizing the worms in microchannels, followed by the fluorescent or electrophysiological recording of neurons' static or dynamic responses before sorting takes place. Behavioral screening has also been achieved by investigating the crawling, swimming, and egg laying responses of an individual or a group of worms to external stimuli, applied controllably inside fluidic microenvironments. Fast and viable delivery of animals to desired interrogation locations on a chip is a critical unit operation in these assays.

A microvalve-controlled "Population Delivery Chip" was developed to rapidly deliver 16 different worm populations, each under 4.7 s, to favorable locations on a microchip to accelerate screening [86]. This technique can be used in assays where a specific population of synchronized worms is needed for downstream high-throughput investigations. In this regard, Rohde et al. [21] developed a microfluidic system for high-speed sorting and subcellular phenotypic screening of worms. Pre-synchronized worms were captured in a microchannel by immobilization via side suction channels and imaged at high resolution. To expand the utility of their platform, the authors developed an interface chip consisting of 96 aspirator tips to connect the main screening device to 96-well plates. This setup could be used to carry out chemical and RNA interference (RNAi) screening, although no such screening was actually demonstrated in the paper. Chung et al. [20] also developed a similar device but with several differences. In their setup, worms could be rapidly immobilized in a channel for high-resolution imaging of neuronal processes by lowering the temperature of the channel using a Peltier cooler. The entire setup was automated and could operate at a speed of several hundred worms per hour. A variation of this device (without the cooling unit) was later combined with a computer-assisted setup for automated and faster genetic screening of mutants [87]. Another paper by Crane et al. [88] also reported development of an automated platform for fast screening of neuronal phenotypes. The setup was designed to load worms, image neurons using GFP markers, process fluorescence data using computer software, and perform sorting based on the analyzed data. A Peltier cooler approach (similar to [20]) was used to transiently immobilize animals for ~10 s in order to capture and process the images. Animals having desired phenotypes were collected in the end. This entire operation was performed without a camera and microscope. Moreover, very little human intervention was required. A throughput of more than 220 worms per hour was demonstrated, which is suitable for large-scale mutant, RNAi, and drug screening for neuronal function.

In certain screening it is desirable to orient worms in a specific manner, which cannot be achieved in the abovementioned devices. Cáceres et al. [89] have demonstrated that orientation along the dorsal–ventral body axis can play an important role in the inspection of morphological features with specific dorsal–ventral alignments. Their device was capable of orienting worms laterally in a U-shaped microchannel with high frequency such that neuronal processes could be screened at the rate of roughly 500 worms an hour. Several new mutants affecting motor neurons were recovered in this study.

The above devices rely on physical or thermal contact with animals in order to immobilize them. Recently, Yan et al. [90] developed an approach in which individual worms with different levels of GFP expression were encapsulated in oil droplets and analyzed fluorescently in an automated setup. The system provided 100% accuracy and an analysis speed of 0.5 s per worm, although resolution was poor.

A limitation of the screening systems discussed so far in this section is that live imaging of transient cellular processes in active animals cannot be performed. Chokshi et al. [91] reported an automated high-speed platform to image neuronal activity using transient calcium imaging in live worms exposed to different odors. The platform could screen tens to hundreds of worms per hour. As a test case, calcium responses were measured in worms of different ages exposed to hyperosmotic stimulus (1 M glycerol). The results revealed that the activity of a specific chemosensory neuron ASH was significantly diminished in older animals. Similar screening may be performed for other odors and neurons.

While fluorescent microscopy based on GFP or GCaMP sensors has led to a number of very useful microfluidic devices for high throughput studies, two recent papers have demonstrated the power of PDMS-based microfluidics in non-invasive electrophysiological readout of neuromuscular activities. These devices could provide direct information about the target of drug action, and are amenable to parallelization. Lockery et al. [92] developed a device (Figure 8A) to record electrophysiological activities of pharynx (termed electropharyngeogram or EPG), a neuromuscular organ involved in feeding. The device consisted of a funnel-shaped channel to clamp the pharynx of a single worm, electrodes at the anterior and posterior sides to record the EPG signal, and side channels for perfusion of chemicals into the device to expose the worm. Although the system was not fully automated, as it required placing worms manually into the inlet port, it could, however, perform recordings of eight worms simultaneously, which is significant compared to existing manual methods [93]. Although no screening was demonstrated, the setup offers a rapid and sensitive way to identify chemicals, e.g., anthelmintics, which may affect the physiology of worms. Another similar device (Figure 8B) for neurobiological studies in *C. elegans*, including neurotoxicology and drug screening, was developed by Hu et al. [94]. Worms were exposed to chemicals and EPG responses were monitored along with neuronal activities using fluorescent probes. Worms could also be recovered, which makes it possible for phenotypic screening. A throughput of 12 worms per hour demonstrated that it is significantly faster than conventional EPG protocols.

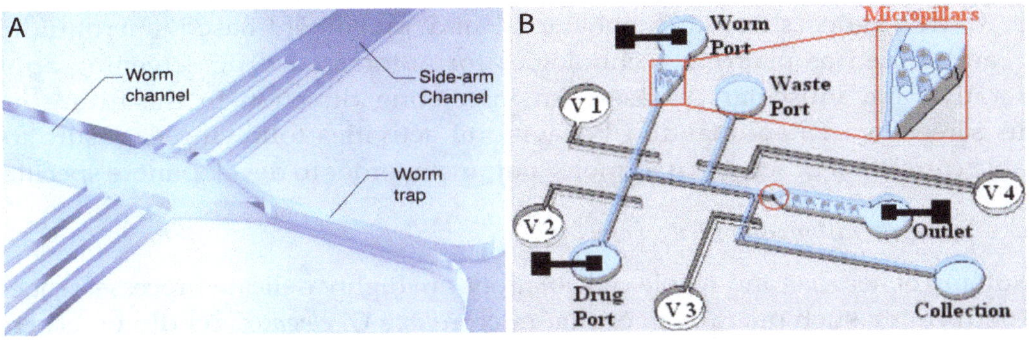

Figure 8. Microfluidic devices for neuromuscular electrophysiological studies on *C. elegans* (Lockery et al. and Hu et al. [92,94]). Panel (**A**) reproduced with permission from the Royal Society of Chemistry. The EPG recording device (panel (**A**)) contains a worm channel and a funnel-shaped trap region. Fluid flows through the side-arm channel. Panel (**B**) shows the neurochip. Blue indicates the layer containing the microfluidic region (for worms and chemicals) and white shows the pneumatic control layer. The red circle contains the trapped worm's head. The red square contains micropillars to correctly orient the worm. V1–4 are valves and the solid black squares are microelectrodes. Refer to respective references for more details.

4.2. *Drug Delivery and Behavioral Screening*

Developmental and physiological changes such as crawling, swimming, and egg laying are widely used readouts in *C. elegans* assays. Screening based on such features has been successfully demonstrated in microfluidic environments.

Chung et al. [95] developed a device that contains parallel arrays of worm chambers to facilitate behavior-based chemical screening at single-animal resolution. A total of 48 chambers, each 1.5 mm diameter, were designed in an 8 (column) × 6 (row) single layer format. A serpentine channel of 500 μm width was used to deliver worms and media to each chamber. All chambers are visible in the field of view, making it possible to monitor individual worm responses following exposure to chemicals. Using this setup, researchers succeeded in measuring changes in behavioral activities, such as body bending and mating, following exposure to drugs and sex pheromones. The embryo-culturing platform of Cornaglia et al. [39] discussed before (Section 2.3) can also be used for fast and automated screening of biological processes and drugs. The setup contains two separate regions for embryo and worm cultures that are connected via a microchannel. The animals could be maintained in the device for extended periods of time and periodically imaged as desired. Up to 20 single embryos could be arrayed automatically upon delivery and monitored simultaneously, making it possible to monitor gene expression and developmental events. It is possible to visualize the entire embryogenesis using this device. The platform was successfully used to study mitochondrial biogenesis and mitochondrial unfolded protein response (UPR) processes.

In addition to egg-laying, locomotion behavior has also been investigated in a medium-to-high throughput manner using microfluidic devices. Three reports have successfully implemented electrotaxis behavior to achieve these goals [69,74,96]. Carr et al. [96] developed a microfluidic device to perform drug screening using electrotactic motion as a readout. The chip contained a reservoir to expose worms to defined chemicals and a microchannel with electrodes across it to perform electrotaxis to investigate their movement responses. Levamisole, an anthelmintic drug, was used as a test case to demonstrate the successful use of the device for *C. elegans* as well as a parasitic nematode *Oesophagotomum dentatum*. In the paper by Salam et al. [74], our group characterized the electrotaxis defects of neurotoxin-treated worms. The neurotoxins caused damage to dopaminergic (DAergic) neurons, resulting in movement defect and serving as a model for the study of Parkinson's disease. To accelerate screening of electrotaxis phenotypes, we recently developed an advanced version of this device [69] by automating the process of single worm loading, stabilizing the fluid flow, exposing the worm to electrical signals, and recording and analyzing the electrotactic response.

Overall, these studies show that behavioral and movement-based microfluidic screening approaches can be used as platform technologies for automated drug screening applications in neurodegenerative and muscular diseases. An interesting direction in the future may involve simultaneous screening of neuronal and behavioural activities following exposure to chemicals. This will enable correlation of sensory and motor outputs in order to develop more specific treatments.

4.3. Microsurgery, Regeneration, and Drug Screening

Nanoaxotomy of neurons and whole cell ablations are highly delicate processes, especially when it comes to performing such operations on the micron-size *C. elegans*. As discussed above, worm manipulation and immobilization can be achieved in an automated, high-throughput, and very precise manner using microfluidics. These assets have made microfluidic devices very effective and useful in laser nanoaxotomy and nerve regeneration studies of *C. elegans*. Guo et al. [97] and Zeng et al. [98] have developed nanoaxotomy chips for fast surgical operations of nerves, in order to study nerve regeneration and perform high-throughput screening. Both setups were designed to trap worms by applying pressure on a flexible PDMS membrane (Figure 9A). This allowed fast operation and recovery of worms after nanosurgery. GFP fluorescence tags were used to locate neurons to perform surgery and monitor the regeneration processes (Figure 9B). A throughput of 60 worms per hour (from time of loading the worm to performing surgery) has been reported [97]. It was shown that axonal regrowth occurs faster in the distal fragment in the absence of anesthetics. Later on, Chung and Lu [99] used their cooling immobilization technique to develop a microfluidic device to rapidly paralyze L1 stage worms in two parallel traps and ablate neuronal cell bodies sequentially. The two-trap system greatly increases the speed of operation. While one trap loads a worm, the other is used to perform the surgery. This is accomplished by an automated setup that moves the stage to align the desired worm with the laser beam. The authors reported a throughput of 110 worms per hour, which was enough to obtain a population of animals for behavioural assays. Samara et al. [100] combined microfluidics and femtosecond laser microsurgery to perform semi-automated loading and single-axon ablation in *C. elegans* at a rate of approximately 180 worms per hour. Their immobilization approach involved first orienting the worm using an array of suction channels and then applying pressure on the top PDMS layer membrane to fully constrain the animal motion. The post-ablated worms were exposed to a small library of chemicals to screen for candidates with regenerative effects. A total of 10 drugs were identified that significantly enhanced the neurite regeneration.

Another device to perform high throughput laser axotomy was developed by Gokce et al. [101]. The device contains a loading chamber for housing up to 250 synchronized stage worms. The chamber is connected to a staging area that allows trapping of single worms for axotomy. The animals were immobilized by a deflecting membrane technique that physically confines them in a small region. The entire process of animal loading, staging, neuron identification, axon ablation, and unloading was fully automated with the help of computer-controlled microfluidic components and image processing tools. An average throughput of about 210 worms per hour was reported for ablation of 300 nm-wide axons at sub-micron resolution.

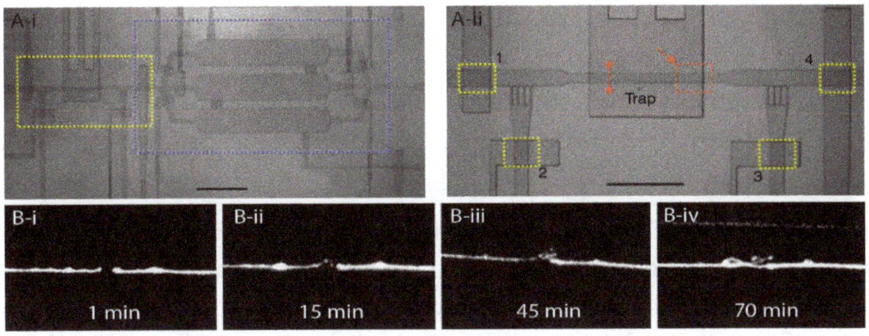

Figure 9. A microfluidic device for (**A**) laser nanoaxotomy of the ALM neuron in *C. elegans* and (**B**) investigation of time-lapse nerve regeneration (Guo et al. [97]). Reproduced with permission from Adapted by permission from Macmillan Publishers Ltd. Nature Methods. On the left (**A-i**) the trap system (**yellow** rectangle) and the three recovery chambers (**blue** rectangle) are indicated. The right panel (**A-ii**) shows a magnified view of the trapping system. The small yellow dotted rectangles show four valves to control worms. Panels (**B-i to B-iv**) show axonal recovery. Branching is visible several minutes after axotomy. By 70 min the nerve has regrown and appears to be reconnected. Refer to the references for more details.

All in all, researchers have successfully developed automated microfluidic systems for rapid surgery of axons and ablation of neuronal cell bodies in *C. elegans* and post-operated manipulations of such animals. The existing technologies are summarized in Table 2 for an easy comparison. The reported throughputs are an order of magnitude faster than the conventional manual methods. These developments are expected to facilitate the study of nerve regeneration and perform drug screening.

Table 2. Overview of microfluidic devices to perform microsurgery in *C. elegans*.

Immobilization Method	Throughput	Operation Performed	References
Pressure-based	60 worm per hour	Neuroaxotomy and nerve regeneration	Guo et al. [97]
Pressure-based	One worm every few seconds	Neuroaxotomy	Zeng et al. [98]
Cooling	120 worms per hour	Cell ablation	Chung and Lu [99]
Suction- and pressure-based	180 worms per hour	Neuroaxotomy and nerve regeneration	Samara et al. [100]
Pressure-based	210 worms per hour	Neuroaxotomy	Gokce et al. [101]

5. Concluding Remarks

Microfluidics offers powerful tools to accelerate fundamental and drug discovery research in *C. elegans*. This review provides a comprehensive summary of major devices and platforms that are enabling applications ranging from the routine handling of animals to fast, automated, and high-throughput screenings. We have primarily focused on the technological aspects of the progress to align with the audience of this journal. The description of experimental data is largely meant to support the usefulness of approaches.

Despite the excellent progress in the above fields, there are a few hurdles for greater acceptance of the technology by mainstream *C. elegans* labs. At present, the technical know-how about microfluidic systems and the infrastructure needed to produce devices is restricted mainly to labs with significant engineering expertise. Other hurdles include the high cost of fabrication, the need for customized hardware and software to process data, and the manual work required to operate many of the devices. Overcoming these barriers will require training the next generation of *C. elegans* scientists to be comfortable with fabricating and operating custom microdevices.

Although virtually all areas of *C. elegans* research have been miniaturized, there are a few where growth has been rapid and impressive. These include in vivo analysis of neuronal function, longitudinal studies of gene expression, and age-related processes. Among the various approaches, a genetically encoded GCaMP Ca^{2+} sensor in combination with pressure-based and temperature-sensitive gel-based immobilization offers powerful means to accelerate in vivo studies of cellular processes at high resolution. The electrotaxis response is also a promising tool to manipulate worms in different ways including sorting, transport, and exercise treatments. We expect to see a lot more progress in these directions, including improved devices for characterization of disease models and gene target identifications that are relevant to humans.

Drug discovery is one frontier where microfluidics holds significant potential and *C. elegans* screening can help in the identification of candidate targets in a rapid and cost-effective manner. To this end, several devices have been developed (e.g., [20,21,91]) that are capable of performing high-throughput assays. The system by Cornaglia et al. [39] is interesting as it allows monitoring of the whole of embryogenesis following chemical exposures and other manipulations. Axonal regeneration and neurodegenerative disease model-based screenings are also expected to yield exciting results in the near future due to the ease of visualizing neuronal morphologies in live animals. We anticipate that high-throughput platforms such as Samara et al. [100], Gokce et al. [101], and Cornaglia et al. [31] will enable new screening to identify drug candidates as well as genetic targets for further validations in higher eukaryotes. Overall, *C. elegans* microfluidics research holds significant promise to advance our understanding of human diseases and to help develop potential treatments.

Acknowledgments: Microfluidics research in our labs has been supported by the CHRP, NSERC, ERA, and CRC funding agencies. We thank Lesley MacNeil and the members of the Gupta lab for their comments on the manuscript. Cory Richman assisted with finalizing some of the images.

Author Contributions: Bhagwati P. Gupta conceived the layout and organized sections of the manuscript. Images were selected by Pouya Rezai. Bhagwati P. Gupta and Pouya Rezai wrote the paper.

References

1. Markaki, M.; Tavernarakis, N. Modeling human diseases in *Caenorhabditis elegans*. *Biotechnol. J.* **2010**, *5*, 1261–1276. [CrossRef] [PubMed]
2. Corsi, A.K. A biochemist's guide to *Caenorhabditis elegans*. *Anal. Biochem.* **2006**, *359*, 1–17. [CrossRef] [PubMed]
3. Emmons, S.W. The beginning of connectomics: A commentary on White et al.(1986)'The structure of the nervous system of the nematode *Caenorhabditis elegans*'. *Philos. Trans. R. Soc. Lond. B Biol. Sci.* **2015**, *370*, 2014039. [CrossRef] [PubMed]
4. White, J.G.; Southgate, E.; Thomson, J.N.; Brenner, S. The structure of the nervous system of the nematode *Caenorhabditis elegans*. *Philos. Trans. R. Soc. Lond. B Biol. Sci.* **1986**, *314*, 1–340. [CrossRef] [PubMed]
5. Silverman, G.A.; Luke, C.J.; Bhatia, S.R.; Long, O.S.; Vetica, A.C.; Perlmutter, D.H.; Pak, S.C. Modeling molecular and cellular aspects of human disease using the nematode *Caenorhabditis elegans*. *Pediatr. Res.* **2009**, *65*, 10–18. [CrossRef] [PubMed]
6. Kaletta, T.; Hengartner, M.O. Finding function in novel targets: *C. elegans* as a model organism. *Nat. Rev. Drug Discov.* **2006**, *5*, 387–398. [CrossRef] [PubMed]
7. Mark, D.; Haeberle, S.; Roth, G.; von Stetten, F.; Zengerle, R. Microfluidic lab-on-a-chip platforms: Requirements, characteristics and applications. *Chem. Soc. Rev.* **2010**, *39*, 1153–1182. [CrossRef] [PubMed]
8. Spiller, D.G.; Wood, C.D.; Rand, D.A.; White, M.R. Measurement of single-cell dynamics. *Nature* **2010**, *465*, 736–745. [CrossRef] [PubMed]
9. Rezai, P.; Siddiqui, A.; Selvaganapathy, P.R.; Gupta, B.P. Electrotaxis of Caenorhabditis elegans in a microfluidic environment. *Lab Chip* **2010**, *10*, 220–226. [CrossRef] [PubMed]
10. Rezai, P.; Salam, S.; Selvaganapathy, P.R.; Gupta, B.P. Electrical sorting of *Caenorhabditis elegans*. *Lab Chip* **2012**, *12*, 1831–1840. [CrossRef] [PubMed]
11. Maniere, X.; Lebois, F.; Matic, I.; Ladoux, B.; Di Meglio, J.M.; Hersen, P. Running worms: *C. elegans* self-sorting by electrotaxis. *PLoS ONE* **2011**, *6*, e16637. [CrossRef] [PubMed]
12. Solvas, X.C.i.; Geier, F.M.; Leroi, A.M.; Bundy, J.G.; Edel, J.B.; DeMello, A.J. High-throughput age synchronisation of *Caenorhabditis elegans*. *Chem. Commun.* **2011**, *47*, 9801–9803. [CrossRef] [PubMed]
13. Han, B.; Kim, D.; Ko, U.H.; Shin, J.H. A sorting strategy for *C. elegans* based on size-dependent motility and electrotaxis in a micro-structured channel. *Lab Chip* **2012**, *12*, 4128–4134. [CrossRef] [PubMed]

14. Ai, X.; Zhuo, W.; Liang, Q.; McGrath, P.T.; Lu, H. A high-throughput device for size based separation of *C. elegans* developmental stages. *Lab Chip* **2014**, *14*, 1746–1752. [CrossRef] [PubMed]
15. Yan, Y.; Ng, L.F.; Ng, L.T.; Choi, K.B.; Gruber, J.; Bettiol, A.A.; Thakor, N.V. A continuous-flow *C. elegans* sorting system with integrated optical fiber detection and laminar flow switching. *Lab Chip* **2014**, *14*, 4000–4006. [CrossRef] [PubMed]
16. Wang, X.; Hu, R.; Ge, A.; Hu, L.; Wang, S.; Feng, X.; Du, W.; Liu, B.F. Highly efficient microfluidic sorting device for synchronizing developmental stages of *C. elegans* based on deflecting electrotaxis. *Lab Chip* **2015**, *15*, 2513–2521. [CrossRef] [PubMed]
17. Dong, L.; Cornaglia, M.; Lehnert, T.; Gijs, M.A. Versatile size-dependent sorting of *C. elegans* nematodes and embryos using a tunable microfluidic filter structure. *Lab Chip* **2016**, *16*, 574–585. [CrossRef] [PubMed]
18. Gabel, C.V.; Gabel, H.; Pavlichin, D.; Kao, A.; Clark, D.A.; Samuel, A.D. Neural circuits mediate electrosensory behavior in *Caenorhabditis elegans*. *J. Neurosci.* **2007**, *27*, 7586–7596. [CrossRef] [PubMed]
19. Aubry, G.; Zhan, M.; Lu, H. Hydrogel-droplet microfluidic platform for high-resolution imaging and sorting of early larval *Caenorhabditis elegans*. *Lab Chip* **2015**, *15*, 1424–1431. [CrossRef] [PubMed]
20. Chung, K.; Crane, M.M.; Lu, H. Automated on-chip rapid microscopy, phenotyping and sorting of *C. elegans*. *Nat. Methods* **2008**, *5*, 637–643. [CrossRef] [PubMed]
21. Rohde, C.B.; Zeng, F.; Gonzalez-Rubio, R.; Angel, M.; Yanik, M.F. Microfluidic system for on-chip high-throughput whole-animal sorting and screening at subcellular resolution. *Proc. Natl. Acad. Sci. USA* **2007**, *104*, 13891–13895. [CrossRef] [PubMed]
22. Ma, H.; Jiang, L.; Shi, W.; Qin, J.; Lin, B. A programmable microvalve-based microfluidic array for characterization of neurotoxin-induced responses of individual *C. elegans*. *Biomicrofluidics* **2009**, *3*, 44114. [CrossRef] [PubMed]
23. Hulme, S.E.; Shevkoplyas, S.S.; McGuigan, A.P.; Apfeld, J.; Fontana, W.; Whitesides, G.M. Lifespan-on-a-chip: Microfluidic chambers for performing lifelong observation of *C. elegans*. *Lab Chip* **2010**, *10*, 589–597. [CrossRef] [PubMed]
24. Wen, H.; Shi, W.; Qin, J. Multiparameter evaluation of the longevity in *C. elegans* under stress using an integrated microfluidic device. *Biomed. Microdevices* **2012**, *14*, 721–728. [CrossRef] [PubMed]
25. Belfer, S.J.; Chuang, H.S.; Freedman, B.L.; Yuan, J.; Norton, M.; Bau, H.H.; Raizen, D.M. Caenorhabditis-in-drop array for monitoring *C. elegans* quiescent behavior. *Sleep* **2013**, *36*, 689–698. [CrossRef] [PubMed]
26. Lionaki, E.; Tavernarakis, N. High-throughput and longitudinal analysis of aging and senescent decline in *Caenorhabditis elegans*. *Methods Mol. Biol.* **2013**, *965*, 485–500. [PubMed]
27. Nghe, P.; Boulineau, S.; Gude, S.; Recouvreux, P.; van Zon, J.S.; Tans, S.J. Microfabricated polyacrylamide devices for the controlled culture of growing cells and developing organisms. *PLoS ONE* **2013**, *8*, e75537.
28. Turek, M.; Besseling, J.; Bringmann, H. Agarose Microchambers for Long-term Calcium Imaging of *Caenorhabditis elegans*. *J. Vis. Exp.* **2015**, *100*, 52742. [CrossRef] [PubMed]
29. Krajniak, J.; Lu, H. Long-term high-resolution imaging and culture of *C. elegans* in chip-gel hybrid microfluidic device for developmental studies. *Lab Chip* **2010**, *10*, 1862–1868. [CrossRef] [PubMed]
30. Wen, H.; Yu, Y.; Zhu, G.; Jiang, L.; Qin, J. A droplet microchip with substance exchange capability for the developmental study of *C. elegans*. *Lab Chip* **2015**, *15*, 1905–1911. [CrossRef] [PubMed]
31. Cornaglia, M.; Krishnamani, G.; Mouchiroud, L.; Sorrentino, V.; Lehnert, T.; Auwerx, J.; Gijs, M.A. Automated longitudinal monitoring of in vivo protein aggregation in neurodegenerative disease *C. elegans* models. *Mol. Neurodegener.* **2016**, *11*, 17. [CrossRef] [PubMed]
32. Kim, N.; Dempsey, C.M.; Zoval, J.V.; Sze, J.Y.; Madou, M.J. Automated microfluidic compact disc (CD) cultivation system of *Caenorhabditis elegans*. *Sens. Actuators B* **2007**, *122*, 511–518. [CrossRef]
33. Gilleland, C.L.; Rohde, C.B.; Zeng, F.; Yanik, M.F. Microfluidic immobilization of physiologically active *Caenorhabditis elegans*. *Nat. Protoc.* **2010**, *5*, 1888–1902. [CrossRef] [PubMed]
34. Kopito, R.B.; Levine, E. Durable spatiotemporal surveillance of *Caenorhabditis elegans* response to environmental cues. *Lab Chip* **2014**, *14*, 764–770. [CrossRef] [PubMed]
35. Chokshi, T.V.; Ben-Yakar, A.; Chronis, N. CO_2 and compressive immobilization of *C. elegans* on-chip. *Lab Chip* **2009**, *9*, 151–157. [CrossRef] [PubMed]
36. Liu, D.; Ranawade, A.; Gupta, B.P.; Selvaganapathy, P.R. A microfluidic device for *C. elegans* immobilization and long term imaging. In preparation.
37. Mondal, S.; Ahlawat, S.; Rau, K.; Venkataraman, V.; Koushika, S.P. Imaging in vivo neuronal transport in genetic model organisms using microfluidic devices. *Traffic* **2011**, *12*, 372–385. [CrossRef] [PubMed]

38. Chuang, H.S.; Chen, H.Y.; Chen, C.S.; Chiu, W.T. Immobilization of the nematode *Caenorhabditis elegans* with addressable light-induced heat knockdown (ALINK). *Lab Chip* **2013**, *13*, 2980–2989. [CrossRef] [PubMed]
39. Cornaglia, M.; Mouchiroud, L.; Marette, A.; Narasimhan, S.; Lehnert, T.; Jovaisaite, V.; Auwerx, J.; Gijs, M.A. An automated microfluidic platform for *C. elegans* embryo arraying, phenotyping, and long-term live imaging. *Sci. Rep.* **2015**, *5*, 10192. [CrossRef] [PubMed]
40. Ghaemi, R.; Tong, J.; Selvaganapathy, P.R.; Gupta, B.P. Microfluidic device for microinjection of *Caenorhabditis elegans*. In Proceedings of the 17th International Conference on Miniaturized Systems for Chemistry and Life Sciences, Freiburg, Germany, 27–31 October 2013; pp. 1821–1823.
41. Ghaemi, R. Microfluidic device for microinjection of *Caenorhabditis elegans*. Available online: https://macsphere.mcmaster.ca/handle/11375/15339 (accessed on 18 July 2016).
42. Zhao, X.; Xu, F.; Tang, L.; Du, W.; Feng, X.; Liu, B.F. Microfluidic chip-based *C. elegans* microinjection system for investigating cell-cell communication in vivo. *Biosens. Bioelectron.* **2013**, *50*, 28–34. [CrossRef] [PubMed]
43. Song, P.; Dong, X.; Liu, X. A microfluidic device for automated, high-speed microinjection of *Caenorhabditis elegans*. *Biomicrofluidics* **2016**, *10*, 011912. [CrossRef] [PubMed]
44. Park, S.; Hwang, H.; Nam, S.W.; Martinez, F.; Austin, R.H.; Ryu, W.S. Enhanced *Caenorhabditis elegans* locomotion in a structured microfluidic environment. *PLoS ONE* **2008**, *3*, e2550. [CrossRef] [PubMed]
45. Parashar, A.; Lycke, R.; Carr, J.A.; Pandey, S. Amplitude-modulated sinusoidal microchannels for observing adaptability in *C. elegans* locomotion. *Biomicrofluidics* **2011**, *5*, 24112. [CrossRef] [PubMed]
46. Johari, S.; Nock, V.; Alkaisi, M.M.; Wang, W. On-chip analysis of *C. elegans* muscular forces and locomotion patterns in microstructured environments. *Lab Chip* **2013**, *13*, 1699–1707. [CrossRef] [PubMed]
47. Qiu, Z.; Tu, L.; Huang, L.; Zhu, T.; Nock, V.; Yu, E.; Liu, X.; Wang, W. An integrated platform enabling optogenetic illumination of *Caenorhabditis elegans* neurons and muscular force measurement in microstructured environments. *Biomicrofluidics* **2015**, *9*, 014123. [CrossRef] [PubMed]
48. Hwang, H.; Barnes, D.E.; Matsunaga, Y.; Benian, G.M.; Ono, S.; Lu, H. Muscle contraction phenotypic analysis enabled by optogenetics reveals functional relationships of sarcomere components in *Caenorhabditis elegans*. *Sci. Rep.* **2016**, *6*, 19900. [CrossRef] [PubMed]
49. Allen, P.B.; Sgro, A.E.; Chao, D.L.; Doepker, B.E.; Scott Edgar, J.; Shen, K.; Chiu, D.T. Single-synapse ablation and long-term imaging in live *C. elegans*. *J. Neurosci. Methods* **2008**, *173*, 20–26. [CrossRef] [PubMed]
50. Lee, H.; Kim, S.A.; Coakley, S.; Mugno, P.; Hammarlund, M.; Hilliard, M.A.; Lu, H. A multi-channel device for high-density target-selective stimulation and long-term monitoring of cells and subcellular features in *C. elegans*. *Lab Chip* **2014**, *14*, 4513–4522. [CrossRef] [PubMed]
51. Bulina, M.E.; Chudakov, D.M.; Britanova, O.V.; Yanushevich, Y.G.; Staroverov, D.B.; Chepurnykh, T.V.; Merzlyak, E.M.; Shkrob, M.A.; Lukyanov, S.; Lukyanov, K.A. A genetically encoded photosensitizer. *Nat. Biotechnol.* **2006**, *24*, 95–99. [CrossRef] [PubMed]
52. Larsch, J.; Ventimiglia, D.; Bargmann, C.I.; Albrecht, D.R. High-throughput imaging of neuronal activity in *Caenorhabditis elegans*. *Proc. Natl. Acad. Sci. USA* **2013**, *110*, E4266–E4273. [CrossRef] [PubMed]
53. Miyawaki, A. Fluorescence imaging of physiological activity in complex systems using GFP-based probes. *Curr. Opin. Neurobiol.* **2003**, *13*, 591–596. [CrossRef] [PubMed]
54. Ardeshiri, R.; Murphy, B.; Zhen, M.; Rezai, P. A Hybrid Microfluidic Device for *C. elegans* On-Demand Orientation and Multidirectional Imaging. In Proceedings of the International Conference on Miniaturized Systems for Chemistry and Life Sciences, Gyengju, Korea, 25–29 October 2015; pp. 707–709.
55. Ahmed, D.; Ozcelik, A.; Bojanala, N.; Nama, N.; Upadhyay, A.; Chen, Y.; Hanna-Rose, W.; Huang, T.J. Rotational manipulation of single cells and organisms using acoustic waves. *Nat. Commun.* **2016**, *7*, 11085. [CrossRef] [PubMed]
56. Chronis, N.; Zimmer, M.; Bargmann, C.I. Microfluidics for in vivo imaging of neuronal and behavioral activity in *Caenorhabditis elegans*. *Nat. Methods* **2007**, *4*, 727–731. [CrossRef] [PubMed]
57. Chalasani, S.H.; Chronis, N.; Tsunozaki, M.; Gray, J.M.; Ramot, D.; Goodman, M.B.; Bargmann, C.I. Dissecting a circuit for olfactory behaviour in *Caenorhabditis elegans*. *Nature* **2007**, *450*, 63–70. [CrossRef] [PubMed]
58. Albrecht, D.R.; Bargmann, C.I. High-content behavioral analysis of *Caenorhabditis elegans* in precise spatiotemporal chemical environments. *Nat. Methods* **2011**, *8*, 599–605. [CrossRef] [PubMed]
59. McCormick, K.E.; Gaertner, B.E.; Sottile, M.; Phillips, P.C.; Lockery, S.R. Microfluidic devices for analysis of spatial orientation behaviors in semi-restrained *Caenorhabditis elegans*. *PLoS ONE* **2011**, *6*, e25710. [CrossRef] [PubMed]

60. Hu, L.; Ye, J.; Tan, H.; Ge, A.; Tang, L.; Feng, X.; Du, W.; Liu, B.F. Quantitative analysis of *Caenorhabditis elegans* chemotaxis using a microfluidic device. *Anal. Chim. Acta* **2015**, *887*, 155–162. [CrossRef] [PubMed]
61. Hu, L.; Wang, J.J.; Feng, X.J.; Du, W.; Liu, B.F. Microfluidic device for analysis of gas-evoked neuronal sensing in *C. elegans*. *Sens. Actuators B Chem.* **2015**, *209*, 109–115. [CrossRef]
62. Vidal-Gadea, A.; Ward, K.; Beron, C.; Ghorashian, N.; Gokce, S.; Russell, J.; Truong, N.; Parikh, A.; Gadea, O.; Ben-Yakar, A.; et al. Magnetosensitive neurons mediate geomagnetic orientation in *Caenorhabditis elegans*. *Elife* **2015**, *4*, e07493. [CrossRef] [PubMed]
63. Sukul, N.C.; Croll, N.A. Influence of Potential Difference and Current on the Electrotaxis of *Caenorhaditis elegans*. *J. Nematol.* **1978**, *10*, 314–317. [PubMed]
64. Caveness, F.E.; Panzer, J.D. Nemic galvanotaxis. *Proc. Helminthol. Soc. Wash.* **1960**, *27*, 73–74.
65. Gupta, S.P. Galvanotactic reaction of infective larvae of Trichostrongylus retortaeformis. *Exp. Parasitol.* **1962**, *12*, 118–119. [CrossRef]
66. Rezai, P.; Salam, S.; Selvaganapathy, P.R.; Gupta, B.P. Microfluidic systems to study the biology of human diseases and identify potential therapeutic targets in *Caenorhabditis elegans*. In *Integrated Microsystems*; Iniewski, K., Ed.; CRC Press: Boca Raton, FL, USA, 2012; pp. 581–608.
67. Rezai, P.; Salam, S.; Selvaganapathy, P.R.; Gupta, B.P. Effect of pulse direct current signals on electrotactic movement of nematodes *Caenorhabditis elegans* and Caenorhabditis briggsae. *Biomicrofluidics* **2011**, *5*, 044116. [CrossRef] [PubMed]
68. Rezai, P.; Siddiqui, A.; Selvaganapathy, P.R.; Gupta, B.P. Behavior of *Caenorhabditis elegans* in alternating electric field and its application to their localization and control. *Appl. Phys. Lett.* **2010**, *96*, 153702. [CrossRef]
69. Liu, D.; Gupta, B.; Selvaganapathy, P.R. An automated microfluidic system for screening *Caenorhabditis elegans* behaviors using electrotaxis. *Biomicrofluidics* **2016**, *10*, 014117. [CrossRef] [PubMed]
70. Leung, M.C.; Williams, P.L.; Benedetto, A.; Au, C.; Helmcke, K.J.; Aschner, M.; Meyer, J.N. *Caenorhabditis elegans*: An emerging model in biomedical and environmental toxicology. *Toxicol. Sci.* **2008**, *106*, 5–28. [CrossRef] [PubMed]
71. Hagerbaumer, A.; Hoss, S.; Heininger, P.; Traunspurger, W. Experimental studies with nematodes in ecotoxicology: An overview. *J. Nematol.* **2015**, *47*, 11–27. [PubMed]
72. Jung, J.; Nakajima, M.; Kojima, M.; Ooe, K.; Fukuda, T. Microchip device for measurement of body volume of *C. elegans* as bioindicator application. *J. Micro Nano Mechatron.* **2011**, *7*, 3–11. [CrossRef]
73. Jung, J.; Nakajima, M.; Tajima, H.; Huang, Q.; Fukuda, T. A microfluidic device for the continuous culture and analysis of *Caenorhabditis elegans* in a toxic aqueous environment. *J. Micromech. Microeng.* **2013**, *23*, 1–8. [CrossRef]
74. Salam, S.; Ansari, A.; Amon, S.; Rezai, P.; Selvaganapathy, P.R.; Mishra, R.K.; Gupta, B.P. A microfluidic phenotype analysis system reveals function of sensory and dopaminergic neuron signaling in *C. elegans* electrotactic swimming behavior. *Worm* **2013**, *2*, e24558. [CrossRef] [PubMed]
75. Salam, S.; Gupta, B.P. Unpublished observations.
76. Zhang, B.; Li, Y.; He, Q.; Qin, J.; Yu, Y.; Li, X.; Zhang, L.; Yao, M.; Liu, J.; Chen, Z. Microfluidic platform integrated with worm-counting setup for assessing manganese toxicity. *Biomicrofluidics* **2014**, *8*, 054110. [CrossRef] [PubMed]
77. Yang, J.; Chen, Z.; Ching, P.; Shi, Q.; Li, X. An integrated microfluidic platform for evaluating in vivo antimicrobial activity of natural compounds using a whole-animal infection model. *Lab Chip* **2013**, *13*, 3373–3382. [CrossRef] [PubMed]
78. Saeki, S.; Yamamoto, M.; Iino, Y. Plasticity of chemotaxis revealed by paired presentation of a chemoattractant and starvation in the nematode *Caenorhabditis elegans*. *J. Exp. Biol.* **2001**, *204*, 1757–1764. [PubMed]
79. Wen, J.Y.; Kumar, N.; Morrison, G.; Rambaldini, G.; Runciman, S.; Rousseau, J.; van der Kooy, D. Mutations that prevent associative learning in *C. elegans*. *Behav. Neurosci.* **1997**, *111*, 354–368. [CrossRef] [PubMed]
80. Zhang, Y.; Lu, H.; Bargmann, C.I. Pathogenic bacteria induce aversive olfactory learning in *Caenorhabditis elegans*. *Nature* **2005**, *438*, 179–184. [CrossRef] [PubMed]
81. Qin, J.; Wheeler, A.R. Maze exploration and learning in *C. elegans*. *Lab Chip* **2007**, *7*, 186–192. [CrossRef] [PubMed]
82. Xian, B.; Shen, J.; Chen, W.; Sun, N.; Qiao, N.; Jiang, D.; Yu, T.; Men, Y.; Han, Z.; Pang, Y.; et al. WormFarm: A quantitative control and measurement device toward automated *Caenorhabditis elegans* aging analysis. *Aging Cell* **2013**, *12*, 398–409. [CrossRef] [PubMed]

83. Li, S.; Stone, H.A.; Murphy, C.T. A microfluidic device and automatic counting system for the study of *C. elegans* reproductive aging. *Lab Chip* **2015**, *15*, 524–531. [CrossRef] [PubMed]
84. Chuang, H.S.; Kuo, W.J.; Lee, C.L.; Chu, I.H.; Chen, C.S. Exercise in an electrotactic flow chamber ameliorates age-related degeneration in *Caenorhabditis elegans*. *Sci. Rep.* **2016**, *6*, 28064. [CrossRef] [PubMed]
85. Hulme, S.E.; Whitesides, G.M. Chemistry and the worm: *Caenorhabditis elegans* as a platform for integrating chemical and biological research. *Angew. Chem. Int. Ed. Engl.* **2011**, *50*, 4774–4807. [CrossRef] [PubMed]
86. Ghorashian, N.; Gokce, S.K.; Guo, S.X.; Everett, W.N.; Ben-Yakar, A. An automated microfluidic multiplexer for fast delivery of *C. elegans* populations from multiwells. *PLoS ONE* **2013**, *8*, e74480. [CrossRef] [PubMed]
87. Crane, M.M.; Chung, K.; Lu, H. Computer-enhanced high-throughput genetic screens of *C. elegans* in a microfluidic system. *Lab Chip* **2009**, *9*, 38–40. [CrossRef] [PubMed]
88. Crane, M.M.; Stirman, J.N.; Ou, C.Y.; Kurshan, P.T.; Rehg, J.M.; Shen, K.; Lu, H. Autonomous screening of *C. elegans* identifies genes implicated in synaptogenesis. *Nat. Methods* **2012**, *9*, 977–980. [CrossRef] [PubMed]
89. Caceres Ide, C.; Valmas, N.; Hilliard, M.A.; Lu, H. Laterally orienting *C. elegans* using geometry at microscale for high-throughput visual screens in neurodegeneration and neuronal development studies. *PLoS ONE* **2012**, *7*, e35037.
90. Yan, Y.; Boey, D.; Ng, L.T.; Gruber, J.; Bettiol, A.; Thakor, N.V.; Chen, C.H. Continuous-flow *C. elegans* fluorescence expression analysis with real-time image processing through microfluidics. *Biosens. Bioelectron.* **2016**, *77*, 428–434. [CrossRef] [PubMed]
91. Chokshi, T.V.; Bazopoulou, D.; Chronis, N. An automated microfluidic platform for calcium imaging of chemosensory neurons in *Caenorhabditis elegans*. *Lab Chip* **2010**, *10*, 2758–2763. [CrossRef] [PubMed]
92. Lockery, S.R.; Hulme, S.E.; Roberts, W.M.; Robinson, K.J.; Laromaine, A.; Lindsay, T.H.; Whitesides, G.M.; Weeks, J.C. A microfluidic device for whole-animal drug screening using electrophysiological measures in the nematode *C. elegans*. *Lab Chip* **2012**, *12*, 2211–2220. [CrossRef] [PubMed]
93. Raizen, D.; Song, B.M.; Trojanowski, N.; You, Y.J. Methods for Measuring Pharyngeal Behaviors. Available online: http://www.ncbi.nlm.nih.gov/books/NBK126648/ (accessed on 13 July 2016).
94. Hu, C.; Dillon, J.; Kearn, J.; Murray, C.; O'Connor, V.; Holden-Dye, L.; Morgan, H. NeuroChip: A microfluidic electrophysiological device for genetic and chemical biology screening of *Caenorhabditis elegans* adult and larvae. *PLoS ONE* **2013**, *8*, e64297. [CrossRef] [PubMed]
95. Chung, K.; Zhan, M.; Srinivasan, J.; Sternberg, P.W.; Gong, E.; Schroeder, F.C.; Lu, H. Microfluidic chamber arrays for whole-organism behavior-based chemical screening. *Lab Chip* **2011**, *11*, 3689–3697. [CrossRef] [PubMed]
96. Carr, J.A.; Parashar, A.; Gibson, R.; Robertson, A.P.; Martin, R.J.; Pandey, S. A microfluidic platform for high-sensitivity, real-time drug screening on *C. elegans* and parasitic nematodes. *Lab Chip* **2011**, *11*, 2385–2396. [CrossRef] [PubMed]
97. Guo, S.X.; Bourgeois, F.; Chokshi, T.; Durr, N.J.; Hilliard, M.A.; Chronis, N.; Ben-Yakar, A. Femtosecond laser nanoaxotomy lab-on-a-chip for in vivo nerve regeneration studies. *Nat. Methods* **2008**, *5*, 531–533. [CrossRef] [PubMed]
98. Zeng, F.; Rohde, C.B.; Yanik, M.F. Sub-cellular precision on-chip small-animal immobilization, multi-photon imaging and femtosecond-laser manipulation. *Lab Chip* **2008**, *8*, 653–656. [CrossRef] [PubMed]
99. Chung, K.; Lu, H. Automated high-throughput cell microsurgery on-chip. *Lab Chip* **2009**, *9*, 2764–2766. [CrossRef] [PubMed]
100. Samara, C.; Rohde, C.B.; Gilleland, C.L.; Norton, S.; Haggarty, S.J.; Yanik, M.F. Large-scale in vivo femtosecond laser neurosurgery screen reveals small-molecule enhancer of regeneration. *Proc. Natl. Acad. Sci. USA* **2010**, *107*, 18342–18347. [CrossRef] [PubMed]
101. Gokce, S.K.; Guo, S.X.; Ghorashian, N.; Everett, W.N.; Jarrell, T.; Kottek, A.; Bovik, A.C.; Ben-Yakar, A. A fully automated microfluidic femtosecond laser axotomy platform for nerve regeneration studies in *C. elegans*. *PLoS ONE* **2014**, *9*, e113917. [CrossRef] [PubMed]

17

Three-Dimensional Fabrication for Microfluidics by Conventional Techniques and Equipment used in Mass Production

Toyohiro Naito [1,*], Makoto Nakamura [1], Noritada Kaji [2,3], Takuya Kubo [1], Yoshinobu Baba [2,3] and Koji Otsuka [1]

[1] Department of Material Chemistry, Graduate School of Engineering, Kyoto University, Katsura, Nishikyo-ku, Kyoto 615-8510, Japan; nakamura.makoto.87a@st.kyoto-u.ac.jp (M.N.); kubo@anchem.mc.kyoto-u.ac.jp (T.K.); otsuka@anchem.mc.kyoto-u.ac.jp (K.O.)

[2] Department of Applied Chemistry, Graduate School of Engineering, Nagoya University, Furo-cho, Chikusa-ku, Nagoya 464-8603, Japan; kaji@apchem.nagoya-u.ac.jp (N.K.); babaymtt@apchem.nagoya-u.ac.jp (Y.B.)

[3] ImPACT Research Center for Advanced Nanobiodevices, Nagoya University, Furo-cho, Chikusa-ku, Nagoya 464-8603, Japan

* Correspondence: naito@anchem.mc.kyoto-u.ac.jp

Academic Editors: Manabu Tokeshi and Kiichi Sato

Abstract: This paper presents a simple three-dimensional (3D) fabrication method based on soft lithography techniques and laminated object manufacturing. The method can create 3D structures that have undercuts with general machines for mass production and laboratory scale prototyping. The minimum layer thickness of the method is at least 4 μm and bonding strength between layers is over 330 kPa. The performance reaches conventional fabrication techniques used for two-dimensionally (2D)-designed microfluidic devices. We fabricated some 3D structures, *i.e.*, fractal structures, spiral structures, and a channel-in-channel structure, in microfluidic channels and demonstrated 3D microfluidics. The fabrication method can be achieved with a simple black light for bio-molecule detection; thus, it is useful for not only lab-scale rapid prototyping, but also for commercial manufacturing.

Keywords: 3D microfluidics; microfabrication

---

1. Introduction

Research in the field of microfluidics is moving from understanding two-dimensionally (2D)-designed flow to three-dimensionally (3D)-designed flow. 2D flow, which is planar flow with no vertical flow, has been studied since the 1990s, and various 2D-designed microfluidic devices such as multiple branching channels [1–3], zigzag shaped channels [3,4], Tesla mixers [5], and deterministic lateral displacement devices [6–8] have been developed. These channels have no changes in their cross-sectional shape and are fabricated by simple soft lithography or an etching process. In some fairly recent reports, Dean flow which is 3D flow formed in a 2D curved microchannel has been applied for particle separation [9,10] and liquid mixing [11,12]. To generate more complex 3D flows in microchannels, 3D-designed structures with configuration changes in a vertical direction have been developed since about 2000. The chaotic mixer reported in 2002 has structures that are grooves similar to rifling in a gun barrel; the structures can generate a 3D twisting flow in the mixer channel [13,14]. The microfluidic baker's transformation (MBT) device that has 2D-designed structures with vertical changes also generate a 3D flow in a microchannel [15,16]. These 3D structures achieved a highly efficient solution mixing in microfluidic channels. However, their configurations were limited by the demolding process to shapes without undercuts such as cuboids or pyramids. Most reports have used 2D or 3D structures without undercuts for microfluidic devices in spite of the research shift from understanding 2D flow to 3D flow due to the difficulties and particularities of the fabrication processes of realizing complex 3D structures unlimited by the demolding process.

To fabricate complex 3D structures with undercuts, 3D microfabrication techniques are required. The complex 3D structures can be fabricated using additive manufacturing such as microstereolithography [17,18] or lamination [19,20]. Microstereolithography employs a liquid UV-curable polymer, an UV laser, and an XYZ stage to build thin layers which are part of a 3D structure. The UV laser patterns the cross sections of a 3D structure on a thin layer with a curing UV-curable polymer. Subsequently, the XYZ stage moves down to the next layer. These two processes are repeated over and over for creating a 3D structure. Although microstereolithography can achieve complex 3D structures, there are some problems: (i) it needs special equipment; (ii) a large amount of UV-curable polymer compared with microstructure is required; and (iii) the method limits the fabrication speed and working area because of the laser scanning method. These are critical issues for the progress of the technology from the basic research phase to the market-related development phase.

Lamination of many thin layers of substrates is a widely used method thanks to the use of conventional fabrication techniques and machines [20–23]. Although this method is possible for the mass production of micro 3D structures, two drawbacks have to be solved. One is that some materials require a cumbersome bonding process to build up layers with chemical, thermal, or both processes. The other is the difficulty of controlling the layer thickness. The layer thickness of the lamination method is not thin enough for creating 3D structures in microfluidic channels.

Many 3D printers based on microstereolithography or lamination techniques have been launched in the last five years. These technologies allow us to create microfluidic devices and peripherals. Some groups have reported 3D-printed microfluidic devices [24–26], and others have applied 3D printers for bioprinting [27,28]. The Lewis group fabricated microfluidic print-heads for 3D printing [29]. 3D printers can save labor for the first stage device fabrication and reduce fabrication limitation in the field of microfluidics. On the other hand, although 3D printing is a powerful technique for rapid prototyping, personalized manufacturing, and distributed manufacturing, it is still less efficient than injection molding and other conventional techniques from the viewpoint of mass production.

In recent years, thiol-ene reaction has attracted a lot of attention as an alternative to poly(methyl methacrylate) (PMMA) or poly(dimethylsiloxane) (PDMS) [30–34]. The thiol-ene reaction-based materials polymerize rapidly, have solvent resistance, have good mechanical properties, and have the possible application of 3D fabrication. We also applied the reaction to fabricate a thermo-pneumatic pump that is composed of several layers [35]. Carlborg *et al.* reported a new thiol-ene reaction-based microfabrication [36]. They characterize mechanical and chemical properties of microfluidic devices based on the thiol-ene reaction and demonstrate a potential to bridge the gap between rapid prototyping and mass production. However, their fabrication technique is limited to 2D-designed microfluidic channels. In this study, we propose a simple rapid prototyping method based on soft lithography without any support material for high throughput 3D fabrication. The method is a combination of injection-molding and lamination, which uses double-sided molding with PDMS molds and a UV-curable adhesive based on thiol-ene reaction. It can be done by conventional equipment for soft lithography and requires no cumbersome bonding and washing processes for multilayer lamination.

2. Materials and Methods

2.1. Device Fabrication

The fabrication process is schematically depicted in Figure 1. There are two main procedures in the fabrication method: the fabrication of PDMS molds by soft lithography and the lamination of thin sheets. The PDMS molds are used for making thin sheets of a commercial UV-curable adhesive,

Norland Optical Adhesive 81 (NOA 81). NOA 81 can be cured rapidly by UV irradiation within a few minutes or even seconds. However, the reaction is inhibited by oxygen, and PDMS has high oxygen permeability and UV transmittance. The NOA 81 in the PDMS molds is cured by UV irradiation, while surfaces of NOA 81 sheets in contact with PDMS remains uncured [31,33]. The uncured surfaces work as bonding layers, which are helpful in piling up thin layers to 3D structures.

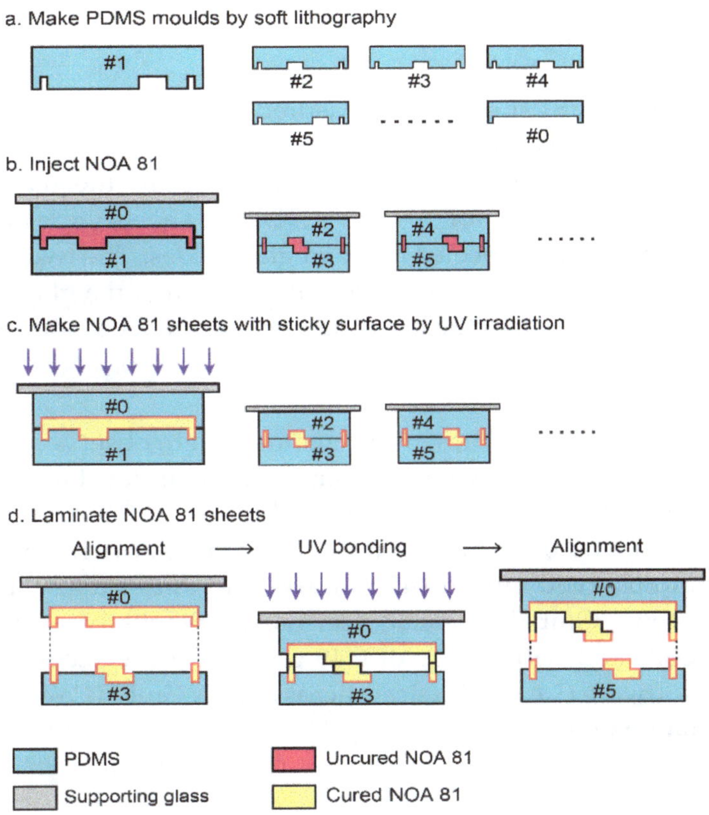

Figure 1. Schematic cross-sectional illustrations of the 3D fabrication process by conventional photolithography with NOA 81. (**a**) Fabricated PDMS molds by soft lithography. The numbers indicate layer order and #0 is a mold for a lid. (**b**) Injection of uncured NOA 81 to the spaces between PDMS molds by capillary force. (**c**) UV irradiation for partially curing NOA 81. (**d**) Lamination of NOA 81 sheets. NOA sheets are aligned after one side of PDMS molds are peeled off. The sheets are bonded by UV irradiation. The alignment and UV bonding processes are repeated.

The sliced images of objective structures along the *Z*-axis were drawn on overhead projector (OHP) films that were used as photomasks. The master molds were made with negative photoresist SU-8 series (Microchem, Tokyo, Japan) to make PDMS molds. Two components of the PDMS kit (Sylgard 184, Dow Corning, Tokyo, Japan) were mixed at a weight ratio of base:curing agent = 10:1. The prepolymer was poured into the SU-8 master molds. Concavities and convexities of the PDMS molds were opposite to those of conventional PDMS microfluidic channels; channels in SU-8 molds were concaves, and channels in PDMS molds were convexes (Figure 1a).

The PDMS molds of the objective 3D structure were divided into pairs to make interspaces. The upper PDMS molds were attached to supporting glass plates such as a cover glass or a glass slide for preventing the upper PDMS molds from dead load deflection. The interspaces were filled with NOA 81 (Norland Products Inc., Cranbury, NJ, USA) by capillary force (Figure 1b). Uncured NOA 81 were polymerized by UV irradiation for 90 s with a simple black light (AS ONE Co., Osaka, Japan) that illuminates 1.65 mW/cm^2 at 365 nm. As mentioned above, the polymerized NOA 81 sheets with a cross-sectional shape of the objective 3D structure have uncured surfaces (Figure 1c). The NOA sheets were bonded via UV irradiation for 30 s after one side of the PDMS molds were peeled off (Figure 1d).

The lamination process was repeated until the top layer. Thin films were aligned with a positioning stage (TR6047-S1, Chuo Precision Industrial Co. Ltd., Tokyo, Japan) under a stereomicroscope (SZ61, Olympus, Tokyo, Japan). The devices with 3D structures were exposed to UV light for 20 min for complete polymerization, because NOA 81 requires 2 J/cm^2 to fully cure, according to the product data sheet. Three kinds of 3D structures—Menger sponges, spiral structures, and a channel in channel structure—were fabricated by using this method.

2.2. Bond Strength Measurement

Bond strength between NOA 81 sheets was measured by a tensile adhesion test with glass slides bonded by NOA 81 sheets. Two PDMS molds without any pattern were put on two glass slides, and uncured NOA 81 was injected into the spaces between the molds and the glass slides. Two glass slides with NOA 81 were exposed to UV light for 90 s. The NOA 81 sheets were cut into 1×1 cm^2 with a utility knife after the PDMS molds were peeled off from glass slides. The two glass slides were bonded with the NOA 81 by UV exposure for 20 min. A weight was hung on a glass slide, and the load was made heavier to reach the point at which the bonding between NOA 81 sheets broke.

2.3. Flow Visualization

The 3D microstructures and 3D fluidics in the channels were observed with a scanning electron microscope (SEM, TM-1000, Hitachi, Tokyo, Japan) and a laser confocal microscope (TCS-STED-CW, Leica Microsystems, Wetzlar, Germany). Aqueous solutions of 0.5 mM fluorescein sodium salt (Sigma-Aldrich, Tokyo, Japan) and 0.1 mM rhodamine B (Sigma-Aldrich, Tokyo, Japan) were used to visualize the 3D flow and observed with a 442-nm excitation laser and a 10×/0.40 lens (HCX PL APO CS, Leica Microsystems GmbH, Wetslar, Germany). Stacks of each confocal X-Y scan of 1024×1024 pixels were collected with a step of 0.49 μm in the Z direction. Z-series images were loaded in to the imaging software (LAS AF, Leica Microsystems GmbH, Wetslar, Germany) and made into vertical cross-sectional images.

3. Results and Discussion

3.1. Characterization of The Method

Pneumatic valve-like structures, which have a membrane clamped with 35-μm-deep and 50-μm-deep chambers, were fabricated to confirm the minimum layer thickness (Figure 2a). Figure 2b,c shows the SEM images of cross-sectional shapes of the membranes. SU-8 3025 was used to make a PDMS mold for a thicker membrane, and SU-8 3005 was used for a thinner membrane. The layer thickness can be controlled by changing the thickness of the SU-8 master molds, and the minimum layer thickness was 4 μm. This performance is higher than that of the high-end commercialized 3D printer with a minimum layer resolution of around 10 μm [26,37–41].

As for the bond strength, two pieces of NOA 81 sheets could maintain the bond against a load of 3.4 kg for a few minutes. The bond between NOA 81 sheets was broken upon loading 3.5 kg. The bond strength with a bonding area of 1 cm^2 against a load of 3.4 kg was about 330 kPa. The average bond strength of the conventional bonding method was from 100 to 500 kPa [42–45]. The bonding between NOA 81 sheets by our method has enough bond strength for using it as a material in microfluidic devices.

The Menger sponge is one of the fractals that are geometric configurations with a self-similar pattern [46]. The basic structure is a cube with holes on every surface in the center. The 20 basic structures are arranged in a way that they form the same configuration. The higher level Menger sponges are made by the Menger sponges of the same regulation one level lower. In this paper, level-1 and level-2 Menger sponges were fabricated. Figure 2d–f shows SEM images of 90-μm level-1 and 810-μm level-2 Menger sponges. They are composed of 30-μm cubes and 270-μm level-1 Menger sponges, respectively. The alignment accuracy is less than 10% of the minimum structure size; however, the accuracy can improve by using bonding aligner. Their unique structure with many undercuts can be observed.

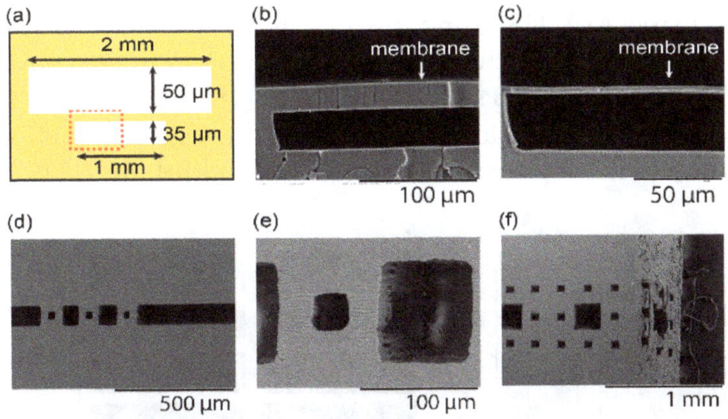

Figure 2. Structures for characterization. (**a**) A schematic illustration of a structure for layer thickness characterization. A red dotted box shows a region for close-up views in (**b**,**c**). (**b**) An SEM image of a membrane made by a PDMS mold with a thickness of 50 μm and (**c**) a membrane made by a PDMS mold with a thickness of 4 μm. (**d**) 90-μm level-1 Menger sponges; (**e**) A close up image of the level-1 Menger sponge. (**f**) 810-μm level-2 Menger sponges from oblique view points.

3.2. Spiral Structure

The spiral stair-like structure has a deforming wall in a channel, as shown in Figure 3a. This wall consists of five layers, and its cross sections change their shape to rotate 180° around a central longitudinal axis of a channel. Figure 3b,c is the top-down view of the spiral structure taken with a SEM and cross-sectional shapes taken with a confocal microscope, respectively. The structure, consisting of five thin NOA 81 films (black parts), shows the changing cross-sectional shapes: a horizontal wall appears at 200 μm from the channel end, changes to a vertical wall at the point of 1000 μm, and the vertical wall keeps rotating until it becomes horizontal at 1800 μm. The thickness of the NOA 81 sheets made with two molds were 119 ± 2.03 μm measured with a digital micrometer (CLM1-15QM, Mitutoyo Corporation, Kanagawa, Japan) when the sheets were made in 60-μm-high SU-8 molds. The change of the structure shape can be smoother by a redesign of the structure (Movie S1) and an increase in lamination layers.

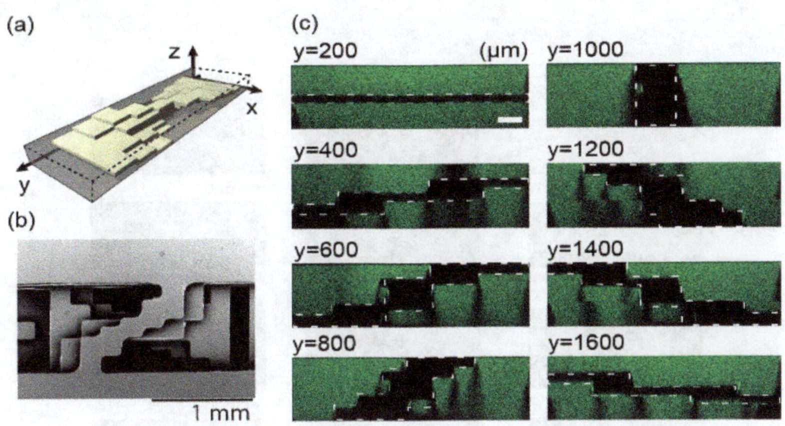

Figure 3. Diagrams of a spiral structure. (**a**) A conceptual image of the structure. (**b**) A SEM image of the structure from top-down view. (**c**) Confocal images of vertical cross sections at every 200 μm in the flow direction of the microchannel filled with fluorescein solution. White dotted lines represent cross-sectional shapes of a five-layer structure. Scale bar is 100 μm.

The spiral structures were embedded in a Y-junction microfluidic channel to visualize the flow in the microchannel (Figure 4a). One cycle of the structure was 2 mm in length and there were five cycles in the channel. Fluorescein solution and water were introduced into the channel with syringe pumps (KD Scientific, Holliston, MA, USA) and the flow in the channel was observed with the confocal microscope. The fluids were mixed with the structure and the flow in the channel is similar to the flow in a chaotic microfluidic mixer [13,14] or a spiral type mixer [18] (Figure 4b,c). The two kinds of liquids were rotated by the change in cross-section structure and they generated a vertical multilayer flow. The multilayer spiral structure has a potential to make and control the 3D flow.

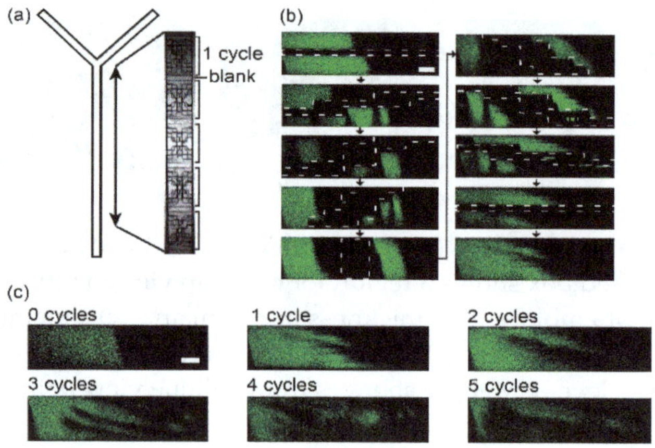

Figure 4. Confocal microscope images of a flow in the spiral structure channel. (**a**) Overhead view of the Y-shaped channel with the 5 spiral structures; (**b**) Confocal microscope images of vertical cross sections of the microchannel at every 200 μm in the flow direction and (**c**) at blanks. The channel was filled with a fluorescein solution (green) and water (dark). Scale bars are 100 μm.

We also made a ten-layer spiral structure in a microfluidic channel, which has two five-layer spiral structures in the channel (Figure 5). Although the lower layer cannot be visualized clearly due to light scattering at the vertical shape of the structure, the double screw structure was observed. The structure has an extruded structure from side to side and from top to bottom, which cannot be fabricated by conventional soft lithography. When the channel was pressurized to introduce a fluorescent reagent with the syringe pump (flow rate: 1 mL/min), there was no liquid leakage from the channel. The same PDMS molds can make another type of ten-layer structure made of the upper and lower five-layer 3D structures.

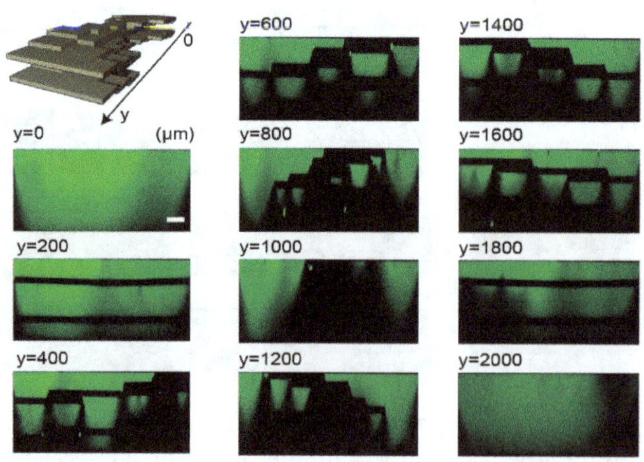

Figure 5. Confocal microscope images of a ten-layer spiral structure at every 200 μm in the flow direction of the microchannel. Scale bar is 100 μm.

3.3. 3D Sheath Device

A 3D sheath device was also made via the fabrication method. The 3D sheath device has a channel with a smaller channel in it, and the channels are connected to different inlets (Figure 6a). The smaller channel is merged with a larger channel to make a sheath flow (Figure 6b,c). The larger channel dimension is 500-μm-wide and 250-μm-deep, and the smaller one is 100-μm-wide and 50-μm-deep. A flow in the merged channel was observed with the confocal microscope at 5 mm from the merged point while a fluorescein solution and rhodamine solution were introduced into the channels from the two inlets with the flow rate of fluorescein as 90 μL/min and rhodamine as 450 μL/min. The fluorescein solution was stably sheathed by the rhodamine solution at least 5 mm in length. The sheathed flow broadened due to decreasing flow rates—fluorescein as 40 μL/min and rhodamine as 210 μL/min.

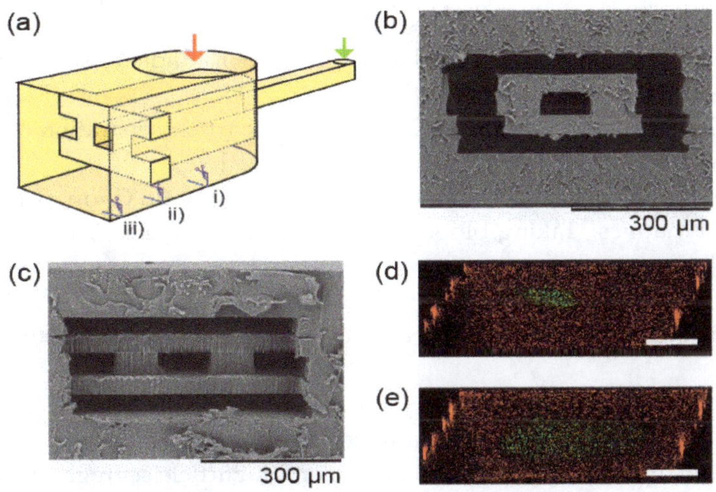

Figure 6. Cross-sectional images of a 3D sheath device. (**a**) Conceptual image of the 3D sheath device. (**b**) A SEM image of a cross section of the device at position (i) in Figure 6a, and (**c**) at position (ii) in Figure 6c. (**d**) A confocal microscope image of the 3D sheath at position (iii) with the flow rate of fluorescein as 90 μL/min and rhodamine as 450 μL/min. (**e**) Fluorescein as 40 μL/min and rhodamine as 210 μL/min. Scale bars are 100 μm.

4. Conclusions

In this paper, we demonstrated a 3D microfabrication method with simple instruments for soft lithography: a black light for detection of biomolecules and a stereomicroscope. The method does not require any robotic action or sacrificial materials [47,48]. The method exhibited enough bond strength for microfluidic devices and thin layer thickness with a high layer resolution. The Menger sponge-like structures, and some 3D microfluidic devices were achieved without limitation by the demolding process needed in conventional molding fabrication. This method still has some problems about the positioning accuracy for manual procedures, which can be solved by instruments and automation techniques for mass production. 3D structures in microfluidic channels grow in importance as approaches to control flow [49,50], produce microlens arrays [51], produce 3D cell culture systems and mimic organs [52–55], and develop multifactor separation [56]. Our fabrication method can meet the requirements of new 3D structures in a lab scale and allow for the commercialization of previously studied prototype devices.

Acknowledgments: This research was partially supported by the Japan Society for the Promotion of Science (JSPS) through its International Training Program (ITP), Institutional Program for Young Researcher Overseas Visits, and Grant-in-Aid for Young Scientists (B) (26790033).

Author Contributions: Toyohiro Naito designed and performed the experiments; Makoto Nakamura made the Menger sponges and characterized the fabrication method; Noritada Kaji and Yoshinobu Baba contributed reagents/materials/microscopes; and Takuya Kubo and Koji Otsuka co-supervised the work and reviewed the manuscript.

References

1. Brody, J.P.; Yager, P.; Goldstein, R.E.; Austin, R.H. Biotechnology at low reynolds numbers. *Biophys. J.* **1996**, *71*, 3430–3441. [CrossRef]
2. Weigl, B.H.; Yager, P. Microfluidic diffusion-based separation and detection. *Science* **1999**, *283*, 346–347. [CrossRef]
3. Kenis, P.J.A.; Ismagilov, R.F.; Takayama, S.; Whitesides, G.M.; Li, S.; White, H.S. Fabrication inside microchannels using fluid flow. *Acc. Chem. Res.* **2000**, *33*, 841–847. [CrossRef] [PubMed]
4. Kenis, P.J.A.; Ismagilov, R.F.; Whitesides, G.M. Microfabrication inside capillaries using multiphase laminar flow patterning. *Science* **1999**, *285*, 83–85. [CrossRef] [PubMed]
5. Hong, C.C.; Choi, J.W.; Ahn, C.H. A novel in-plane passive microfluidic mixer with modified tesla structures. *Lab Chip* **2004**, *4*, 109–113. [CrossRef] [PubMed]
6. Huang, L.R.; Cox, E.C.; Austin, R.H.; Sturm, J.C. Continuous particle separation through deterministic lateral displacement. *Science* **2004**, *304*, 987–990. [CrossRef] [PubMed]
7. Davis, J.A.; Inglis, D.W.; Morton, K.J.; Lawrence, D.A.; Huang, L.R.; Chou, S.Y.; Sturm, J.C.; Austin, R.H. Deterministic hydrodynamics: Taking blood apart. *Proc. Natl. Acad. Sci. USA* **2006**, *103*, 14779–14784. [CrossRef] [PubMed]
8. Morton, K.J.; Loutherback, K.; Inglis, D.W.; Tsui, O.K.; Sturm, J.C.; Chou, S.Y.; Austin, R.H. Crossing microfluidic streamlines to lyse, label and wash cells. *Lab Chip* **2008**, *8*, 1448–1453. [CrossRef] [PubMed]
9. Yoon, D.H.; Ha, J.B.; Bahk, Y.K.; Arakawa, T.; Shoji, S.; Go, J.S. Size-selective separation of micro beads by utilizing secondary flow in a curved rectangular microchannel. *Lab Chip* **2009**, *9*, 87–90. [CrossRef] [PubMed]
10. Di Carlo, D.; Edd, J.F.; Humphry, K.J.; Stone, H.A.; Toner, M. Particle segregation and dynamics in confined flows. *Phys. Rev. Lett.* **2009**, *102*, 094503. [CrossRef] [PubMed]
11. Sudarsan, A.P.; Ugaz, V.M. Multivortex micromixing. *Proc. Natl. Acad. Sci. USA* **2006**, *103*, 7228–7233. [CrossRef] [PubMed]
12. Di Carlo, D.; Irimia, D.; Tompkins, R.G.; Toner, M. Continuous inertial focusing, ordering, and separation of particles in microchannels. *Proc. Natl. Acad. Sci. USA* **2007**, *104*, 18892–18897. [CrossRef] [PubMed]
13. Stroock, A.D.; Dertinger, S.K.W.; Ajdari, A.; Mezić, I.; Stone, H.A.; Whitesides, G.M. Chaotic mixer for microchannels. *Science* **2002**, *295*, 647–651. [CrossRef] [PubMed]
14. Villermaux, E.; Stroock, A.D.; Stone, H.A. Bridging kinematics and concentration content in a chaotic micromixer. *Phys. Rev. E* **2008**, *77*, 015301. [CrossRef] [PubMed]
15. Yasui, T.; Omoto, Y.; Osato, K.; Kaji, N.; Suzuki, N.; Naito, T.; Watanabe, M.; Okamoto, Y.; Tokeshi, M.; Shamoto, E.; *et al.* Microfluidic baker's transformation device for three-dimensional rapid mixing. *Lab Chip* **2011**, *11*, 3356–3360. [CrossRef] [PubMed]
16. Yasui, T.; Omoto, Y.; Osato, K.; Kaji, N.; Suzuki, N.; Naito, T.; Okamoto, Y.; Tokeshi, M.; Shamoto, E.; Baba, Y. Confocal microscopic evaluation of mixing performance or three-dimensional microfluidic mixer. *Anal. Sci.* **2012**, *28*, 57–59. [CrossRef] [PubMed]
17. Melchels, F.P.; Feijen, J.; Grijpma, D.W. A review on stereolithography and its applications in biomedical engineering. *Biomaterials* **2010**, *31*, 6121–6130. [CrossRef] [PubMed]
18. Bertsch, A.; Heimgartner, S.; Cousseau, P.; Renaud, P. Static micromixers based on large-scale industrial mixer geometry. *Lab Chip* **2001**, *1*, 56–60. [CrossRef] [PubMed]
19. Liu, R.H.; Stremler, M.A.; Sharp, K.V.; Olsen, M.G.; Santiago, J.G.; Adrian, R.J.; Aref, H.; Beebe, D.J. Passive mixing in a three-dimensional serpentine microchannel. *J. Microelectromech. Syst.* **2000**, *9*, 190–197. [CrossRef]
20. Zhang, M.; Wu, J.; Wang, L.; Xiao, K.; Wen, W. A simple method for fabricating multi-layer pdms structures for 3d microfluidic chips. *Lab Chip* **2010**, *10*, 1199–1203. [CrossRef] [PubMed]
21. Unger, M.A.; Chou, H.-P.; Thorsen, T.; Scherer, A.; Quake, S.R. Monolithic microfabricated valves and pumps by multilayer soft lithography. *Science* **2000**, *288*, 113–116. [CrossRef] [PubMed]

22. Baek, J.Y.; Park, J.Y.; Ju, J.I.; Lee, T.S.; Lee, S.H. A pneumatically controllable flexible and polymeric microfluidic valve fabricated via *in situ* development. *J. Micromech. Microeng.* **2005**, *15*, 1015–1020. [CrossRef]
23. Mosadegh, B.; Kuo, C.H.; Tung, Y.C.; Torisawa, Y.S.; Bersano-Begey, T.; Tavana, H.; Takayama, S. Integrated elastomeric components for autonomous regulation of sequential and oscillatory flow switching in microfluidic devices. *Nat. Phys.* **2010**, *6*, 433–437. [CrossRef] [PubMed]
24. Shallan, A.I.; Smejkal, P.; Corban, M.; Guijt, R.M.; Breadmore, M.C. Cost-effective three-dimensional printing of visibly transparent microchips within minutes. *Anal. Chem.* **2014**, *86*, 3124–3130. [CrossRef] [PubMed]
25. Bishop, G.W.; Satterwhite, J.E.; Bhakta, S.; Kadimisetty, K.; Gillette, K.M.; Chen, E.; Rusling, J.F. 3D-printed fluidic devices for nanoparticle preparation and flow-injection amperometry using integrated prussian blue nanoparticle-modified electrodes. *Anal. Chem.* **2015**, *87*, 5437–5443. [CrossRef] [PubMed]
26. Chan, H.N.; Chen, Y.; Shu, Y.; Chen, Y.; Tian, Q.; Wu, H. Direct, one-step molding of 3D-printed structures for convenient fabrication of truly 3d pdms microfluidic chips. *Microfluid. Nanofluid.* **2015**, *19*, 9–18. [CrossRef]
27. Kolesky, D.B.; Truby, R.L.; Gladman, A.S.; Busbee, T.A.; Homan, K.A.; Lewis, J.A. 3D bioprinting of vascularized, heterogeneous cell-laden tissue constructs. *Adv. Mater* **2014**, *26*, 3124–3130. [CrossRef] [PubMed]
28. Markstedt, K.; Mantas, A.; Tournier, I.; Martinez Avila, H.; Hagg, D.; Gatenholm, P. 3D bioprinting human chondrocytes with nanocellulose-alginate bioink for cartilage tissue engineering applications. *Biomacromolecules* **2015**, *16*, 1489–1496. [CrossRef] [PubMed]
29. Hardin, J.O.; Ober, T.J.; Valentine, A.D.; Lewis, J.A. Microfluidic printheads for multimaterial 3d printing of viscoelastic inks. *Adv Mater* **2015**, *27*, 3279–3284. [CrossRef] [PubMed]
30. Satyanarayana, S.; Karnik, R.N.; Majumdar, A. Stamp-and-stick room-temperature bonding technique for microdevices. *J. Microelectromech. Syst.* **2005**, *14*, 392–399. [CrossRef]
31. Bartolo, D.; Degre, G.; Nghe, P.; Studer, V. Microfluidic stickers. *Lab Chip* **2008**, *8*, 274–279. [CrossRef] [PubMed]
32. Natali, M.; Begolo, S.; Carofiglio, T.; Mistura, G. Rapid prototyping of multilayer thiolene microfluidic chips by photopolymerization and transfer lamination. *Lab Chip* **2008**, *8*, 492–494. [CrossRef] [PubMed]
33. Arayanarakool, R.; Le Gac, S.; van den Berg, A. Low-temperature, simple and fast integration technique of microfluidic chips by using a uv-curable adhesive. *Lab Chip* **2010**, *10*, 2115–2121. [CrossRef] [PubMed]
34. Sikanen, T.M.; Lafleur, J.P.; Moilanen, M.-E.; Zhuang, G.; Jensen, T.G.; Kutter, J.P. Fabrication and bonding of thiol-ene-based microfluidic devices. *J. Micromech. Microeng.* **2013**, *23*, 037002. [CrossRef]
35. Naito, T.; Arayanarakool, R.; Le Gac, S.; Yasui, T.; Kaji, N.; Tokeshi, M.; van den Berg, A.; Baba, Y. Temperature-driven self-actuated microchamber sealing system for highly integrated microfluidic devices. *Lab Chip* **2013**, *13*, 452–458. [CrossRef] [PubMed]
36. Carlborg, C.F.; Haraldsson, T.; Oberg, K.; Malkoch, M.; van der Wijngaart, W. Beyond pdms: Off-stoichiometry thiol-ene (OSTE) based soft lithography for rapid prototyping of microfluidic devices. *Lab Chip* **2011**, *11*, 3136–3147. [CrossRef] [PubMed]
37. Lifton, V.; Lifton, G.; Simon, S. Options for additive rapid prototyping methods (3D printing) in mems technology. *Rapid Prototyp. J.* **2014**, *20*, 403–412. [CrossRef]
38. Farzadi, A.; Solati-Hashjin, M.; Asadi-Eydivand, M.; Abu Osman, N.A. Effect of layer thickness and printing orientation on mechanical properties and dimensional accuracy of 3D printed porous samples for bone tissue engineering. *PLoS ONE* **2014**, *9*, e108252. [CrossRef] [PubMed]
39. Headley, D.B.; DeLucca, M.V.; Haufler, D.; Pare, D. Incorporating 3d-printing technology in the design of head-caps and electrode drives for recording neurons in multiple brain regions. *J. Neurophysiol.* **2015**, *113*, 2721–2732. [CrossRef] [PubMed]
40. Lee, M.P.; Cooper, G.J.; Hinkley, T.; Gibson, G.M.; Padgett, M.J.; Cronin, L. Development of a 3D printer using scanning projection stereolithography. *Sci. Rep.* **2015**, *5*, 9875. [CrossRef] [PubMed]
41. Melenka, G.W.; Schofield, J.S.; Dawson, M.R.; Carey, J.P. Evaluation of dimensional accuracy and material properties of the makerbot 3d desktop printer. *Rapid Prototyp. J.* **2015**, *21*, 618–627. [CrossRef]
42. Bhattacharya, S.; Datta, A.; Berg, J.M.; Gangopadhyay, S. Studies on surface wettability of poly(dimethyl) siloxane (pdms) and glass under oxygen-plasma treatment and correlation with bond strength. *J. Microelectromech. Syst.* **2005**, *14*, 590–597. [CrossRef]
43. Chow, W.W.Y.; Lei, K.F.; Shi, G.; Li, W.J.; Huang, Q. Microfluidic channel fabrication by pdms-interface bonding. *Smart Mater. Struct.* **2006**, *15*, 112–116. [CrossRef]

44. Eddings, M.A.; Johnson, M.A.; Gale, B.K. Determining the optimal PDMS-PDMS bonding technique for microfluidic devices. *J. Micromech. Microeng.* **2008**, *18*, 067001. [CrossRef]
45. Kim, J.; Surapaneni, R.; Gale, B.K. Rapid prototyping of microfluidic systems using a pdms/polymer tape composite. *Lab Chip* **2009**, *9*, 1290–1293. [CrossRef] [PubMed]
46. Herrmann, R. A fractal approach to the dark silicon problem: A comparison of 3D computer architectures—Standard slices *versus* fractal menger sponge geometry. *Chaos Solitons Fractals* **2015**, *70*, 38–41. [CrossRef]
47. Gelber, M.K.; Bhargava, R. Monolithic multilayer microfluidics via sacrificial molding of 3D-printed isomalt. *Lab Chip* **2015**, *15*, 1736–1741. [CrossRef] [PubMed]
48. Parekh, D.P.; Ladd, C.; Panich, L.; Moussa, K.; Dickey, M.D. 3D printing of liquid metals as fugitive inks for fabrication of 3D microfluidic channels. *Lab Chip* **2016**. [CrossRef] [PubMed]
49. Huang, S.-H.; Tan, W.-H.; Tseng, F.-G.; Takeuchi, S. A monolithically three-dimensional flow-focusing device for formation of single/double emulsions in closed/open microfluidic systems. *J. Micromech. Microeng.* **2006**, *16*, 2336–2344. [CrossRef]
50. King, P.H.; Jones, G.; Morgan, H.; de Planque, M.R.; Zauner, K.P. Interdroplet bilayer arrays in millifluidic droplet traps from 3D-printed moulds. *Lab Chip* **2014**, *14*, 722–729. [CrossRef] [PubMed]
51. Kim, S.; Yeo, E.; Kim, J.H.; Yoo, Y.-E.; Choi, D.-S.; Yoon, J.S. Geometry modulation of microlens array using spin coating and evaporation processes of photoresist mixture. *Int. J. Precis. Eng. Manuf. Green Technol.* **2015**, *2*, 231–235. [CrossRef]
52. Huh, D.; Leslie, D.C.; Matthews, B.D.; Fraser, J.P.; Jurek, S.; Hamilton, G.A.; Thorneloe, K.S.; McAlexander, M.A.; Ingber, D.E. A human disease model of drug toxicity–induced pulmonary edema in a lung-on-a-chip microdevice. *Sci. Transl. Med.* **2012**, *4*, 159ra147. [CrossRef] [PubMed]
53. Kim, H.J.; Huh, D.; Hamilton, G.; Ingber, D.E. Human gut-on-a-chip inhabited by microbial flora that experiences intestinal peristalsis-like motions and flow. *Lab Chip* **2012**, *12*, 2165–2174. [CrossRef] [PubMed]
54. Torisawa, Y.S.; Spina, C.S.; Mammoto, T.; Mammoto, A.; Weaver, J.C.; Tat, T.; Collins, J.J.; Ingber, D.E. Bone marrow-on-a-chip replicates hematopoietic niche physiology *in vitro*. *Nat. Methods* **2014**, *11*, 663–669. [CrossRef] [PubMed]
55. Stucki, A.O.; Stucki, J.D.; Hall, S.R.; Felder, M.; Mermoud, Y.; Schmid, R.A.; Geiser, T.; Guenat, O.T. A lung-on-a-chip array with an integrated bio-inspired respiration mechanism. *Lab Chip* **2015**, *15*, 1302–1310. [CrossRef] [PubMed]
56. Wong, C.C.; Liu, Y.; Wang, K.Y.; Rahman, A.R. Size based sorting and patterning of microbeads by evaporation driven flow in a 3D micro-traps array. *Lab Chip* **2013**, *13*, 3663–3667. [CrossRef] [PubMed]

Permissions

All chapters in this book were first published in MDPI; hereby published with permission under the Creative Commons Attribution License or equivalent. Every chapter published in this book has been scrutinized by our experts. Their significance has been extensively debated. The topics covered herein carry significant findings which will fuel the growth of the discipline. They may even be implemented as practical applications or may be referred to as a beginning point for another development.

The contributors of this book come from diverse backgrounds, making this book a truly international effort. This book will bring forth new frontiers with its revolutionizing research information and detailed analysis of the nascent developments around the world.

We would like to thank all the contributing authors for lending their expertise to make the book truly unique. They have played a crucial role in the development of this book. Without their invaluable contributions this book wouldn't have been possible. They have made vital efforts to compile up to date information on the varied aspects of this subject to make this book a valuable addition to the collection of many professionals and students.

This book was conceptualized with the vision of imparting up-to-date information and advanced data in this field. To ensure the same, a matchless editorial board was set up. Every individual on the board went through rigorous rounds of assessment to prove their worth. After which they invested a large part of their time researching and compiling the most relevant data for our readers.

The editorial board has been involved in producing this book since its inception. They have spent rigorous hours researching and exploring the diverse topics which have resulted in the successful publishing of this book. They have passed on their knowledge of decades through this book. To expedite this challenging task, the publisher supported the team at every step. A small team of assistant editors was also appointed to further simplify the editing procedure and attain best results for the readers.

Apart from the editorial board, the designing team has also invested a significant amount of their time in understanding the subject and creating the most relevant covers. They scrutinized every image to scout for the most suitable representation of the subject and create an appropriate cover for the book.

The publishing team has been an ardent support to the editorial, designing and production team. Their endless efforts to recruit the best for this project, has resulted in the accomplishment of this book. They are a veteran in the field of academics and their pool of knowledge is as vast as their experience in printing. Their expertise and guidance has proved useful at every step. Their uncompromising quality standards have made this book an exceptional effort. Their encouragement from time to time has been an inspiration for everyone.

The publisher and the editorial board hope that this book will prove to be a valuable piece of knowledge for researchers, students, practitioners and scholars across the globe.

List of Contributors

Eleonora Bolli
Department of Industrial Engineering, University of Rome "Tor Vergata", Via del Politectico 1, 00133 Rome, Italy
Institute for the Study of Nanostructured Materials, ISMN–CNR, Monterotondo Stazione, 00015 Rome, Italy

Saulius Kaciulis and Alessio Mezzi
Institute for the Study of Nanostructured Materials, ISMN–CNR, Monterotondo Stazione, 00015 Rome, Italy

Jin Tao, Qiankun Zhang, Wei Pang, Hao Zhang, Xuexin Duan, Daihua Zhang and Jing Liu
State Key Laboratory of Precision Measurement Technology and Instruments, School of Precision Instruments and Opto-Electronics Engineering, Tianjin University, Tianjin 300072, China

Yunfeng Xiao
State Key Laboratory for Mesoscopic Physics, School of Physics, Peking University, Beijing 100044, China

Xiaoying Li
Key Laboratory of Optoelectronics Information Technology, School of Precision Instrument and Opto-Electronics Engineering, Tianjin University, Tianjin 300072, China

Pei Yao
School of Materials Science and Engineering, Tianjin University, Tianjin 300072, China

Milad Navaei, Alireza Mahdavifar and Peter J. Hesketh
School of Mechanical Engineering, Georgia Institute of Technology, Atlanta, GA 30324, USA

Jean-Marie D. Dimandja
Department of Chemistry, Spelman College, Atlanta, GA 30314, USA

Gary McMurray
Food Processing Center, Georgia Tech Research Institute (GTRI), Atlanta, GA 30318, USA

Yumi Mukouyama
Center for International Research on Integrative Biomedical Systems (CIBiS), Institute of Industrial Science (IIS), The University of Tokyo, 4-6-1 Komaba, Meguro-ku, Tokyo 153-8505, Japan

Yuya Morimoto, Shohei Habasaki and Shoji Takeuchi
Center for International Research on Integrative Biomedical Systems (CIBiS), Institute of Industrial Science (IIS), The University of Tokyo, 4-6-1 Komaba, Meguro-ku, Tokyo 153-8505, Japan
Takeuchi biohybrid Innovation Project, Exploratory Research for Advanced Technology (ERATO), Japan Science and Technology (JST), Komaba Open Laboratory (KOL) Room M202, 4-6-1, Komaba, Meguro-ku, Tokyo 153-8904, Japan

Sahana Sarkar, Stanley C. S. Lai and Serge G. Lemay
MESA+ Institute for Nanotechnology, University of Twente, P.O. Box 217, 7500 AE Enschede, The Netherlands

Xiangchen Che and Long Que
Department of Electrical and Computer Engineering, Iowa State University; Ames, IA 50011, USA

Jacob Nuhn
Department of Chemical and Biological Engineering, Iowa State University; Ames, IA 50011, USA

Ian Schneider
Department of Chemical and Biological Engineering, Iowa State University; Ames, IA 50011, USA
Department of Genetics, Development and Cell Biology, Iowa State University; Ames, IA 50011, USA

Cilong Yu, Xiang Qian, Quan Yu and Kai Ni
Division of Advanced Manufacturing, Graduate School at Shenzhen, Tsinghua University, Shenzhen 518055, China

Yan Chen
Shenzhen Institutes of Advanced Technology, Chinese Academy of Sciences, Shenzhen 518055, China

Xiaohao Wang
Division of Advanced Manufacturing, Graduate School at Shenzhen, Tsinghua University, Shenzhen 518055, China
The State Key Laboratory of Precision Measurement Technology and Instruments, Tsinghua University, Beijing 100084, China

Lei Liu, Zhenguo Jiang, Syed Rahman and Md. Itrat Bin Shams
Department of Electrical Engineering, University of Notre Dame, 275 Fitzpatrick, Notre Dame, IN 46556, USA

Benxin Jing
Department of Chemical Engineering and Material Science, Wayne State University, 5050 Anthony Wayne Dr., Detroit, MI 48202, USA

Akash Kannegulla and Li-Jing Cheng
School of Electrical Engineering and Computer Science, Oregon State University, Corvallis, OR 97330, USA

Saeed Mohammadi
Graduate School of Chemical Sciences and Engineering, Hokkaido University, Kita 13 Nishi 8, Kita-ku, Sapporo 060-8628, Japan

Lori Shayne Alamo Busa
Graduate School of Chemical Sciences and Engineering, Hokkaido University, Kita 13 Nishi 8, Kita-ku, Sapporo 060-8628, Japan
Physical Sciences Department, Nueva Vizcaya State University, Bayombong, Nueva Vizcaya 3700, Philippines

Masatoshi Maeki, Akihiko Ishida and Hirofumi Tani
Division of Applied Chemistry, Faculty of Engineering, Hokkaido University, Kita 13 Nishi 8, Kita-ku, Sapporo 060-8628, Japan

Manabu Tokeshi
Division of Applied Chemistry, Faculty of Engineering, Hokkaido University, Kita 13 Nishi 8, Kita-ku, Sapporo 060-8628, Japan
ImPACT Research Center for Advanced Nanobiodevices, Nagoya University, Furo-cho, Chikusa-ku, Nagoya 464-8603, Japan
Innovative Research Center for Preventive Medical Engineering, Nagoya University, Furo-cho, Chikusa-ku, Nagoya 464-8601, Japan
Institute of Innovation for Future Society, Nagoya University, Furo-cho, Chikusa-ku, Nagoya 464-8601, Japan
Division of Applied Chemistry, Faculty of Engineering, Hokkaido University, Sapporo 060-8628, Japan

Chayakom Phurimsak, Mark D. Tarn and Nicole Pamme
Department of Chemistry, University of Hull, Cottingham Road, Hull, HU6 7RX, UK

Yutaka Kazoe
Department of Hemolysis and Apheresis, Graduate School of Medicine, The University of Tokyo, 7-3-1 Hongo, Bunkyo, Tokyo 113-8656, Japan

Ippei Yamashiro, Kazuma Mawatari and Takehiko Kitamori
Deparment of Applied Chemistry, Graduate School of Engineering, The University of Tokyo, 7-3-1 Hongo, Bunkyo, Tokyo 113-8656, Japan

Ko-Ichiro Miyamoto, Takuya Sato and Minami Abe
Department of Electronic Engineering, Tohoku University, 6-6-05 Aza-Aoba, Aramaki, Aoba-ku, Sendai 980-8579, Japan

Torsten Wagner
Institute of Nano- and Biotechnologies, Aachen University of Applied Sciences, Heinrich-Mußmann-Str. 1, Jülich 52428, Germany

Michael J. Schöning
Institute of Nano- and Biotechnologies, Aachen University of Applied Sciences, Heinrich-Mußmann-Str. 1, Jülich 52428, Germany
Peter-Grünberg Institute (PGI-8), Research Centre Jülich, Jülich 52425, Germany

Tatsuo Yoshinobu
Department of Electronic Engineering, Tohoku University, 6-6-05 Aza-Aoba, Aramaki, Aoba-ku, Sendai 980-8579, Japan
Department of Biomedical Engineering, Tohoku University, 6-6-05 Aza-Aoba, Aramaki, Aoba-ku, Sendai 980-8579, Japan

Keisuke Tenda, Riki Ota, Kentaro Yamada, Terence G. Henares, Koji Suzuki and Daniel Citterio
Department of Applied Chemistry, Keio University, 3-14-1 Hiyoshi, Kohoku-ku, Yokohama 223-8522, Kanagawa, Japan

Yaxiaer Yalikun
Department of Mechanical Engineering, Osaka University, 2-1 Yamadaoka, Suita, Osaka 565-0871, Japan
Laboratory for Integrated Biodevice, Quantitative Biology Center, RIKEN, 1-3 Yamadaoka, Suita, Osaka 565-0871, Japan

Yasunari Kanda
Division of Pharmacology, National Institute of Health Sciences, 1-18-1 Kamiyoga, Setagaya, Tokyo 158-8501, Japan

Keisuke Morishima
Department of Mechanical Engineering, Osaka University, 2-1 Yamadaoka, Suita, Osaka 565-0871, Japan
Global Center for Advanced Medical Engineering and Informatics, Osaka University, 2-2 Yamadaoka, Suita, Osaka 565-0871, Japan

Jumpei Morikawa
Department of Applied Chemistry, Graduate School of Engineering, Nagoya University, Furo-cho, Chikusa-ku, Nagoya 464-8603, Japan

Takao Yasui
Department of Applied Chemistry, Graduate School of Engineering, Nagoya University, Furo-cho, Chikusa-ku, Nagoya 464-8603, Japan
ImPACT Research Center for Advanced Nanobiodevices, Nagoya University, Furo-cho, Chikusa-ku, Nagoya 464-8603, Japan
Japan Science and Technology Agency (JST), PRESTO, 4-1-8 Honcho, Kawaguchi, Saitama 332-0012, Japan

Kazuo Tsubota
Department of Ophthalmology, School of Medicine, Keio University, Tokyo 160-8582, Japan

Yoshinobu Baba
Department of Applied Chemistry, Graduate School of Engineering, Nagoya University, Furo-cho, Chikusa-ku, Nagoya 464-8603, Japan
ImPACT Research Center for Advanced Nanobiodevices, Nagoya University, Furo-cho, Chikusa-ku, Nagoya 464-8603, Japan
Health Research Institute, National Institute of Advanced Industrial Science and Technology (AIST), Takamatsu 761-0395, Japan

Bhagwati P. Gupta
Department of Biology, McMaster University, Hamilton, ON L8S 4K1, Canada

Pouya Rezai
Department of Mechanical Engineering, York University, Toronto, ON M3J 1P3, Canada

Toyohiro Naito, Makoto Nakamura, Takuya Kubo and Koji Otsuka
Department of Material Chemistry, Graduate School of Engineering, Kyoto University, Katsura, Nishikyo-ku, Kyoto 615-8510, Japan

Noritada Kaji
Department of Applied Chemistry, Graduate School of Engineering, Nagoya University, Furo-cho, Chikusa-ku, Nagoya 464-8603, Japan
ImPACT Research Center for Advanced Nanobiodevices, Nagoya University, Furo-cho, Chikusa-ku, Nagoya 464-8603, Japan

Index

www.ingramcontent.com/pod-product-compliance
Lightning Source LLC
LaVergne TN
LVHW082201200926
848484LV00008B/2197